M. Chernick

Michael Chernick
15 Quail Drive
Holland, PA 18966

# *Almost Sure Convergence*

# Probability and Mathematical Statistics

*A Series of Monographs and Textbooks*

*Editors*   **Z. W. Birnbaum**   **E. Lukacs**
*University of Washington*   *Bowling Green State University*
*Seattle, Washington*   *Bowling Green, Ohio*

# *Almost Sure Convergence*

WILLIAM F. STOUT

*Department of Mathematics*
*University of Illinois at Urbana-Champaign*
*Urbana, Illinois*

1974

ACADEMIC PRESS    New York   San Francisco   London
*A Subsidiary of Harcourt Brace Jovanovich, Publishers*

ACADEMIC PRESS, INC.
111 Fifth Avenue, New York, New York 10003

*United Kingdom Edition published by*
ACADEMIC PRESS, INC. (LONDON) LTD.
24/28 Oval Road, London NW1


Library of Congress Cataloging in Publication Data

Stout, William F
        Almost sure convergence.

        (Probability and mathematical statistics)
        Bibliography: p.
        1.    Random variables.    2.    Sequences (Mathematics)
3.    Partial sums (Series)    I.    Title.
QA273.5.S76              519.2′4              73-2075
ISBN 0–12–672750–3


AMS(MOS) 1970 Subject Classifications: 60F15, 60G45, 60G50, 60-02


PRINTED IN THE UNITED STATES OF AMERICA

# Contents

# Chapter 3  Stability and Moments

# Chapter 4  Stability of Weighted Sums of Random Variables

# Chapter 5  Exponential Inequalities, the Law of the Iterated Logarithm, and Related Results

# Chapter 6  Recurrence of $\{S_n, n \geq 1\}$ and Related Results

# Preface

Almost sure behavior of partial sums of random variables, the subject of this book, has enjoyed both a rich classical period and a recent resurgence of research activity. This is nicely illustrated by the law of the iterated logarithm for partial sums, the subject of Chapter 5: Attempts to sharpen Borel's [1909] strong law of large numbers culminated in Khintchine's [1924] law of the iterated logarithm for Bernoulli random variables. The Kolmogorov [1929] and the Hartman–Wintner [1941] extensions of Khintchine's result to large classes of independent random variables were milestones in the classical theory. Levy's (1937, see [1954]) law of the iterated logarithm for martingales, an important class of dependent random variables, was another major advance in the classical theory.

The modern period for the law of the iterated logarithm was started by Strassen ([1964], [1965], [1966]) with his discovery of almost sure invariance principles, his deep functional law of the iterated logarithm, and his converse to the Hartman–Wintner law of the iterated logarithm (this last occurring, remarkably, a quarter century after the Hartman–Wintner result). As Chapter 5 indicates, one of the characteristics of the modern period has been an emphasis on laws of the iterated logarithm for dependent random variables.

Because of the rich history and current interest in the law of the iterated logarithm and in other areas of almost sure behavior, it seems desirable to have a monograph which treats almost sure behavior in a systematic and unified manner. This book presents such a treatment of the law of the iterated logarithm and of four other major varieties of almost sure behavior: almost sure convergence of partial sums (Chapter 2), almost sure stability of partial sums (Chapter 3), almost sure stability of weighted partial sums (Chapter 4), and recurrence of partial sums (Chapter 6).

Subdivision into major topics within chapters is usually done on the basis of dependence structure. For example, the law of the iterated logarithm for partial sums of independent random variables, which is developed in Sections 5.1–5.3 of Chapter 5, is such a major topic. The treatment of each major topic starts with more elementary and usually classical theorems and proceeds with more advanced and often recently established theorems.

Since the book's viewpoint is probabilistic, certain mathematically important topics, although logically fitting under the umbrella of almost sure behavior, are omitted because they are not probabilistic in nature. Examples of such excluded topics are the almost everywhere convergence of Fourier series and the pointwise ergodic theory of operators in $L_p$ spaces. Topics of probabilistic interest have often been excluded as well, thereby keeping the book from being too long.

The book was developed from a course on almost sure behavior given by the author. This course was given as a sequel to the basic graduate level probability course at the University of Illinois. The book contains ample material for a one semester course. It assumes familiarity with basic real analysis and basic measure theory (such as provided by Royden's "Real Analysis") and with basic measure theoretic probability (such as provided by Ash [1972, Chapters 5–8] or Chung [1968]). The book should also prove useful for independent study, either to be read systematically or to be used as a reference.

A glossary of symbols and conventions is provided for the reader's convenience. This should be especially helpful when the book is used as a reference rather than systematically read. Results are numbered by chapter and section. For instance, Example 2.8.11 is the eleventh of the examples in Section 8 of Chapter 2. Chapters are somewhat independent; hence each may be read without extensive reference to previous chapters. Exercises are included to help the reader develop a working familiarity with the subject and to provide additional information on the subject. Numerous references are given to stimulate further reading.

I especially thank R. J. Tomkins for reading the entire manuscript, making many valuable suggestions, and spotting numerous errors. I thank R. B. Ash, J. L. Doob, and W. Philipp for reading portions of the manuscript and making valuable suggestions. Carolyn Bloemker's excellent typing was greatly appreciated.

The writing of this book was partially supported by the National Science Foundation.

# Glossary of Symbols
# and Conventions

| | |
|---|---|
| $(\Omega, \mathscr{F}, P)$ | Underlying probability space |
| $EX, E(X)$ | Expectation of $X$ |
| $\{X_i, i \geq 1\}$ | Basic sequence of random variables |
| $\{S_n, n \geq 1\}$ | $S_n = \sum_{i=1}^{n} X_i$ |
| $A^c$ | Complement of $A$ |
| $A = B$ | $P[A \cap B^c] + P[A^c \cap B] = 0$, unless explicitly stated otherwise |
| $A \triangle B$ | $(A \cap B^c) \cup (A^c \cap B)$ |
| $A \subset B$ | $P[A \cap B^c] = 0$ |
| $A$ implies $B$ | $A \subset B$ |
| $a_n \to a$ | $\lvert a \rvert = \infty$ not allowed, unless explicitly stated otherwise |
| On $A$, ... | $A$ implies ... |
| $\text{cov}(X, Y)$ | Covariance of $X$ and $Y$ |
| $\text{var}(Y)$ | Variance of $Y$ |
| $R_n$ | $n$-dimensional Euclidean space |
| $R$ | One-dimensional Euclidean space |
| $R_\infty$ | Infinite-dimensional Euclidean space |
| $\mathscr{C}_n$ | Borel sets of $R_n$ |
| $\mathscr{C}_\infty$ | Borel sets of $R_\infty$ |
| $\mathscr{B}\{Y_\alpha, \alpha \in A\}$ | $\sigma$ field generated by $\{Y_\alpha, \alpha \in A\}$ |
| $\mathscr{B}_\infty$ | $\mathscr{B}\{X_1, X_2, \ldots\}$ |
| Occurrence | See parenthetical remark, p. 1 |
| a.s. | Almost surely |
| $\{\mathscr{F}_n, n \geq 1\}$ adapted to $\{Y_n, n \geq 1\}$ | $\mathscr{F}_n$ increasing $\sigma$ fields, $\subset \mathscr{F}$, $Y_n$ is $\mathscr{F}_n$ measurable |
| $\{Y_n, \mathscr{F}_n, n \geq 1\}$ adapted stochastic sequence | Same as above |
| $\mathscr{F}_0$ | $\{\varnothing, \Omega\}$ |
| $I(A)$ | Indicator function of event $A$ |
| i.o. | Infinitely often |
| $A_n$ i.o. | $\bigcap_{k=1}^{\infty} \bigcup_{n=k}^{\infty} A_n$ |
| $\mu(X)$ | Median of $X$ |
| $X^+$ | $\max(X, 0)$ |

| | |
|---|---|
| $X^-$ | $-\min(X, 0)$ |
| $[x]$ | Greatest integer function of $x$ |
| $P_X$ | Probability measure induced on $(R_1, \mathscr{C}_1)$ by $X$ |
| $F_X$ | Distribution function of $X$ |
| $\alpha(X)$ | Constant in definition of generalized Gaussian random variable |
| $a_n \sim b_n$ | $\lim a_n/b_n = 1$ |
| $a_n = o(b_n)$ | $\lim a_n/b_n = 0$ |
| $a_n = O(b_n)$ | $\lim \sup a_n/b_n < \infty$ |
| $\sum_{i=n+1}^{n} (\cdot)$ | $= 0$ |
| $L_p$ | Banach space of functions with $p$th absolute moment |
| $[X \in A]$ | $[\omega \mid X(\omega) \in A]$ |
| $\equiv$ | Indicates definition; e.g., $h(x) \equiv x^2$ |
| $\log_2 x$ | $\log \log x$ |

# CHAPTER 1

# Introduction

## 1.1. Delineation of the Subject

The purpose of this chapter is to introduce the reader to the subject and content of this book in some detail. The basic setting throughout the book is that of a probability space $(\Omega, \mathscr{F}, P)$ with a sequence of random variables, henceforth referred to as the basic sequence and denoted by $\{X_i, i \geq 1\}$, defined on $(\Omega, \mathscr{F}, P)$. Let $S_n = \sum_{i=1}^{n} X_i$ for $n \geq 1$. $\{S_n, n \geq 1\}$ is referred to as the sequence of partial sums.

In this book we shall study those events whose occurrence is determined by the values of infinitely many $S_n$. (The use of the word "occurrence," although most convenient, can cause confusion. Recall that in probabilistic terminology each $\omega \in \Omega$ is an "outcome" of an "experiment." Given a particular outcome $\omega \in \Omega$, the occurrence of an event $A$ simply means that $\omega \in A$. Thus occurrence simply expresses membership in a set.) This remark should be made precise: For each $n \geq 1$, let $\mathscr{B}_n$ be the $\sigma$ field generated by $\{X_i, 1 \leq i \leq n\}$. Thus $\mathscr{B}_n$ consists of all events of the form $[(X_1, X_2, \ldots, X_n) \in B]$ as $B$ ranges over the Borel sets of $R_n$, $R_n$ denoting an $n$-dimensional Euclidean space. Let $\mathscr{B}_\infty$ be the $\sigma$ field generated by the $\mathscr{B}_n$. Thus $\mathscr{B}_\infty$ consists of all events of the form $[(X_1, X_2, \ldots, X_n, \ldots) \in B]$ as $B$ ranges over the Borel sets of $R_\infty$, $R_\infty$ denoting an infinite-dimensional Euclidean space. Of course, $\mathscr{B}_\infty$ can also be described as consisting of all events of the form $[(S_1, S_2, \ldots, S_n, \ldots) \in B]$ as $B$ ranges over the Borel sets of $R_\infty$. From the viewpoint of the book this last is a more natural way to view $\mathscr{B}_\infty$. Events of $\mathscr{B}_\infty$ are of two types:

**Definition 1.1.1.** An event which is a member of $\mathscr{B}_n$, for some finite $n$, is called a weak event. An event which is a member of $\mathscr{B}_\infty$ and which is not a weak event is called a strong event. ∎

1

The occurrence of a weak event is determined by the values of only finitely many random variables of the basic sequence, whereas the occurrence of a strong event is determined by the values of infinitely many random variables of the basic sequence (equivalently, by the values of infinitely many random variables of the sequence of partial sums). For example, for each integer $n$, $[S_n \geq 0]$ is a weak event because, for a given $\omega \in \Omega$, the question of whether or not $\omega \in [S_n \geq 0]$ can be decided if the values of $X_1(\omega), X_2(\omega), \ldots, X_n(\omega)$ are known. Similarly, $[S_n \geq 0$ for infinitely many $n]$ is a strong event since, for a given $\omega \in \Omega$, for no finite $n$ does knowing the values of $X_1(\omega), X_2(\omega), \ldots, X_n(\omega)$ determine whether $\omega \in [S_n \geq 0$ for infinitely many $n]$.

Rephrasing the statement beginning the preceeding paragraph in the light of Definition 1.1.1, in this book we study strong events. Given a strong event $A$ of interest, the most important problem is to find conceptually and computationally simple conditions which imply that $P(A) = 1$ (or 0). In addition, we often obtain important results in which a strong event $A$ can satisfy $0 < P(A) < 1$. Then it is an important problem to find events $B$ of interest for which $B = A$, $A \subset B$, and $B \subset A$. For example, (letting $EY$ denote the expectation of a random variable $Y$) if the $X_i$ are martingale differences with $E \sup |X_i| < \infty$, then

$$[\sup S_n < \infty] = [S_n \text{ converges}]$$

can be proven. (Here and throughout the book, convergence means convergence to a finite limit and equality of two events means that two events contain the same outcomes with the possible exception of a null event.)

We will look at the probabilities of weak events only when the results obtained are useful in studying the probabilities of strong events. For instance, that will be our only interest in the central limit theorem. In studying the probabilities of strong events two types of assumptions are usually made concerning the basic sequence. First, the random variables of the basic sequence satisfy a dependence relationship, such as being orthogonal, being independent, being Markovian, being martingale differences, etc. Second, the random variables satisfy an absolute moment condition such as

$$\sum_{i=1}^{\infty} EX_i^2 < \infty, \qquad \sum_{i=1}^{\infty} E(X_i - EX_i)^2/i^2 < \infty,$$

$$E \sup |X_i| < \infty, \qquad \sup E|S_n| < \infty,$$

etc. A third type of assumption sometimes imposed is that the random variables of the basic sequence satisfy a stationarity condition such as being weakly stationary, identically distributed, strictly stationary, etc.

These types of assumptions combine to impose bounds on probabilities of weak events. Since the weak events form a field which generates the $\sigma$ field of strong events $\mathscr{B}_\infty$, it follows by the Carathéodory extension theorem of measure theory that specification of the probabilities of the weak events uniquely determines the probabilities of the strong events. Typically, weak and general assumptions such as those discussed above impose bounds on the probabilities of certain weak events, which, in turn, imply that certain strong events of interest occur with probability zero or one.

For example, Kolmogorov's strong law of large numbers for independent random variables states that the $X_i$ being independent and

$$\sum_{i=1}^{\infty} E(X_i - EX_i)^2/i^2 < \infty$$

together imply that

$$P[(S_n - ES_n)/n \to 0 \qquad \text{as} \qquad n \to \infty] = 1.$$

## 1.2. A Brief Chapter by Chapter Outline of Topics Covered

It seems appropriate to outline briefly the chapter by chapter organization of the book, touching on certain highlights of each chapter. In order to accomplish this, knowledge of some advanced probability concepts is assumed (e.g., martingales, Markov processes, mixing sequences). Statements concerning unfamiliar concepts should be passed over with the knowledge that such concepts will be fully developed at the appropriate places in the book.

In Chapter 2, the event $A = [S_n \text{ converges}]$ is studied. Of particular importance is the almost sure convergence of $S_n$ (that is, $P(A) = 1$). A variety of dependence structures for the basic sequence are analyzed in Chapter 2, these structures usually implying that the $X_i$ are orthogonal. As successively more restrictive assumptions are made concerning the dependence structure of the basic sequence, stronger and more specific results concerning the almost sure convergence of $S_n$ are established: Suppose $ES_n^2 < \infty$ for each $n \geq 1$. Orthogonality alone guarantees that

$$\sum_{i=1}^{\infty} (\log i)^2 \, EX_i^2 < \infty$$

implies $S_n$ converges almost surely. Under the further restriction that $\{S_n, n \geq 1\}$ is a martingale,

$$\sum_{i=1}^{\infty} EX_i^2 < \infty \qquad (\text{indeed } \sup E \,|\, S_n \,| < \infty)$$

implies that $S_n$ converges almost surely. Under the still further restriction that the $X_i$ are independent with $EX_i = 0$ for each $i \geq 1$, Kolmogorov's three series theorem characterizes the almost sure convergence of $S_n$ in terms of the convergence of three numerical series. A major part of Chapter 2 is devoted to this progressive restricting of the assumptions concerning the dependence structure. The case of independence, for which most of the major results are classical, and the martingale case, for which many of the major results are the consequence of recent research, both receive particular emphasis in Chapter 2. In the martingale case questions concerning local convergence are carefully examined. For example,

$$\left[ \sum_{i=1}^{\infty} E(X_i^2 \,|\, X_1, X_2, \ldots, X_{i-1}) < \infty \right] = [S_n \text{ converges}]$$

is shown under the assumption that the $X_i$ are martingale differences with $E \sup X_i^2 < \infty$. The question of the almost sure convergence of $S_n$ being implied by absolute moment conditions on the $S_n$ without any particular dependence structure assumed for the basic sequence is analyzed also. Several applications to real analysis are made in Chapter 2. For example, several results about the almost everywhere convergence of Haar series are shown to follow easily from martingale convergence results.

Even if a sequence of random variables $\{T_n, n \geq 1\}$ is almost surely divergent (that is, $P[T_n \text{ converges as } n \to \infty] = 0$) it is easy to show that there exists constants $a_n \to \infty$ and $b_n$ such that $T_n$ is stabilized; that is $(T_n - b_n)/a_n$ converges almost surely to zero. In Chapter 3, the stability of $S_n$ is studied when $S_n$ is almost surely divergent. The case when the $X_i$ are independent and identically distributed is studied first: As successively higher absolute moments of $X_1$ are assumed finite, sequences $\{a_n, n \geq 1\}$ converging successively slower to infinity are shown to stabilize $S_n$. According to Kolmogorov's strong law of large numbers for independent identically distributed random variables, if $E \,|\, X_1 \,| < \infty$, then

$$(S_n - nEX_1)/n \to 0 \qquad \text{almost surely.}$$

If, in addition, $E \,|\, X_1 \,|^p < \infty$ for some $1 < p < 2$, then

$$(S_n - nEX_1)/n^{1/p} \to 0 \qquad \text{almost surely.}$$

Let var $Y$ denote the variance of a random variable $Y$. If, in addition, $EX_1^2 < \infty$, then

$$\limsup \frac{(S_n - nEX_1)}{[2n \, \mathrm{var}(X_1) \log \log(n \, \mathrm{var} \, X_1)]^{1/2}} \leq 1 \qquad \text{almost surely.}$$

Stability is also studied when the $X_i$ are independent but not necessarily identically distributed, when the $X_i$ are martingale differences, when the $X_i$ are strictly stationary, when the $X_i$ are mixing, when the $S_n$ are Markovian, and when the $S_n$ are restricted by absolute moment conditions without any particular dependence structure assumed for the basic sequence. One such important result is the pointwise ergodic theorem: If $\{X_i, i \geq 1\}$ is strictly stationary and ergodic with $E \,|\, X_1 \,| < \infty$, then

$$(S_n - nEX_1)/n \to 0 \qquad \text{almost surely.}$$

A second important result is given by: If $\{X_i, i \geq 1\}$ is a martingale difference sequence with

$$\sum_{i=1}^{\infty} E \,|\, X_i \,|^p / i^{1+p/2} < \infty \qquad \text{for some} \quad p \geq 2,$$

then

$$S_n/n \to 0 \qquad \text{almost surely.}$$

A third important result is given by: If $\{X_i, i \geq 1\}$ is an independent sequence with $E \,|\, X_i \,|^{1+\delta} \leq K < \infty$ for some $\delta > 0$ and all $i \geq 1$, then

$$(S_n - ES_n)/n \to 0 \qquad \text{almost surely.}$$

As discussed in the preceding paragraphs, the almost sure convergence of $(S_n - b_n)/a_n$ to zero is studied in Chapter 3. The more general problem of the almost sure convergence of centered weighted sums $T_n = \sum_{k=1}^{\infty} a_{nk} X_k$ $-b_n$ to zero is studied in Chapter 4. (The $a_{nk}$ are the weights and the $b_n$ are the centering constants.) The purpose is to prove almost sure convergence results for as broad a class of coefficient matrices $\{a_{nk}\}$ as possible. This is in contrast to Chapter 3 where $\{a_{nk}\}$ is always assumed to have a specific structure (for example, $a_{nk} = n^{-1}$ for $k \leq n$, $a_{nk} = 0$ for $k > n$). The case when the $X_i$ are independent and identically distributed is studied most, but the cases when the $X_i$ are independent but not necessarily identically distributed, when the $X_i$ are martingale differences, when the $X_i$ are strictly stationary, and when the $X_i$ are strongly multiplicative receive attention too. The following result is typical of Chapter 4: Let

$$\sum_{k=1}^{\infty} a_{nk}^2 \leq Cn^{-\delta} \qquad \text{for some} \quad C < \infty, \quad \delta > 0 \quad \text{and all} \quad n \geq 1,$$

$$a_{nk}^2 \leq Ck^{-1} \qquad \text{for some} \quad C < \infty \quad \text{and all} \quad n \geq 1, \quad k \geq 1,$$

and the $X_i$ be independent identically distributed with $EX_1 = 0$ and $EX_1^2 < \infty$. Then

$$\sum_{k=1}^{\infty} a_{nk}X_k \to 0 \qquad \text{almost surely as} \quad n \to \infty.$$

In the special case of weighted averages ($T_n = \sum_{i=1}^{n} a_i X_i / \sum_{i=1}^{n} a_i$) with strictly positive weights, more precise results are shown. For example, $\{X_i, i \geq 1\}$ independent identically distributed with $E \mid X_1 \mid \log^+ \mid X_1 \mid < \infty$, $EX_1 = 0$, and $\{a_i, i \geq 1\}$ uniformly bounded implies $T_n \to 0$ almost surely.

In Chapter 5, the law of the iterated logarithm and closely related results are studied. The case when the $X_i$ are independent is studied most. For this case the classical exponential inequalities approach is used to derive Kolmogorov's well-known law of the iterated logarithm. Strassen's almost sure invariance principle is used to derive the Hartman–Wintner law of the iterated logarithm. One of the most interesting results of Chapter 5 is this Hartman–Wintner law of the iterated logarithm combined with its converse due to Strassen: Let the $X_i$ be independent identically distributed. Then

$$\limsup \frac{S_n}{(2n \, c \log \log n)^{1/2}} = 1 \qquad \text{almost surely for some} \quad 0 < c < \infty$$

if and only if $EX_1 = 0$ and $0 < EX_1^2 = c < \infty$. This delicate result about the magnitude of the asymptotic fluctuations of the $S_n$ is typical of the results of Chapter 5 in that it contains remarkably precise information and is quite difficult to prove. Besides the case of independence, the law of the iterated logarithm is also studied when the $X_i$ are martingale differences, mixing, and strongly multiplicative. In the case of mixing sequences, a central limit theorem with an error estimate is used to derive the needed probability inequalities. This, along with Kolmogorov's and Strassen's approach referred to above, is one of the major approaches for deriving laws of the iterated logarithm. A solution to the problem of finding necessary and sufficient conditions for the strong law of large numbers in the case of independence is given also; this is a by-product of the study of the exponential inequalities mentioned above.

Given a sequence of positive constants $\{a_n, n \geq 1\}$, let $T_n = \sum_{i=1}^{n} X_i/a_n$ for each $n \geq 1$. In Chapter 6 the recurrence of $\{T_n, n \geq 1\}$ and related results are studied. The most important problem considered is the determination of the recurrent states of $\{S_n, n \geq 1\}$. That is, for which $c \in R_1$ does

$$P[S_n \in (c - \varepsilon, c + \varepsilon) \text{ for infinitely many } n] = 1$$

hold for each $\varepsilon > 0$? This question is studied when the $X_i$ are independent identically distributed and more generally when $\{S_n, n \geq 1\}$ is Markovian with stationary transition probabilities. The question is also studied when the $X_i$ are independent identically distributed random vectors taking values in an $n$-dimensional lattice. An important result of Chapter 6 is that the $X_i$ independent identically distributed nonlattice random variables with $EX_1 = 0$ implies that each real number $c$ is a recurrent state of $\{S_n, n \geq 1\}$. Assuming that the $X_i$ are independent identically distributed, the determination of the recurrent states of $\{T_n, n \geq 1\}$ for $a_n = n^{\alpha}$, $0 < \alpha \leq 1$ fixed, is studied. For example, $EX_1 = 0$ and $a_n = n^{1/2}$ implies that $-\infty$ and $\infty$ are recurrent states of $\{T_n, n \geq 1\}$. That is,

$$\limsup S_n/n^{1/2} = \infty \qquad \text{and} \qquad \liminf S_n/n^{1/2} = -\infty$$

almost surely. Assuming that the $X_i$ are independent identically distributed, certain questions concerning the amount of time that $\{S_n, n \geq 1\}$ spends in various subsets of $R_1$ are also studied.

## 1.3.  Methodology

Certain remarks about the methodology used in proving results about the occurrence of strong events seems appropriate. Typically, proofs tend to be a mixture of two kinds of analyses. First, magnitudes of probabilities are estimated using mostly elementary techniques of classical real analysis. The inequalities of Chebyshev, Holder, and Jensen, integration and summation by parts, splitting integrals into pieces, Taylor series expansions, etc., are heavily drawn upon. Second, the clever use of certain probabilistic and measure-theoretic techniques translate these estimates of probabilities into statements about the probability of occurrence of strong events of interest. The Borel–Cantelli lemma, truncation of random variables, centering at means or medians, stopping rule techniques, etc., are heavily drawn upon. Combinatorial arguments are sometimes useful. Complex analysis and functional analysis, although occasionally used, do not play a major role. One of the keys to understanding the choice of methodology is that probabilities are seldom computed but rather estimated. Indeed, the generality of typical hypotheses (for example, $X_i$ independent identically distributed with $EX_1 = 0$) prohibits the computation of probabilities of weak events.

In many proofs a major step is the establishment of a maximal inequality. For example, $\{X_i, i \geq 1\}$ orthogonal implies that

$$E[\max_{i \leq n} S_i^2] \leq (\log 4n/\log 2)^2 \sum_{i=1}^{n} EX_i^2 \qquad \text{for all} \quad n \geq 1,$$

an inequality which plays a major role in the analysis of the orthogonal case.

Often certain techniques become associated with a particular dependence structure. For example, stopping rule techniques play a major role in the study of the local convergence of martingales and truncation plays a major role in problems where the $X_i$ are independent.

It is a major purpose of this book to stress methods of proof as well as present interesting results concerning almost sure behavior.

## 1.4. Applications to Fields outside Probability

Besides being of intrinsic interest to probabilists, the results presented in the book have applications to number theory, real analysis, and statistics. In this section we sketch three such applications.

The most famous example of application to number theory is the result that except for a set of Lebesque measure zero, all real numbers in the unit interval are *normal* in the sense that they have decimal expansions in which the digits $0, 1, \ldots, 9$ occur with equal limiting relative frequency. This follows immediately from Kolmogorov's strong law of large numbers for independent identically distributed random variables.

Certain probabilistic results concerning almost sure convergence have application in real analysis to the almost everywhere convergence of certain orthogonal series of real functions. For example, since the successive partial sums of a Haar series form a martingale satisfying a certain regularity condition, certain rather deep results concerning the almost everywhere convergence of Haar series follow rather easily from the study of martingale convergence presented in Chapter 2. Results in probability theory (such as the law of the iterated logarithm) sometimes suggest results for certain classical types of orthogonal series even when the probabilistic proofs do not carry over. This is true of lacunary trigonometric series since intuitively a sequence of lacunary trigonometric functions seems similar to a sequence of independent identically distributed random variables.

In statistics there are certain situations where almost sure convergence seems a more relevant concept than convergence in probability. Consider a physician who treats patients with a drug having the same unknown cure probability of $p$ for each patient. The physician is willing to continue use of the drug as long as no superior drug is found. Along with administering the drug, he estimates the cure probability from time to time by dividing the number of cures up to that point in time by the number of patients treated. If $n$ is the number of patients treated, denote this estimating random variable by $\bar{X}_{(n)}$. Suppose the physician wishes to estimate $p$ within a prescribed tolerance $\varepsilon > 0$. He asks whether he will ever reach a point in time such that with high probability, all subsequent estimates will fall within $\varepsilon$ of $p$. That is, he wonders for prescribed $\delta > 0$ whether there exists an integer $N$ such that

$$P[\max_{n \geq N} | \bar{X}_{(n)} - p | \leq \varepsilon] \geq 1 - \delta.$$

The weak law of large numbers says only that

$$P[| \bar{X}_{(n)} - p | \leq \varepsilon] \to 1 \qquad \text{as} \qquad n \to \infty$$

and hence does not answer his question. It is only by the strong law of large numbers that the existence of such an $N$ is indeed guaranteed.

The major emphasis of the book is on almost sure behavior results for their own intrinsic probabilistic flavor. However, applications to fields outside probability are often presented. This is especially so when these applications are in some sense natural, such as the three applications sketched above.

# Pointwise Convergence
# of Partial Sums

## 2.1. Application of the Borel–Cantelli Lemma and the Chebyshev Inequality to Almost Sure Convergence

Given the basic sequence $\{X_i, i \geq 1\}$, let $S_n = \sum_{i=1}^{n} X_i$ define $\{S_n, n \geq 1\}$. $\{S_n, n \geq 1\}$ will hereafter be referred to as the sequence of partial sums. In Chapter 2 the occurrence of the event

$$[S_n \text{ converges as } n \to \infty]$$

is studied. The Borel–Cantelli lemma plays a fundamental role in this study as well as throughout the book.

Events $\{A_n, n \geq 1\}$ are said to occur infinitely often if

$$\limsup A_n = \bigcap_{k=1}^{\infty} \bigcup_{n=k}^{\infty} A_n$$

occurs. $\limsup A_n$ is often denoted by

$$[A_n \text{ i.o.}],$$

where i.o. denotes infinitely often. Let $\{T_n, n \geq 1\}$ and $T$ be random variables. Recall that $T_n$ is said to converge almost surely to $T$ if

$$P[T_n \to T \text{ as } n \to \infty] = 1.$$

Almost sure convergence is denoted by $T_n \to T$ a.s. $T_n$ is said to converge in probability to $T$ if

$$P[|T_n - T| > \varepsilon] \to 0 \qquad \text{as} \quad n \to \infty$$

10

for each $\varepsilon > 0$. Recall that the Borel–Cantelli lemma states that

$$\sum_{n=1}^{\infty} P[A_n] < \infty$$

implies that

$$P[A_n \text{ i.o.}] = 0.$$

If the $A_n$ are independent, then the partial converse to the Borel–Cantelli lemma states that

$$\sum_{n=1}^{\infty} P[A_n] = \infty$$

implies that

$$P[A_n \text{ i.o.}] = 1.$$

(It is an instructive exercise to prove the Borel–Cantelli lemma and its partial converse.)

**Theorem 2.1.1.**   Let $\{T_n, n \geq 1\}$ be a sequence of random variables such that

$$\sum_{n=1}^{\infty} P[\,|\,T_n\,| > \varepsilon] < \infty$$

for each $\varepsilon > 0$. Then

$$T_n \to 0 \qquad \text{a.s.}$$

***Proof.***   By hypothesis, for each $k \geq 1$,

$$\sum_{n=1}^{\infty} P[\,|\,T_n\,| > 2^{-k}] < \infty.$$

Hence, by the Borel–Cantelli lemma, for each $k \geq 1$, $|\,T_n\,| \leq 2^{-k}$ for all $n$ sufficiently large, except on a null event $N_k$. It follows that

$$T_n(\omega) \to 0 \qquad \text{for all} \quad \omega \notin \bigcup_{k=1}^{\infty} N_k.$$

Since $\bigcup_{k=1}^{\infty} N_k$ is a null event, $T_n \to 0$ a.s. follows.   ∎

**Theorem 2.1.2.**   Let $\{T_n, n \geq 1\}$ be a sequence of random variables. Suppose

$$\sum_{n=1}^{\infty} P[\,|\,T_n\,| > \varepsilon_n] < \infty$$

for positive constants $\varepsilon_n \to 0$. Then

$$T_n \to 0 \qquad \text{a.s.}$$

**Proof.** Immediate from the Borel–Cantelli lemma applied to events $[|T_n| > \varepsilon_n], n \geq 1$. ∎

One might suspect that the converse of Theorem 2.1.2 holds. Example 2.1.1 below shows this is not the case.

**Example 2.1.1.** Let $\Omega = [0, 1]$, $\mathscr{F}$ be the Borel sets of $[0, 1]$ and $P$ be Lebesgue measure. Let

$$T_n(\omega) = \begin{cases} 0 & \text{if} \quad 0 \leq \omega \leq 1 - n^{-1} \\ 1 & \text{if} \quad 1 - n^{-1} < \omega \leq 1. \end{cases}$$

Clearly $T_n \to 0$ a.s. But note that for any positive $\varepsilon_n \to 0$ we have

$$P[|T_n| > \varepsilon_n] = n^{-1}$$

for $n$ sufficiently large. Thus

$$\sum_{n=1}^{\infty} P[|T_n| > \varepsilon_n] = \infty. \quad ∎$$

**Corollary 2.1.1.** Suppose $T_n \to 0$ in probability. Then there exists positive integers $n_k \to \infty$ such that

$$T_{n_k} \to 0 \qquad \text{a.s. as} \quad k \to \infty.$$

**Proof.** $P[|T_n| > \varepsilon] \to 0$ for each $\varepsilon > 0$. Thus there exists positive integers $n_k \to \infty$ such that

$$P[|T_{n_k}| > 2^{-k}] \leq 2^{-k}$$

for each $k \geq 1$. Thus

$$\sum_{k=1}^{\infty} P[|T_{n_k}| > 2^{-k}] < \infty$$

and hence $T_{n_k} \to 0$ a.s. follows from Theorem 2.1.2. ∎

Example 2.1.2 illustrates the basic fact that convergence in probability does not imply convergence almost surely.

***Example 2.1.2.***   Let $\Omega = (0, 1]$, $\mathscr{F}$ be the Borel sets of $(0, 1]$ and $P$ be Lebesgue measure. Let

$$T_{nk}(\omega) = \begin{cases} 1 & \text{if } (k-1)/n < \omega \leq k/n \\ 0 & \text{otherwise} \end{cases}$$

for $1 \leq k \leq n$, $n \geq 1$. Then the sequence $T_{11}, T_{21}, T_{22}, T_{31}, T_{32}, T_{33}, \ldots$ is easily seen to converge in probability but not almost surely.  ∎

It is clear from Theorem 2.1.1 and Theorem 2.1.2 that upper bounds for $P[|\,T_n\,| > \varepsilon]$ should be useful in trying to show that $T_n \to 0$ a.s. The Chebyshev inequality (or Markov inequality, as it is sometimes called) provides such bounds. This simple inequality and more sophisticated inequalities of a similar nature play a major role in Chapter 2 as well as throughout the book. Recall that for a random variable $X$, the Chebyshev inequality says

$$P[|\,X\,| > \varepsilon] \leq E\,|\,X\,|/\varepsilon$$

for each $\varepsilon > 0$. Two other useful versions which follow immediately are

$$P[|\,X - EX\,| > \varepsilon] \leq \operatorname{var} X/\varepsilon^2$$

and

$$P[X > \varepsilon] \leq \exp(-t\varepsilon)E\exp(tX)$$

for each $\varepsilon > 0$ and real $t$.

**Theorem 2.1.3.**   Suppose

$$\sum_{n=1}^{\infty} E\,|\,T_n\,|^p < \infty$$

for some $p > 0$. Then

$$T_n \to 0 \qquad \text{a.s.}$$

***Proof.***

$$\sum_{n=1}^{\infty} E\,|\,T_n\,|^p < \infty \qquad \text{implies} \qquad E\sum_{n=1}^{\infty} |\,T_n\,|^p < \infty,$$

implying that $\sum_{n=1}^{\infty} |\,T_n\,|^p < \infty$ a.s. and hence that $T_n \to 0$ a.s.  ∎

Section 2.1 illustrates the close connection between bounds on absolute moments and almost sure behavior. The Chebyshev inequality is the simplest of many inequalities which establish a connection between magnitudes of absolute moments and magnitudes of probabilities. The Borel–Cantelli lemma is one of the most used results for translating bounds for the magnitudes of probabilities of weak events into statements about almost sure behavior. In particular, the Borel–Cantelli lemma is very useful in establishing almost sure convergence.

**Exercise 2.1.1.** Prove Theorem 2.1.3 by means of the Chebyshev inequality and the Borel–Cantelli lemma. ∎

**Exercise 2.1.2.** Let $\{X_n, n \geq 1\}$ be identically distributed

(i)  Suppose $E\,|\,X_1\,| < \infty$. Prove that $X_n/n \to 0$ a.s. Hint: Show that $E\,|\,X_1\,| < \infty$ implies that $\sum_{i=1}^{\infty} P[|\,X_1\,| > i] < \infty$.

(ii)  Let $0 \leq a_n \to \infty$. Show that there exists $\{X_i, i \geq 1\}$ such that $\sum_{i=1}^{\infty} P[|\,X_i\,| > a_i] = \infty$ and $X_n/a_n \to 0$ a.s. ∎

## 2.2.  Almost Sure Convergence of Subsequences of $\{S_n, n \geq 1\}$ When the $X_i$ Are Orthogonal

**Definition 2.2.1.**  The basic sequence $\{X_i, i \geq 1\}$ is said to be orthogonal if $EX_i^2 < \infty$ for all $i \geq 1$ and

$$EX_iX_j = 0 \qquad \text{for all} \quad i \neq j.$$

The basic sequence is said to be uncorrelated if $EX_i^2 < \infty$ for all $i \geq 1$ and

$$E(X_iX_j) = (EX_i)(EX_j)$$

for all $i \neq j$. ∎

If $EX_i = 0$ for all $i \geq 1$, it is immediate that the $X_i$ are orthogonal if and only if the $X_i$ are uncorrelated; that is, if the $X_i$ have mean zero, the theory of orthogonal random variables coincides with the theory of uncorrelated random variables. Centering uncorrelated random variables at their means preserves the property of being uncorrelated.

Intuitively, uncorrelated random variables are random variables which exhibit no stochastic linear dependence. Thus probabilistic intuition is a guide to the understanding of uncorrelated random variables. Because of this, it is sometimes helpful to view the study of orthogonal random variables centered at means as being the study of the more intuitively familiar (to a probabilist) uncorrelated random variables. One should of course keep in mind that many systems of functions studied in real analysis such as the trigonometric system are orthogonal and hence uncorrelated if they have zero means.

In the remainder of Section 2.2, the basic sequence is assumed to be orthogonal.

**Lemma 2.2.1.**  (Pythagorean relation).

$$E(S_n - S_m)^2 = \sum_{i=m+1}^{n} EX_i^2$$

for all $m < n$.

***Proof.***

$$(S_n - S_m)^2 = \sum_{i=m+1}^{n} X_i^2 + 2 \sum_{m+1 \le i < j \le n} X_i X_j.$$

The result then follows by taking expectations in the above equality, using the definition of orthogonality.  ∎

$E(S_n - S_m)^2 = \sum_{i=m+1}^{n} EX_i^2$ for all $m < n$ is equivalent to the definition of orthogonality. (Prove this.) Thus orthogonality relates the second moments of the partial sums $\sum_{i=m+1}^{n} X_i$ to the second moments of the individual $X_i$.

**Lemma 2.2.2.**  Suppose $\sum_{i=1}^{\infty} EX_i^2 < \infty$. Then there exists a random variable $S$ such that

$$ES^2 < \infty \qquad \text{and} \qquad E(S_n - S)^2 \to 0.$$

(For obvious reasons we often denote $S$ by $\sum_{i=1}^{\infty} X_i$.)

***Proof.***  Let $m < n$ be positive integers.

$$E(S_n - S_m)^2 = \sum_{i=m+1}^{n} EX_i^2 \le \sum_{i=m+1}^{\infty} EX_i^2 \to 0$$

as $m \to \infty$ by the Pythagorean relation and by hypothesis. Hence $\{S_n, n \ge 1\}$ is a Cauchy sequence with respect to the $L_2$ norm. Hence, by completeness of $L_2$, there exists a random variable $S$ with $ES^2 < \infty$ and $E(S_n - S)^2 \to 0$.  ∎

Lemma 2.2.2 will be used to identify the candidate for the limit in almost sure convergence of $S_n$ when $\sum_{i=1}^{\infty} EX_i^2 < \infty$.

If

$$\sum_{i=1}^{\infty} EX_i^2 < \infty, \tag{2.2.1}$$

then by Lemma 2.2.2, there exists a random variable $S$ such that $E(S_n - S)^2 \to 0$. By the Chebyshev inequality, $S_n \to S$ in probability. Hence by Corollary 2.1.1, there exists positive integers $n_k \to \infty$ for which $S_{n_k} \to S$ a.s. as $k \to \infty$. An examination of the proof of Corollary 2.1.1 shows that we have no information about the density of $\{n_k, k \geq 1\}$ in the integers. The question arises whether a stronger hypothesis than Eq. (2.2.1) might imply the existence of positive integers $n_k \to \infty$ such that

$$S_{n_k} \to S \qquad \text{a.s. as} \quad k \to \infty$$

and the $n_k$ form not too sparse a subsequence of the integers.

**Theorem 2.2.1.** Let $\{b_n, n \geq 1\}$ be a positive sequence increasing monotonely to infinity for which $\sum_{n=1}^{\infty} b_n EX_n^2 < \infty$. For each $k \geq 1$, let $n_k$ be the first integer $n$ such that $b_n \geq k$. Then $S_{n_k}$ converges almost surely.

The proof of Theorem 2.2.1 depends on a simple result concerning infinite series of positive numbers. We isolate this result as a lemma in order to emphasize its role.

**Lemma 2.2.3.** Let $\{a_n, n \geq 1\}$ and $\{b_n, n \geq 1\}$ be positive sequences with $b_n \uparrow \infty$ and $\sum_{n=1}^{\infty} a_n b_n < \infty$. For each $k \geq 1$, let $n_k$ be the smallest integer $n \geq 1$ such that $b_n \geq k$. Then $\sum_{k=1}^{\infty} \sum_{n=n_k}^{\infty} a_n < \infty$.

**Proof of the Lemma.** Adopt the convention that $\sum_{k=j}^{j-1} a_k = 0$.

$$\sum_{k=1}^{\infty} \sum_{n=n_k}^{\infty} a_n = \sum_{k=1}^{\infty} k \sum_{n=n_k}^{n_{k+1}-1} a_n = \sum_{k=1}^{\infty} \sum_{n=n_k}^{n_{k+1}-1} (k/b_n) b_n a_n$$

$$\leq \sum_{k=1}^{\infty} \sum_{n=n_k}^{n_{k+1}-1} a_n b_n = \sum_{n=n_1}^{\infty} a_n b_n < \infty. \quad \blacksquare$$

**Proof of the Theorem.** $\sum_{n=1}^{\infty} b_n EX_n^2 < \infty$ and $0 < b_n \to \infty$ implies that $\sum_{n=1}^{\infty} EX_n^2 < \infty$. Hence by Lemma 2.2.2, there exists a random variable $S = \sum_{i=1}^{\infty} X_i$ such that $E(S_n - S)^2 \to 0$ as $n \to \infty$. By Theorem 2.1.3 (essentially the Borel–Cantelli lemma) it suffices to prove that $\sum_{k=1}^{\infty} E(S_{n_k} - S)^2 < \infty$. Using $E(S_n - S)^2 \to 0$ and the Pythagorean relation,

$$E(S - S_{n_k})^2 = \lim_{n \to \infty} E(S_n - S_{n_k})^2 = \sum_{n=n_k+1}^{\infty} EX_n^2.$$

Hence it suffices to prove that $\sum_{k=1}^{\infty} \sum_{n=n_k+1}^{\infty} EX_n^2 < \infty$. But this follows from Lemma 2.2.3 with $a_n = EX_n^2$.  ∎

**Corollary 2.2.1.**  Suppose $\sum_{n=1}^{\infty} nEX_n^2 < \infty$. Then $S_n$ converges almost surely.

*Proof.*  The proof is immediate from Theorem 2.2.1 with $b_n = n$.  ∎

## 2.3.  The Rademacher–Menchoff Fundamental Convergence Theorem

Throughout Section 2.3, the basic sequence is assumed to be orthogonal. Since the proof of Corollary 2.2.1 is rather coarse, it seems reasonable to expect that Corollary 2.2.1 can be improved. We will do so using a general method for establishing almost sure convergence, the method of sub-sequences:

**Lemma 2.3.1.**  Suppose there exists a random variable $T$ and a positive integer subsequence $n_k \uparrow \infty$ such that

$$T_{n_k} \to T \qquad \text{a.s.} \tag{2.3.1}$$

and

$$\max_{n_{k-1} < n \leq n_k} |T_n - T_{n_{k-1}}| \to 0 \qquad \text{a.s. as } k \to \infty. \tag{2.3.2}$$

Then $T_n \to T$ a.s.

*Proof.*  Without loss of generality, suppose that the $n_k$ are strictly increasing. For a given positive integer $n$, choose $n_k$ such that $n_{k-1} < n \leq n_k$.

$$|T_n - T| \leq |T_{n_{k-1}} - T| + |T_n - T_{n_{k-1}}|$$

$$\leq |T_{n_{k-1}} - T| + \max_{n_{k-1} < n \leq n_k} |T_n - T_{n_{k-1}}|.$$

Since the final sum in the above inequality converges to zero almost surely as $k$ approaches infinity by Eq. (2.3.1) and (2.3.2), it follows that $T_n \to T$ a.s. as $n \to \infty$.  ∎

Thus, one can try to prove $S_n \to S$ a.s. by choosing judiciously a positive integer subsequence $\{n_k, k \geq 1\}$ such that Eqs. (2.3.1) and (2.3.2) hold. Theorem 2.2.1 suggests a way to achieve Eq. (2.3.1). What about Eq.

(2.3.2)? If we could conclude that

$$\sum_{k=1}^{\infty} E[\max_{n_{k-1} < n \leq n_k} |S_n - S_{n_{k-1}}|^2] < \infty,$$

by Theorem 2.1.3, Eq. (2.3.2) would follow. If this approach is to yield decent results, we must find a good upper bound for $E[\max_{n_{k-1} < n \leq n_k} |S_n - S_{n_{k-1}}|^2]$.

**Theorem 2.3.1.** (Fundamental maximal inequality for partial sums of orthogonal random variables.)

$$E[\max_{1 \leq i \leq n} S_i^2] \leq \left[\frac{\log(4n)}{\log 2}\right]^2 \sum_{i=1}^{n} EX_i^2 \qquad \text{for each} \quad n \geq 1.$$

**Remark.** Note that $S_i^2 \leq i \sum_{j=1}^{i} X_j^2$ follows from the Cauchy–Schwarz inequality. Hence

$$\max_{1 \leq i \leq n} S_i^2 \leq n \sum_{j=1}^{n} X_j^2$$

and, therefore

$$E[\max_{1 \leq i \leq n} S_i^2] \leq n \sum_{j=1}^{n} EX_j^2.$$

Theorem 2.3.1, due to Menchoff [1923], improves this primitive inequality considerably.

**Proof.** The proof is combinatorial and requires a rather forbidding notation. The first step consists in proving the inequality for $n = 2^k$, $k$ being an arbitrary positive integer. To avoid obscuring the basic idea we present the proof for $k = 6$ only. Let $X_{r,s} = \sum_{i=r+1}^{s} X_i$ for $0 \leq r < s \leq 2^6$. Consider the following collections of $X_{r,s}$:

$$\{X_{0,64}\}$$
$$\{X_{0,32}, X_{32,64}\}$$
$$\{X_{0,16}, X_{16,32}, X_{32,48}, X_{48,64}\}$$
$$\{X_{0,8}, \ldots, \qquad X_{56,64}\}$$
$$\{X_{0,4}, \ldots, \qquad X_{60,64}\}$$
$$\{X_{0,2}, \ldots, \qquad X_{62,64}\}$$
$$\{X_{0,1}, \ldots, \qquad X_{63,64}\}$$

There are $k + 1 = 7$ collections. Choose $1 \leq i \leq 2^6$. We expand $S_i$, choosing the terms of this expansion from the collections above and choosing the minimal possible number of terms in the expansion. It is clear that at most one term is required from each collection. As an example,

$$X_{0,61} = X_{0,32} + X_{32,48} + X_{48,56} + X_{56,60} + X_{60,61}.$$

Thus each expansion then has at most $k + 1 = 7$ terms in it. (Actually there are at most $k = 6$ terms, but it is convenient to ignore this.) Represent the expansion of $S_i$ by

$$S_i = \sum_{j=1}^{h} X_{i_{j-1}, i_j} \qquad (h \leq 7).$$

Then

$$S_i^2 \leq 7 \sum_{j=1}^{h} (X_{i_{j-1}, i_j})^2$$

follows by the Cauchy–Schwarz inequality. Now

$$\sum_{j=1}^{h} X_{i_{j-1}, i_j}^2 \leq X_{0,64}^2 + (X_{0,32}^2 + X_{32,64}^2) + (X_{0,16}^2 + X_{16,32}^2 + X_{32,48}^2$$
$$+ X_{48,64}^2) + \cdots + (X_{0,1}^2 + X_{1,2}^2 + \cdots + X_{63,64}^2).$$

Thus (the key step)

$$\max_{1 \leq i \leq 2^6} S_i^2 \leq 7[X_{0,64}^2 + (X_{0,32}^2 + X_{32,64}^2) + (X_{0,16}^2 + X_{16,32}^2$$
$$+ X_{32,48}^2 + X_{48,64}^2) + \cdots + (X_{0,1}^2 + X_{1,2}^2 + \cdots + X_{63,64}^2)].$$

There are $k + 1 = 7$ parenthetical expressions inside the square brackets. By orthogonality,

$$E[X_{0,64}^2] = E[X_{0,32}^2 + X_{32,64}^2] = \cdots$$
$$= E[X_{0,1}^2 + X_{1,2}^2 + \cdots + X_{63,64}^2] = \sum_{i=1}^{2^6} EX_i^2.$$

Hence

$$E[\max_{i \leq 2^6} S_i^2] \leq 7 \cdot 7 \sum_{i=1}^{2^6} EX_i^2.$$

Using a suitable notation, the above argument generalized to arbitrary $k \geq 1$ *yields*

$$E[\max_{i \leq 2^k} S_i^2] \leq (k + 1)^2 \sum_{i=1}^{2^k} EX_i^2. \qquad (2.3.3)$$

Given an $n$ such that $n \neq 2^k$ for every $k \geq 1$, choose $k$ such that $2^{k-1} < n < 2^k$ and redefine $X_i = 0$ if $n < i \leq 2^k$. By Eq. (2.3.3),

$$E[\max_{i \leq n} S_i^2] \leq (k + 1)^2 \sum_{i=1}^{n} EX_i^2$$

follows trivially. Since $2^{k-1} < n$ implies $(k + 1)^2 \leq [\log(4n)/\log 2]^2$, the desired result follows. ∎

**Exercise 2.3.1.** Develop a rigorous proof of Theorem 2.3.1 by replacing the argument for $k = 6$ by an argument for arbitrary $k$. Modify the proof slightly and show that $\log 4n$ can be replaced by $\log 2n$. ∎

We now state and prove the Rademacher–Menchoff fundamental convergence theorem for orthogonal random variables.

**Theorem 2.3.2.** Suppose

$$\sum_{n=1}^{\infty} (\log^2 n)EX_n^2 < \infty.$$

Then $S_n$ converges almost surely.

**Proof.** We use the method of subsequences. $\sum_{n=1}^{\infty} \log^2 nEX_n^2 < \infty$ implies that

$$\sum_{n=1}^{\infty} (\log n/\log 2)EX_n^2 < \infty.$$

Hence by Theorem 2.2.1, $S_{2^k}$ converges almost surely. By Lemma 2.3.1, it suffices to show that

$$\max_{2^{k-1} < n \leq 2^k} | S_n - S_{2^{k-1}} | \to 0 \qquad \text{a.s.}$$

as $k \to \infty$. By Theorem 2.1.3 it suffices to prove that

$$\sum_{k=1}^{\infty} E[\max_{2^{k-1} < n \leq 2^k} (S_n - S_{2^{k-1}})^2] < \infty.$$

By the fundamental maximal inequality

$$E[\max_{2^{k-1} < n \leq 2^k} (S_n - S_{2^{k-1}})^2] \leq \left\{ \frac{\log[4(2^k - 2^{k-1})]}{\log 2} \right\}^2 \sum_{i=2^{k-1}+1}^{2^k} EX_i^2$$

$$= (\log 2^{k+1}/\log 2)^2 \sum_{i=2^{k-1}+1}^{2^k} EX_i^2.$$

Thus it suffices to show that

$$\sum_{k=1}^{\infty} (\log 2^{k-1})^2 \sum_{i=2^{k-1}+1}^{2^k} EX_i^2 < \infty.$$

$$\sum_{k=1}^{\infty} (\log 2^{k-1})^2 \sum_{i=2^{k-1}+1}^{2^k} EX_i^2 \leq \sum_{k=1}^{\infty} \sum_{i=2^{k-1}+1}^{2^k} (\log i)^2 \, EX_i^2$$

$$= \sum_{i=2}^{\infty} (\log i)^2 EX_i^2 < \infty. \quad \blacksquare$$

The following theorem of Menchoff [1923] shows that Theorem 2.3.2 is sharp.

**Theorem 2.3.3.** If $\{b_n, n \geq 1\}$ is an increasing positive sequence such that $b_n = o(\log^2 n)$, then there exist orthogonal random variables $\{Y_i, i \geq 1\}$ such that

$$\sum_{n=1}^{\infty} b_n EY_n^2 < \infty$$

and $\sum_{i=1}^{\infty} Y_i$ diverges almost surely.

**Proof.** We omit the involved proof (see the work of Alexitz [1961, p. 88]).  ∎

**Exercise 2.3.2.** For many orthogonal systems (such as $X_i$ martingale differences) the maximal inequality of Theorem 2.3.1 can be improved to

$$E[\max_{m \leq i \leq n} (S_i - S_m)^2] \leq C \sum_{i=m+1}^{n} EX_i^2 \qquad \text{for} \quad n > m \geq 0. \tag{2.3.4}$$

Assuming Eq. (2.3.4) and $\sum_{i=1}^{\infty} EX_i^2 < \infty$ prove that there exists $S$ such that $S_n \to S$ a.s.  ∎

Theorem 2.3.2 is the basic result in the study of the almost everywhere convergence of arbitrary orthogonal series of real functions. Alexitz's "Convergence Problems of Orthogonal Series" [1961] is a good reference for this study. Garsia's "Topics in Almost Everywhere Convergence" [1970] is another good reference.

We indicate briefly the correspondence between Theorem 2.3.2 and this study. Let $\Omega$ be a Borel subset of $R_1$ ($\Omega$ is usually an interval or $R_1$ itself) and let $\mathscr{F}$ be the Borel subsets of $\Omega$. Let $\mu$ be a finite measure defined on $(\Omega, \mathscr{F})$ ($\mu$ is often a Lebesgue measure). Let $P = c\mu$ where $c$ is chosen to make $P$ a probability. Then measurable square integrable real valued functions $\{f_i, i \geq 1\}$ defined on $\Omega$ are said to be orthogonal if

$$\int_\Omega f_i(x)f_j(x)\,d\mu(x) = 0 \qquad \text{for all} \quad i \neq j.$$

An obvious translation of Theorem 2.3.2 yields:

**Corollary 2.3.1.** Let $\{f_i, i \geq 1\}$ be an orthogonal sequence of real functions defined on $\Omega$ with

$$\int_\Omega f_i^2(x)\,d\mu(x) = 1 \qquad \text{for} \quad i \geq 1$$

and let $c_n$ be constants. If

$$\sum_{n=1}^\infty c_n^2(\log n)^2 < \infty,$$

then $S_n = \sum_{i=1}^n c_i f_i$ converges almost surely; that is, almost everywhere with respect to $\mu$.

**Proof.** Using probabilistic notation, let $X_i = c_i f_i$. Then $EX_i^2 = c_i^2 Ef_i^2 = c_i^2$. Apply Theorem 2.3.2. ∎

As an example we apply Corollary 2.3.1 to the trigonometric system.

**Example 2.3.1.** [Alexitz, 1961]. Let $\Omega = [0, 1]$ and $\mu$ be a Lebesgue measure. Choose any sequence of positive integers $\{n_i, i \geq 2\}$ strictly increasing to infinity. Let

$$S_n = \sum_{i=2}^n \cos(2\pi n_i x)/[i(\log i)^{3+\varepsilon}]^{1/2}$$

for some $\varepsilon > 0$. Note that

$$\sum_{i=2}^\infty [i(\log i)^{3+\varepsilon}]^{-1}(\log i)^2 < \infty$$

and that

$$\{\cos 2\pi n_2 x, \cos 2\pi n_3 x, \ldots, \cos 2\pi n_i x, \ldots\}$$

is a sequence of orthogonal functions, with each function square integrable to 1. Hence, applying Corollary 2.3.1, $S_n$ converges almost surely. ∎

Note that in the above example none of the special properties of the trigonometric system other than orthogonality and square integrability to

1 were used. For given $a_i$ and $b_i$ let

$$S_n = \sum_{i=1}^{n} a_i \sin(2\pi ix) + \sum_{i=1}^{n} b_i \cos(2\pi ix)$$

for $n \geq 1$. If

$$\sum_{i=1}^{\infty} (a_i^2 + b_i^2) < \infty,$$

then by a very deep analysis Carleson [1966] proves that $S_n$ converges almost everywhere. This extremely important result includes and sharpens the above example. Garsia's monograph on almost everywhere convergence [1970, Chapter 4] looks at the problem of almost everywhere convergence of trigonometric series.

## 2.4. Convergence of $\{S_n, n \geq 1\}$ Assuming Only Moment Restrictions

Orthogonality was used in the proof of the fundamental convergence theorem (Theorem 2.3.2) in order to obtain a bound for $E(\sum_{i=r+1}^{s} X_i)^2$ in terms of the $EX_i^2$, namely

$$E\left( \sum_{i=r+1}^{s} X_i \right)^2 = \sum_{i=r+1}^{s} EX_i^2.$$

How essential is this particular bound; that is, how essential is orthogonality? Serfling [1970a,b] provides an interesting answer to this question and at the same time gives an alternate proof of the fundamental maximal inequality (Theorem 2.3.1). The results are somewhat involved to state; however, their wide applicability makes the effort worthwhile. In Section 2.4, the basic sequence is *not* assumed to be orthogonal.

For each $a \geq 0$ and $n \geq 1$ let $F_{a,n}$ be the joint distribution function of $X_{a+1}, \ldots, X_{a+n}$, that is,

$$F_{a,n}(x_1, x_2, \ldots, x_n) = P[X_{a+1} \leq x_1, X_{a+2} \leq x_2, \ldots, X_{a+n} \leq x_n]$$

for each $(x_1, x_2, \ldots, x_n) \in R_n$. For each $a \geq 0$ and $n \geq 1$ let

$$M_{a,n} = \max_{a < k \leq n} \left| \sum_{i=a+1}^{a+k} X_i \right|.$$

We will deal with nonnegative functionals defined on the collection of joint distribution functions. That is, $g$ is such a functional if $g$ is a function with the collection of joint distribution functions as its domain and $[0, \infty)$ as its range. For example, $g(F_{a,n}) = \sum_{i=a+1}^{a+n} EX_i^2$ is such a functional.

Our first step is to look at Serfling's generalization of the fundamental maximal inequality for orthogonal random variables [1970a].

**Theorem 2.4.1.** Suppose $g$ is a functional defined on the joint distribution functions such that

$$g(F_{a,k}) + g(F_{a+k,m}) \leq g(F_{a,k+m}) \tag{2.4.1}$$

for all $1 \leq k < k + m$ and $a \geq 0$,

$$E\left(\sum_{i=a+1}^{a+n} X_i\right)^2 \leq g(F_{a,n}) \tag{2.4.2}$$

for all $n \geq 1$ and $a \geq 0$. Then

$$EM_{a,n}^2 \leq [\log(2n)/\log 2]^2 g(F_{a,n}) \tag{2.4.3}$$

for all $n \geq 1$ and $a \geq 0$.

**Proof.** Fix $a \geq 0$. We proceed by induction on $n$. For $n = 1$, Eq. (2.4.3) is immediate by Eq. (2.4.2). Assume Eq. (2.4.3) holds for all $n < N$ and all $a \geq 0$, taking $N$ to be even. We seek an upper bound for $M_{a,N}^2$ and thus try to bound $(\sum_{i=a+1}^{a+n} X_i)^2$ for each $n \leq N$ as a first step. There are two cases: either $1 \leq n \leq N/2$ or $N/2 < n \leq N$.

If $1 \leq n \leq N/2$,

$$\left(\sum_{i=a+1}^{a+n} X_i\right)^2 \leq M_{a,N/2}^2.$$

If $N/2 < n \leq N$,

$$\left(\sum_{i=a+1}^{a+n} X_i\right)^2 = \left(\sum_{i=a+1}^{a+N/2} X_i + \sum_{i=a+N/2+1}^{a+n} X_i\right)^2$$

$$= \left(\sum_{i=a+1}^{a+N/2} X_i\right)^2 + 2\left(\sum_{i=a+1}^{a+N/2} X_i\right)\left(\sum_{i=a+N/2+1}^{a+n} X_i\right) + \left(\sum_{i=a+N/2+1}^{a+n} X_i\right)^2$$

$$\leq M_{a,N/2}^2 + 2\left|\sum_{i=a+1}^{a+N/2} X_i\right| M_{a+N/2,N/2} + M_{a+N/2,N/2}^2.$$

Hence

$$M_{a,N}^2 \leq M_{a,N/2}^2 + 2\left|\sum_{i=a+1}^{a+N/2} X_i\right| M_{a+N/2,N/2} + M_{a+N/2,N/2}^2.$$

Taking expectations on both sides and applying the induction hypotheses, we obtain

$$EM_{a,N}^2 \le (\log N/\log 2)^2 \, g(F_{a,N/2}) + 2E\left(\left|\sum_{i=a+1}^{a+N/2} X_i\right| M_{a+N/2,N/2}\right)$$
$$+ (\log N/\log 2)^2 \, g(F_{a+N/2,N/2}). \tag{2.4.4}$$

$$2E\left(\left|\sum_{i=a+1}^{a+N/2} X_i\right| M_{a+N/2,N}/2\right) \le 2E^{1/2}\left(\sum_{i=a+1}^{a+N/2} X_i\right)^2 E^{1/2}(M_{a+N/2,N/2}^2)$$
$$\le 2(\log N/\log 2)g^{1/2}(F_{a,N/2})g^{1/2}(F_{a+N/2,N/2})$$
$$\le (\log N/\log 2)[g(F_{a,N/2}) + g(F_{a+N/2,N/2})],$$

using the Cauchy–Schwarz inequality, Eq. (2.4.2), the induction hypotheses, and the elementary inequality $2xy \le x^2 + y^2$. Combining this with Eq. (2.4.4) yields

$$EM_{a,N}^2 \le \left[\frac{\log N}{\log 2} + \left(\frac{\log N}{\log 2}\right)^2\right][g(F_{a,N/2}) + g(F_{a+N/2,N/2})].$$

Using Eq. (2.4.1) and the fact that $(\log N)(\log 2) + (\log N)^2 \le (\log 2N)^2$, the desired result follows for $N$ even.

For $N$ odd, proceeding as above, and considering the two cases $1 \le n \le (N+1)/2$ and $(N+1)/2 < n \le N$ yields the desired result. Hence by induction the result holds for all $a \ge 0$ and $n \ge 1$. ∎

Note that taking

$$g(F_{a,k}) = \sum_{i=a+1}^{a+k} EX_i^2$$

shows that the fundamental maximal inequality (indeed a slight improvement of it) for orthogonal random variables is an immediate corollary of Theorem 2.4.1. A special case of Theorem 2.4.1 is due to Billingsley [1968, p. 102]. There, $g(F_{a,n}) = (\sum_{i=a+1}^{a+n} u_i)^\alpha$ for arbitrary $u_i \ge 0$ and $\alpha \ge 1$ is assumed in Eq. (2.4.2). Billingsley's Section 12 contains several other maximal equalities of interest.

Theorem 2.4.1 will now be used to obtain a widely applicable convergence result in the same manner in which the fundamental maximal inequality was used to derive the fundamental convergence theorem for orthogonal random variables.

**Theorem 2.4.2.** Let $g$ and $h$ be functionals defined on the joint distribution functions. Suppose Eq. (2.4.1) and Eq. (2.4.2) hold. (Recall that

$$g(F_{a,k}) + g(F_{a+k,m}) \le g(F_{a,k+m}) \tag{2.4.1}$$

for all $1 \leq k < k + m$ and $a \geq 0$ and

$$E\left(\sum_{i=a+1}^{a+n} X_i\right)^2 \leq g(F_{a,n}) \tag{2.4.2}$$

for all $n \geq 1$ and $a \geq 0$.) Suppose that, in addition,

$$h(F_{a,k}) + h(F_{a+k,m}) \leq h(F_{a,k+m}) \tag{2.4.5}$$

for all $1 \leq k < k + m$ and $a \geq 0$,

$$h(F_{a,n}) \leq K < \infty \tag{2.4.6}$$

for all $n \geq 1$ and $a \geq 0$, and

$$g(F_{a,n}) \leq K h(F_{a,n})/\log^2(a + 1) \tag{2.4.7}$$

for all $n \geq 1$ and $a > 0$.

Then $S_n$ converges almost surely.

**Proof.**   We use the method of subsequences. By Eqs. (2.4.2), (2.4.7), and (2.4.6),

$$E\left(\sum_{i=a+1}^{a+n} X_i\right)^2 \leq g(F_{a,n}) \leq \frac{K h(F_{a,n})}{\log^2(a + 1)} \leq \frac{K^2}{\log^2(a + 1)} \to 0$$

as $a \to \infty$. Hence $\{S_n, n \geq 1\}$ is a Cauchy sequence with respect to the $L_2$ norm. Hence, by the completeness of $L_2$, there exists a random variable $S$ with $ES^2 < \infty$ and $E(S_n - S)^2 \to 0$.

We show that $S_{2^k} \to S$ a.s. By the Chebyshev inequality,

$$P[|S - S_{2^k}| > \varepsilon] \leq E(S - S_{2^k})^2/\varepsilon^2.$$

Thus, (Theorem 2.1.3, again) it suffices to show that $\sum_{k=1}^{\infty} E(S - S_{2^k})^2 < \infty$.

$$E(S - S_{2^k})^2 = \lim_{n \to \infty} E(S_n - S_{2^k})^2$$

$$\leq \limsup_{n \to \infty} g(F_{2^k, n-2^k})$$

$$\leq K \limsup_{n \to \infty} h(F_{2^k, n-2^k})/\log^2(2^k + 1)$$

$$\leq K^2/\log^2(2^k + 1),$$

using Eqs. (2.4.2), (2.4.7), and (2.4.6), again. Thus

$$\sum_{k=1}^{\infty} E(S - S_{2^k})^2 \leq K^2 \sum_{k=1}^{\infty} \log^{-2}(2^k + 1) < \infty,$$

establishing $S_{2^k} \to S$ a.s.

By the method of subsequences (Lemma 2.3.1), it suffices to prove

$$\max_{2^{k-1} < n \leq 2^k} |S_n - S_{2^{k-1}}| \to 0 \qquad \text{a.s. as} \qquad k \to \infty$$

and hence to prove that

$$\sum_{k=1}^{\infty} E[\max_{2^{k-1} < n \leq 2^k} (S_n - S_{2^{k-1}})^2] < \infty.$$

By Eqs. (2.4.3) and (2.4.7),

$$\sum_{k=1}^{\infty} E[\max_{2^{k-1} < n \leq 2^k} (S_n - S_{2^{k-1}})^2] \leq \sum_{k=1}^{\infty} \left(\frac{\log 2^k}{\log 2}\right)^2 g(F_{2^{k-1}, 2^{k-1}})$$

$$\leq K \sum_{k=1}^{\infty} \left(\frac{\log 2^k}{\log 2}\right)^2 \frac{h(F_{2^{k-1}, 2^{k-1}})}{\log^2(2^{k-1} + 1)}$$

$$\leq KC \sum_{k=1}^{\infty} h(F_{2^{k-1}, 2^{k-1}}) \quad \text{for some } C < \infty.$$

By Eq. (2.4.5) and (2.4.6),

$$\sum_{k=1}^{\infty} h(F_{2^{k-1}, 2^{k-1}}) \leq K < \infty,$$

establishing the theorem. ∎

*Exercise 2.4.1.* Show that the fundamental convergence theorem for orthogonal series is a special case of Theorem 2.4.2. ∎

We now consider an application of Theorem 2.4.2. Recall that the correlation of two random variables $X$ and $Y$ is defined by

$$\text{cor}(X, Y) = \text{cov}(X, Y)/\{(\text{var } X)^{1/2}(\text{var } Y)^{1/2}\},$$

where $\text{cov}(X, Y)$ is the covariance of $X$ and $Y$. Thus, $-1 \leq \text{cor}(X, Y) \leq 1$ is just a restatement of the Cauchy–Schwarz inequality. In an introductory probability course examples are given and theorems are established which build the intuitive idea that correlation measures the degree of linear dependence. In particular, if $|\,\text{cor}(X, Y)\,| = 1$, then $X$ and $Y$ can be shown to be linearly related in a deterministic manner, that is, there exists $a \neq 0$ and $b$ real such that $Y = aX + b$ a.s. If $0 < \text{cor}(X, Y) < 1$, then there exists a line of strictly positive slope such that $(X, Y)$ has a joint distribution which tends to concentrate its mass around the line. The closer $\text{cor}(X, Y)$ is to 1, the more pronounced is this concentration of mass. If $-1 < \text{cor}(X, Y) < 0$, the same remarks apply to a line with strictly negative slope. If $\text{cor}(X, Y) = 0$, such as when $X$ and $Y$ are independent, then there does not exist a line with nonzero slope having the above property. These remarks perhaps help in interpreting the following corollary.

**Corollary 2.4.1.** Suppose there exists nonnegative constants $\{\varrho_i, i \geq 0\}$ such that $0 \leq \varrho_i \leq 1$ and

$$E[X_i X_j] \leq \varrho_{j-i}(EX_i^2 EX_j^2)^{1/2} \qquad \text{for all} \quad j \geq i > 0, \qquad (2.4.8)$$

$$\sum_{i=0}^{\infty} \varrho_i < \infty, \qquad (2.4.9)$$

and

$$\sum_{i=1}^{\infty} (\log^2 i) EX_i^2 < \infty. \qquad (2.4.10)$$

Then $S_n$ converges almost surely.

**Remark.** Note that $\varrho_{j-i}$ is merely an upper bound for the correlation between $X_i$ and $X_j$ if we assume, in addition, that $EX_i = 0$ for $i \geq 1$. Thus, when $EX_i = 0$ for $i \geq 1$, Eqs. (2.4.8) and (2.4.9) imply that random variables far apart in the basic sequence exhibit very little correlation.

**Proof.** We apply Theorem 2.4.2 by finding functionals $g$ and $h$ satisfying Eqs. (2.4.1), (2.4.2), (2.4.5), (2.4.6), and (2.4.7). Fix $a \geq 0$ and $n \geq 1$. By estimating $E(\sum_{i=a+1}^{a+n} X_i)^2$ first, we motivate the choice of $g$.

$$E\left( \sum_{i=a+1}^{a+n} X_i \right)^2 = E\left[ \sum_{i=a+1}^{a+n} X_i^2 + 2 \sum_{a+1 \leq i < j \leq a+n} X_i X_j \right]$$

$$\leq \sum_{i=a+1}^{a+n} EX_i^2 + 2 \sum_{a+1 \leq i < j \leq a+n} \varrho_{j-i} E^{1/2} X_i^2 E^{1/2} X_j^2$$

$$= \sum_{i=a+1}^{a+n} EX_i^2 + 2 \sum_{i=a+1}^{a+n-1} \sum_{j=i+1}^{a+n} \varrho_{j-i} E^{1/2} X_i^2 E^{1/2} X_j^2$$

by Eq. (2.4.8).

Rearranging terms, we obtain

$$2 \sum_{i=a+1}^{a+n-1} \sum_{j=i+1}^{a+n} \varrho_{j-i} E^{1/2} X_i^2 E^{1/2} X_j^2 = 2 \sum_{k=1}^{n-1} \varrho_k \sum_{i=a+1}^{a+n-k} E^{1/2} X_i^2 E^{1/2} X_{k+i}^2$$

$$\leq \sum_{k=1}^{n-1} \varrho_k \sum_{i=a+1}^{a+n-k} (EX_i^2 + EX_{k+i}^2)$$

$$\leq 2 \sum_{k=1}^{n-1} \varrho_k \sum_{i=a+1}^{a+n} EX_i^2.$$

Combining,

$$E\left( \sum_{i=a+1}^{a+n} X_i \right)^2 \leq 2\left( \sum_{i=a+1}^{a+n} EX_i^2 \right)\left( 1 + \sum_{k=1}^{n-1} \varrho_k \right)$$

$$\leq 2\left( \sum_{i=a+1}^{a+n} EX_i^2 \right)\left( \sum_{k=0}^{n} \varrho_k \right).$$

Motivated by this, we define

$$g(F_{a,n}) = 2\left( \sum_{i=a+1}^{a+n} EX_i^2 \right)\left( \sum_{k=0}^{n} \varrho_k \right).$$

Let

$$h(F_{a,n}) = 2 \sum_{i=a+1}^{a+n} (\log^2 i) EX_i^2 \left( \sum_{k=0}^{n} \varrho_k \right).$$

(As a first choice for $h$, Eq. (2.4.7) suggests $h(F_{a,n}) = g(F_{a,n}) \log^2(a+1)$. But then Eq. (2.4.5) may not be satisfied. As defined above, $h$ is one obvious way to modify this first choice so that Eq. (2.4.5) is satisfied and Eq. (2.4.6) and (2.4.7) remain true.) Equations (2.4.1), (2.4.2), (2.4.5), (2.4.6), and (2.4.7) are easily verified. Thus, the corollary is established by the application of Theorem 2.4.2. ∎

**Example 2.4.1.**  Let $\{X_i, i \geq 1\}$ be jointly normally distributed with

$$EX_i = 0 \qquad \text{and} \qquad EX_i^2 = (i \log^4 i)^{-1}$$

for $i \geq 1$ and

$$EX_i X_j / (EX_i^2 EX_j^2)^{1/2} = e^{-|i-j|}$$

for all $i \neq j$. It can be shown that such a distribution for a stochastic sequence is possible since $e^{-|u|}$ is a nonnegative definite function. Then, by Corollary 2.4.1, $S_n$ converges almost surely since

$$\sum_{i=0}^{\infty} e^{-|i|} < \infty \qquad \text{and} \qquad \sum_{i=1}^{\infty} \log^2 i (i \log^4 i)^{-1} < \infty.$$

The assumption of normality here is used only to guarantee the existence of the stochastic sequence. Almost sure convergence follows for any stochastic sequence with the first and second moment structure of the example. Since the $X_i$ do not satisfy any of the standard dependence assumptions (independence, orthogonality, etc.) a result like Corollary 2.4.1 is useful in this example.  ▌

## 2.5.  Martingale Difference Sequences

The next topic to be studied is the convergence of $S_n$ when the basic sequence is a martingale difference sequence. The martingale difference sequences with finite second moments form a subclass of the orthogonal sequences that we have studied in Sections 2.2–2.4. A martingale difference sequence can be viewed as a mathematical model for a game of chance that is a fair game in a certain precise way. By game we mean a sequence of trials for which money is won or lost on each trial.

**Definition 2.5.1.**   Let $\{Y_n, n \geq 1\}$ be a stochastic sequence (that is, a sequence of random variables) and $\{\mathscr{F}_n, n \geq 1\}$ an increasing sequence of $\sigma$ fields with $\mathscr{F}_n \subset \mathscr{F}$ for each $n \geq 1$. If $Y_n$ is $\mathscr{F}_n$ measurable for each $n \geq 1$ the $\sigma$ fields $\{\mathscr{F}_n, n \geq 1\}$ are said to be adapted to the sequence $\{Y_n, n \geq 1\}$ and $\{Y_n, \mathscr{F}_n, n \geq 1\}$ is said to be an adapted stochastic sequence.   ▌

Throughout the book $\mathscr{F}_0$ will denote the trivial $\sigma$ field $(\varnothing, \Omega)$.

**Definition 2.5.2.**   If $\{Y_n, \mathscr{F}_n, n \geq 1\}$ is an adapted stochastic sequence with

$$E[Y_n \mid \mathscr{F}_{n-1}] = 0 \qquad \text{a.s. for each } \ n \geq 2,$$

then $\{Y_n, \mathscr{F}_n, n \geq 1\}$ is called a martingale difference sequence.   ▌

Equivalently, an adapted stochastic sequence $\{Y_n, \mathscr{F}_n, n \geq 1\}$ is a martingale difference sequence if

$$\int_B Y_n \, dP = 0 \qquad \text{for each } \ B \in \mathscr{F}_{n-1} \ \text{ and } \ n \geq 2.$$

This equivalence is immediate from the definition of conditional expectation.

**Exercise 2.5.1.**  Let $\{Y_n, \mathscr{F}_n, n \geq 1\}$ be a martingale difference sequence and let $U_n = \sum_{i=1}^{n} Y_i$ for $n \geq 1$.

(i)  Show that

$$E[U_n \mid \mathscr{F}_{n-1}] = U_n \qquad \text{a.s. for each} \quad n \geq 2. \qquad (2.5.1)$$

(ii)  Suppose also that $EY_i^2 < \infty$ for each $i \geq 1$ and show that $\{Y_i, \mathscr{F}_i, i \geq 1\}$ is an orthogonal sequence. (Thus the square integrable martingale difference sequences form a subclass of the orthogonal sequences as remarked in the first paragraph of Section 2.5.) ∎

**Definition 2.5.3.**  An adapted sequence $\{U_n, \mathscr{F}_n, n \geq 1\}$ satisfying Eq. (2.5.1) is called a martingale. ∎

**Exercise 2.5.2.**  Let $\{U_n, \mathscr{F}_n, n \geq 1\}$ be a martingale. Let $U_0 = 0$ and $Y_n = U_n - U_{n-1}$ for $n \geq 1$. Show that $\{Y_n, \mathscr{F}_n, n \geq 1\}$ is a martingale difference sequence. ∎

Whether results are stated for martingales or for martingale differences is a matter of convenience. In stating results about martingales or martingale differences, it is sometimes convenient to suppress any mention of $\sigma$ fields. Thus the statement that $\{Y_i, i \geq 1\}$ is a martingale difference sequence means that for some choice of $\sigma$ fields $\{\mathscr{F}_i, i \geq 1\}$, $\{Y_i, \mathscr{F}_i, i \geq 1\}$ is a martingale difference sequence. Suppose that $\{Y_i, \mathscr{F}_i, i \geq 1\}$ is a martingale difference sequence. Letting $\mathscr{G}_i$ be the $\sigma$ field generated by $\{Y_1, Y_2, \ldots, Y_i\}$ for each $i \geq 1$, it follows that

$$E[Y_{i+1} \mid \mathscr{G}_i] = E\{E[Y_{i+1} \mid \mathscr{F}_i] \mid \mathscr{G}_i\} = 0 \qquad \text{a.s. for each} \quad i \geq 2.$$

Thus $\{Y_i, \mathscr{F}_i, i \geq 1\}$ a martingale difference sequence always implies that $\{Y_i, \mathscr{G}_i, i \geq 1\}$ is a martingale difference sequence.

Let $\{Y_n, n \geq 1\}$ be a stochastic sequence meant to serve as a mathematical model for a game of chance. That is, $Y_n$ represents the amount of money won on trial $n$. One way of defining the game as fair is to assume

$$E[Y_n \mid Y_{n-1}, Y_{n-2}, \ldots, Y_1] = 0 \qquad \text{a.s.}$$

for each $n \geq 2$ and that $EY_1 = 0$. That is, for each trial $n \geq 1$ the expected amount won on trial $n$ is zero *regardless* of the past evolution of the sequence. Let $\mathscr{G}_n$ be the $\sigma$ field generated by $\{Y_1, Y_2, \ldots, Y_n\}$ for each $n \geq 1$. Then the above definition of a fair game merely says that

$$\{Y_n, \mathscr{G}_n, n \geq 1\}$$

is a martingale difference sequence with $EY_1 = 0$. It is an interesting philosophical and practical question as to whether the martingale difference assumption is really a good description of fairness. We will not discuss this. Thinking of a martingale difference sequence as a mathematical description of a fair game is certainly an intuitive aid to the study of martingale theory. It provides us with many examples of martingale difference sequences.

**Example 2.5.1.** Consider repeated tosses of a fair coin. If a gambler bets $x$ dollars on trial $n$, he wins $x$ dollars if the coin is heads and $-x$ dollars if the coin comes up tails. Suppose the gambler has infinite capital. The gambler's strategy is to bet 1 dollar on trial 1 and for each $n \geq 1$, to double his bet on trial $n + 1$ if he loses on trial $n$ and to quit at trial $n$ (that is, bet 0 dollars on trial $n + 1$ and all succeeding trials) if he wins on trial $n$. Letting $\{Y_n, n \geq 1\}$ be the sequence of his winnings, we clearly have

$$P[Y_n \in A \mid Y_{n-1}, Y_{n-2}, \ldots, Y_1] = P[Y_n \in A \mid Y_{n-1}] \qquad \text{a.s.}$$

for each $n \geq 2$ (that is, $\{Y_n, n \geq 1\}$ is a Markov sequence—a dependence structure to be studied in detail later). Also

$$P[Y_n = 2^{n-1} \mid Y_{n-1} = -2^{n-2}] = \tfrac{1}{2}$$

and

$$P[Y_n = -2^{n-1} \mid Y_{n-1} = -2^{n-2}] = \tfrac{1}{2}$$

for each $n \geq 2$. Further

$$P[Y_n = 0 \mid Y_{n-1} = 0] = P[Y_n = 0 \mid Y_{n-1} = 2^{n-2}] = 1$$

for each $n \geq 2$. Let $\mathscr{G}_n$ be the $\sigma$ field generated by $\{Y_1, Y_2, \ldots, Y_n\}$ for each $n \geq 1$. It is immediate that $\{Y_n, \mathscr{G}_n, n \geq 1\}$ is a martingale difference sequence. Of course, the gambler's fate is inevitable since with probability one he must eventually flip a head with probability one. Thus he will go home one dollar richer than he came. Thus the gambler wins one dollar with probability one by playing this "fair" game. ▮

It is interesting to note that (as remarked to the author by R. J. Tomkins) in nonmathematical usage a "martingale" denotes the precise gambling system of this example (see, for example, Webster's New Collegiate Dictionary).

Suppose in Example 2.5.1 that both the gambler and his adversary (the "house") have infinite capital. Let $\{Z_i, i \geq 1\}$ denote the gambler's winnings with a new strategy substituted for that of the example. It is interesting to ask whether there exists a strategy for the gambler which guarantees that $\sum_{i=1}^{n} Y_i \to \infty$ a.s. We will see (Theorem 2.10.4) that the answer is no.

**Exercise 2.5.3.**  Consider the game of Example 2.5.1. Fix an integer $n \geq 1$. Under the assumption of infinite capital for both the gambler and the house, devise a strategy for the gambler that guarantees that he goes home $n$ dollars richer. Devise a strategy such that his expected take home winnings are infinite. ▮

The martingale assumption is rather weak. Thus there are many and diverse examples of martingales and as a result martingale theory is widely applied, extending far beyond the modeling of fair games. We mention several examples to give some feeling for this diversity.

One obvious but important example is that a sequence of independent mean zero random variables $\{Y_i, i \geq 1\}$ forms a martingale difference sequence $\{Y_i, \mathcal{G}_i, i \geq 1\}$, $\mathcal{G}_i$ denoting the $\sigma$ field generated by $\{Y_1, Y_2, \ldots, Y_i\}$ for each $i \geq 1$.

As will be developed later in Chapter 2, the partial sums of a Haar series form a martingale.

Let $\{Y_i, \mathcal{F}_i, n \geq 1\}$ be any adapted stochastic sequence with $E \mid Y_i \mid < \infty$ for each $i \geq 1$. Then

$$\{Y_i - E[Y_i \mid \mathcal{F}_{i-1}], \mathcal{F}_{i-1}, i \geq 1\}$$

is trivially a martingale difference sequence. This centering device of P. Levy is sometimes very useful. It essentially reduces the study of the behavior of an arbitrary stochastic sequence with finite first moments to the study of the behavior of a martingale difference sequence and the study of the behavior of a sequence of conditional expectations.

Lebesgue's theory of differentiation can be based on the theory of martingale convergence:

**Example 2.5.2.**  Let $0 = x_0^{(n)} < x_1^{(n)} < \cdots < x_n^{(n)} = 1$ define a set of increasingly finer partitions becoming dense in $[0, 1]$. Let $f(\cdot)$ be a measurable real valued function of bounded variation defined on $[0, 1]$. Let

$$A_1^{(n)} = [0, x_1^{(n)}]$$

and

$$A_i^{(n)} = (x_{i-1}^{(n)}, x_i^{(n)}] \qquad \text{for} \quad 2 \leq i \leq n \quad \text{and each} \quad n \geq 1.$$

For each $n \geq 1$ and $0 \leq x \leq 1$ let

$$T_n(x) = \frac{f(x_i^{(n)}) - f(x_{i-1}^{(n)})}{x_i^{(n)} - x_{i-1}^{(n)}} \qquad \text{when} \quad x \in A_i^{(n)}$$

for some $1 \leq i \leq n$. For each $n \geq 1$, let $\mathscr{F}_n$ be the $\sigma$ field generated by unions of the $A_i^{(n)}$. Let $\Omega = [0, 1]$, $\mathscr{F}$ be the Borel sets of $[0, 1]$, and $P$ be a Lebesgue measure. A notationally messy but simple calculation shows that

$$\int_A T_n(x)\, dx = \int_A T_{n-1}(x)\, dx$$

for each $A \in \mathscr{F}_{n-1}$ and each $n \geq 2$. Thus $\{T_n, \mathscr{F}_n, n \geq 1\}$ is a martingale. $T_n$ can be shown to converge almost surely. This result is helpful in establishing the existence of the derivative of $f$ almost everywhere. ∎

**Example 2.5.3.**   Let $T$ be an integrable random variable and $\{\mathscr{F}_n, n \geq 1\}$ an increasing sequence of $\sigma$ fields with $\mathscr{F}_n \subset \mathscr{F}$ for each $n \geq 1$. Let

$$T_n = E[T \mid \mathscr{F}_n]$$

for each $n \geq 1$. Since

$$E[T_n \mid \mathscr{F}_{n-1}] = E\{E[T \mid \mathscr{F}_n] \mid \mathscr{F}_{n-1}\} = T_{n-1} \qquad \text{a.s.}$$

for $n \geq 2$, $\{T_n, \mathscr{F}_n, n \geq 1\}$ is a martingale. Let $\ldots, \mathscr{F}_n, \mathscr{F}_{n+1}, \ldots, \mathscr{F}_{-1}$ be an increasing sequence of $\sigma$ fields with $\mathscr{F}_n \subset \mathscr{F}$ for each integer $n \leq -1$. Let

$$T_n = E[T \mid \mathscr{F}_n]$$

for each $n \leq -1$. Then $\{T_n, \mathscr{F}_n, n \leq -1\}$ is easily seen to be a martingale. (Letting $U_n = T_{-n}$ and $\mathscr{G}_n = \mathscr{F}_{-n}$ for each $n \leq -1$, $(U_m, m \geq 1)$ is an example of a "reverse martingale"; that is, a martingale in the ordinary sense provided the index set $\{m : m \geq 1\}$ is ordered in reverse from the usual ordering.) ∎

**Example 2.5.4.**   (The Polya urn scheme).   Let an urn contain $b$ black and $r$ red balls. Let $T_0 = b/(b + r)$. At each drawing a ball is drawn at random, its color is noted and $a$ balls of that color are added to the urn. Let $b_n$ be the number of black balls and $a_n$ be the number of red balls after the $n$th drawing. Let $T_n$ be the proportion of black balls after the $n$th drawing.

$$E[T_n \mid T_{n-1}, \ldots, T_1] = E[T_n \mid T_{n-1}] \qquad \text{a.s.}$$

$$E\left[T_n \mid T_{n-1} = \frac{b_{n-1}}{b_{n-1} + r_{n-1}}\right]$$

$$= \frac{b_{n-1} + a}{b_{n-1} + a + r_{n-1}} \frac{b_{n-1}}{b_{n-1} + r_{n-1}} + \frac{b_{n-1}}{b_{n-1} + r_{n-1} + a} \frac{r_{n-1}}{.b_{n-1} + r_{n-1}}$$

$$= \frac{b_{n-1}}{b_{n-1} + r_{n-1}} = T_{n-1}.$$

Thus $\{T_n, n \geq 1\}$ is a martingale.  ∎

**Example 2.5.5.**  (Likelihood ratio).  Let $\{Y_i, i \geq 1\}$ be a stochastic sequence for which the joint density of $\{Y_1, Y_2, \ldots, Y_n\}$ is either the function $p_n$ for every $n \geq 1$ or is the function $q_n$ for every $n \geq 1$. The statistical problem is to decide whether $\{p_n, n \geq 1\}$ or $\{q_n, n \geq 1\}$ is the correct set of densities. Let

$$T_n = q_n(Y_1, Y_2, \ldots, Y_n)/p_n(Y_1, Y_2, \ldots, Y_n)$$

for each $n \geq 1$. The $T_n$ are known as likelihood ratios. If the $p_n$ are correct, it seems plausible that the observed $Y_i$ will, when substituted into $T_n$, make $T_n$ approach zero as $n$ gets large. Likewise, if the $q_n$ are correct, it seems plausible that the observed $Y_i$ will make $T_n$ approach infinity as $n$ gets large. In this way, $T_n$, for large $n$, can be observed to decide whether the $p_n$ or the $q_n$ seem to be the correct densities. Suppose, in order to avoid problems, that the $p_n$ are strictly positive. For each $n \geq 1$, let $\mathscr{G}_n$ be the $\sigma$ field generated by $\{Y_1, Y_2, \ldots, Y_n\}$. Then under the assumption that the $p_n$ are the correct densities, $\{T_n, \mathscr{G}_n, n \geq 1\}$ can be shown to be a martingale:

$$E[T_n \mid Y_{n-1} = y_{n-1}, Y_{n-2} = y_{n-2}, \ldots, Y_1 = y_1]$$

$$= \int_{-\infty}^{\infty} \frac{q_n(y_1, y_2, \ldots, y_{n-1}, y)}{p_n(y_1, y_2, \ldots, y_{n-1}, y)} \frac{p_n(y_1, y_2, \ldots, y_{n-1}, y)}{p_{n-1}(y_1, y_2, \ldots, y_{n-1})} \, dy$$

$$= \frac{q_{n-1}(y_1, y_2, \ldots, y_{n-1})}{p_{n-1}(y_1, y_2, \ldots, y_{n-1})}$$

by cancellation and integration. Thus $E[T_n \mid Y_{n-1}, Y_{n-2}, \ldots, Y_1] = T_{n-1}$ a.s., showing that $\{T_n, \mathscr{G}_n, n \geq 1\}$ is a martingale. This fact is helpful in studying the behavior of $T_n$. For further discussion, see the work of Doob [1953, p. 348]. ∎

Consideration of games which are possibly favorable instead of merely fair suggests the following:

**Definition 2.5.4.** An adapted stochastic sequence $\{Y_n, \mathscr{F}_n, n \geq 1\}$ is said to be a submartingale difference sequence if for each $n \geq 2$,

$$E[Y_n \mid \mathscr{F}_{n-1}] \geq 0 \qquad \text{a.s.}$$

An adapted stochastic sequence $\{T_n, \mathscr{F}_n, n \geq 1\}$ is called a submartingale if for each $n \geq 2$

$$E[T_n \mid \mathscr{F}_{n-1}] \geq T_{n-1} \qquad \text{a.s.} \quad ∎$$

Note that a martingale difference sequence is trivially a submartingale difference sequence and that a martingale is trivially a submartingale. Note also that $\{T_n, n \geq 1\}$ defined by $T_n = \sum_{i=1}^{n} Y_i$ for each $n \geq 1$ is a submartingale if and only if $\{Y_i, i \geq 1\}$ is a submartingale difference sequence.

If $\{Y_i, i \geq 1\}$ is a positive sequence of random variables (that is, $P[Y_i \geq 0] = 1$ for each $i \geq 1$) then, trivially, $\{Y_i, i \geq 1\}$ is a submartingale difference sequence. If $\{Y_i, i \geq 1\}$ are independent with $EY_i \geq 0$ for each $i \geq 1$, then $\{Y_i, i \geq 1\}$ is a submartingale difference sequence.

## 2.6.  Preservation of the Martingale and Submartingale Structure

We will study the effect of three types of transformations on martingales and submartingales. Martingales will be transformed into martingales or, at least, into submartingales. Submartingales will be transformed into submartingales.

Transformation by a convex function is considered first.

**Theorem 2.6.1.**  (i)  If $\Phi$ is a continuous convex nondecreasing real valued function defined on $R_1$ and $\{U_n, n \geq 1\}$ is a submartingale with $E \mid \Phi(U_n) \mid < \infty$ for each $n \geq 1$, then $\{\Phi(U_n), n \geq 1\}$ is a submartingale.

(ii)  If $\Phi$ is a continuous convex real valued function defined on $R_1$ and $\{U_n, n \geq 1\}$ is a martingale with $E \mid \Phi(U_n) \mid < \infty$ for each $n \geq 1$, then $\{\Phi(U_n), n \geq 1\}$ is a submartingale.

**Proof.** (i) Fix $n \geq 2$. According to the conditional version of the convexity inequality for expectations, $E[\Phi(U_n) \mid \mathscr{F}_{n-1}] \geq \Phi(E[U_n \mid \mathscr{F}_{n-1}])$ a.s. Using the submartingale assumption and the fact that $\Phi$ is nondecreasing, the desired result follows.

The proof of (ii) is essentially the same and omitted. ∎

**Corollary 2.6.1.** (i)   Fix $p \geq 1$. If $\{U_n, n \geq 1\}$ is a martingale with $E \mid U_n \mid^p < \infty$ for $n \geq 1$ then $\{\mid U_n \mid^p, n \geq 1\}$ is a submartingale.

(ii) If $\{U_n, n \geq 1\}$ is a submartingale then $\{U_n^+, n \geq 1\}$ is a submartingale.

**Proof.** Immediate from Theorem 2.6.1. ∎

If $\{U_n, n \geq 1\}$ is a martingale with $U_n$ representing a gambler's accumulated winnings at trial $n$ of a nonending game, we might wish to build into the mathematical structure the possibility of the gambler stopping play at some finite time $t$ with winnings $U_t$. This *optional stopping* preserves the martingale property in a way made precise below.

**Definition 2.6.1.** Let $\{\mathscr{F}_n, n \geq 1\}$ be an increasing sequence of $\sigma$ fields with $\mathscr{F}_n \subset \mathscr{F}$ for each $n \geq 1$. A function $t$, defined on $\Omega$, is called a stopping rule if $t$ has range $\{1, 2, \ldots, n, \ldots, \infty\}$, $t$ is measurable, and

$$(t = n) \in \mathscr{F}_n \qquad \text{for each} \quad 1 \leq n < \infty.$$

(Occasionally 0 will be adjoined to the range of $t$ with $(t = 0) \in \mathscr{F}_0$ required.) ∎

**Remark.** Recall that a random variable must be finite almost surely. Since $P[t = \infty] > 0$ is allowed, $t$ thus may not be a random variable. The term "stopping rule" is not universally used. Some authors use "stopping time," for example. Whether $P[t = \infty] > 0$ is permitted varies from paper to paper and has to be closely watched.

Given a stopping rule $t$ and a martingale $\{U_n, \mathscr{F}_n, n \geq 1\}$, let

$$U_n^{(t)}(\omega) = U_{\min(t(\omega), n)}(\omega)$$

define the sequence $\{U_n^{(t)}, n \geq 1\}$. That is, for each $n \geq 1$, on the set $(t = n)$ the defined sequence is

$$U_1, U_2, \ldots, U_{n-1}, U_n, U_n, \ldots.$$

Let $\mathscr{F}_n^{(t)} = \mathscr{B}(U_i^{(t)}, i \leq n)$. Here and throughout the book $\mathscr{B}(X_\alpha, \alpha \in A)$ denotes the $\sigma$ field generated by $\{X_\alpha, \alpha \in A\}$.

**Exercise 2.6.1.** Prove that $\{U_n^{(t)}, n \geq 1\}$ is a stochastic sequence. ∎

It is now convenient to introduce a new notation. Throughout the book for a given $A \in \mathscr{F}$, the indicator random variable $I(A)$ is defined by

$$I(A) = \begin{cases} 1 & \text{if} \quad \omega \in A \\ 0 & \text{if} \quad \omega \notin A. \end{cases}$$

One of the most useful tools in establishing convergence results for martingales is the fact that optional stopping preserves the martingale structure.

**Theorem 2.6.2.** (i) If $(U_n, \mathscr{F}_n, n \geq 1)$ is a martingale and $t$ a stopping rule, then $\{U_n^{(t)}, \mathscr{F}_n^{(t)}, n \geq 1\}$ is a martingale.

(ii) If $(U_n, \mathscr{F}_n, n \geq 1)$ is a submartingale and $t$ is a stopping rule, then $\{U_n^{(t)}, \mathscr{F}_n^{(t)}, n \geq 1\}$ is a submartingale.

**Proof.** We prove (i) only, the proof of (ii) being almost the same. We first must show that $E \mid U_n^{(t)} \mid < \infty$ for each $n \geq 1$. Let $Y_i = U_i - U_{i-1}$ for $i \geq 1$ with $U_0 = 0$.

$$\mid U_n^{(t)} \mid = \left| \sum_{i=1}^{n} Y_i I(t \geq i) \right| \leq \sum_{i=1}^{n} \mid Y_i \mid.$$

Since $E \mid Y_i \mid < \infty$ for each $i \geq 1$, $E \mid U_n^{(t)} \mid < \infty$ for each $n \geq 1$. In order to establish that $\{U_n^{(t)}, n \geq 1\}$ is a martingale, it must be shown that

$$\int_B U_n^{(t)} dP = \int_B U_{n-1}^{(t)} dP$$

for each $B \in \mathscr{F}_{n-1}^{(t)}$ and each $n \geq 2$. Choose such a $B$ and $n$.

$$\int_B U_n^{(t)} dP = \int_{B \cap (t \geq n)} U_n^{(t)} dP + \int_{B \cap (t < n)} U_n^{(t)} dP$$

$$= \int_{B \cap (t \geq n)} U_n dP + \int_{B \cap (t < n)} U_{n-1}^{(t)} dP \qquad (2.6.1)$$

since $U_n = U_n^{(t)}$ on $t \geq n$ and $U_n^{(t)} = U_{n-1}^{(t)}$ on $t < n$. On the event $[t \geq n]$,

$$(U_1^{(t)}, U_2^{(t)}, \ldots, U_{n-1}^{(t)}) = (U_1, U_2, \ldots, U_{n-1}).$$

Hence, there exists an $(n-1)$-dimensional Borel set $C$ such that $B \cap [t \geq n]$

$= [(U_1, U_2, \ldots, U_{n-1}) \in C] \cap [t \geq n]. \ (t \geq n) \in \mathscr{F}_{n-1}.$ Hence $B \cap (t \geq n)$
$\in \mathscr{F}_{n-1}.$ Thus, by the martingale assumption,

$$\int_{B \cap (t \geq n)} U_n \, dP = \int_{B \cap (t \geq n)} U_{n-1} \, dP.$$

Combining this with Eq. (2.6.1) yields

$$\int_B U_n^{(t)} \, dP = \int_{B \cap (t \geq n)} U_{n-1} \, dP + \int_{B \cap (t < n)} U_{n-1}^{(t)} \, dP$$

$$= \int_{B \cap (t \geq n)} U_{n-1}^{(t)} \, dP + \int_{B \cap (t < n)} U_{n-1}^{(t)} \, dP$$

$$= \int_B U_{n-1}^{(t)} \, dP,$$

establishing the theorem. ∎

**Remarks.**  If $\{U_n, n \geq 1\}$ is a martingale and $t$ is a stopping rule, then

$$EU_1 = EU_n = EU_n^{(t)} \qquad \text{for each} \quad n \geq 1$$

follows from Theorem 2.6.2 (i) since $U_1^{(t)} = U_1$ a.s. However, even if $t$ is almost surely finite it is *not* necessarily true that $EU_1 = EU_t$. This misunderstanding has sometimes led to serious errors. As a simple example, let $\{Y_i, i \geq 1\}$ be independent identically distributed with $P[Y_1 = 1] = P[Y_1 = -1] = \frac{1}{2}$ (coin tossing model). Let $U_n = \sum_{i=1}^n Y_i$ for each $n \geq 1$ define the martingale $\{U_n, n \geq 1\}$. Let $t$ be the first integer $n \geq 1$ such that $U_n \geq 1$. Clearly $EU_t = 1$ yet $EU_1 = 0$. Interestingly $EU_n^{(t)} = 0$ for all $n \geq 1$.

**Exercise 2.6.2.**  Let $\{Y_i, \mathscr{F}_i, i \geq 1\}$ be a martingale difference sequence with $EY_1 = 0$ and $t$ be a stopping rule satisfying $t \leq M$ a.s. for some $M < \infty$. Prove that

$$E \sum_{i=1}^t Y_i = 0.$$

($Et < \infty$ is actually sufficient for the result to hold.) ∎

Besides the optional stopping discussed above, *optional sampling* of a martingale (or submartingale) is possible: Let $1 \leq t_1 \leq t_2 \leq \cdots \leq t_n$ $\leq \cdots$ be a sequence of stopping rules and $\{U_n, \mathscr{F}_n, n \geq 1\}$ a martingale. Does it follow that $U_{t_1}, U_{t_2}, \ldots, U_{t_n}, \ldots$ is also a martingale? The coin tossing example just discussed shows that the answer in general is no. For a rather comprehensive set of conditions for which the answer is yes, the reader is referred to Doob's text [1953, p. 302].

A third transformation which preserves (indeed *creates*) the martingale structure is Levy's centering combined with truncation. Recall that given a random variable $X$ and a constant $C > 0$ the random variable $XI(|X| \leq C)$ is said to be truncated at $C$. Note that truncated random variables have all absolute moments finite. If $\{Y_n, \mathscr{F}_n, n \geq 1\}$ is any adapted stochastic sequence and $\{c_n, n \geq 1\}$ any choice of positive constants then

$$\{Y_n I(|Y_n| \leq c_n) - E[Y_n I(|Y_n| \leq c_n) \mid \mathscr{F}_{n-1}], n \geq 1\}$$

is a martingale difference sequence with all absolute moments finite. Truncation and centering will often be a useful technique.

### 2.7. The Upcrossings Inequality and Basic Convergence Theorem for Submartingales

The proof of the Rademacher–Menchoff fundamental convergence theorem for orthogonal random variables depends on a maximal inequality whose proof is combinatorial in nature. The proof of the basic convergence theorem for submartingales depends on a different sort of inequality whose proof is also somewhat combinatorial in nature, an "upcrossings" inequality. Lemma 2.7.1 separates out the combinatorial nature of this upcrossings inequality.

Let $N$ be a positive integer, let $0 = u_0, u_1, u_2, \ldots, u_N$ be real numbers, and let $a < b$ define an interval $[a, b]$. We look at "crossings" of this interval and define crossing times $\tau_1, \ldots, \tau_N$. Let

$\tau_1 = $ smallest $i \geq 1$ such that $u_i \leq a$ if such an $i$ exists,

$\tau_2 = $ smallest $i \geq \tau_1$ such that $u_i \geq b$ if such an $i$ exists,

$\tau_3 = $ smallest $i \geq \tau_2$ such that $u_i \leq a$ if such an $i$ exists,

$\vdots$

$\tau_n = $ last such definable $\tau_i$ (let $n = 0$ if no $\tau$'s are yet defined).

Let $\tau_{n+1} = \tau_{n+2} = \cdots = \tau_N = N + 1$. We will now define $d_N$ as the algebraic sum of the lengths of the downcrossings after the first upcrossing, including possibly the length of a partial downcrossing after the last upcrossing. Let $M = N - 1$ if $N$ even and $M = N$ if $N$ odd. Let

$$i_k = \begin{cases} 1 & \text{if} \quad \tau_2 < k \le \tau_3, \tau_4 < k \le \tau_5, \ldots, \tau_{M-1} < k \le \tau_M \\ 0 & \text{otherwise} \end{cases}$$

for each $1 \le k \le N$. Let

$$d_N = \sum_{k=1}^{N} i_k(u_k - u_{k-1}) \qquad \text{and} \qquad h_N = [n/2],$$

noting that $h_N$ is the number of upcrossings.

**Lemma 2.7.1.**

$$d_N \le (a - b)h_N + (u_N - a)^+. \tag{2.7.1}$$

***Proof.*** There are three cases to consider.

(i)   Assume $\tau_2 = N + 1$ (no upcrossings). In this case, $d_N = 0$ and $h_N = 0$, establishing Eq. (2.7.1).

(ii)   Assume $h_N > 0$ and $n$ odd. (There exists an upcrossing and the last crossing is a downcrossing.) In this case

$$d_N \le (a - b)h_N,$$

using the fact that the number of upcrossings $h_N$ equals the number of downcrossings after the first upcrossing and that the length of each down-crossing is less than or equal to $a - b$. Thus Eq. (2.7.1) holds.

(iii)   Assume $h_N > 0$ and $n$ even (last crossing an upcrossing). In this case

$$d_N \le (a - b)(h_N - 1) + u_N - u_{\tau_n},$$

using the fact that the number of downcrossings after the first upcrossing equals the number of upcrossings minus one, that is, $h_N - 1$.

$$(a - b)(h_N - 1) + (u_N - u_{\tau_n}) = (a - b)h_N + (u_N - u_{\tau_n} + b) - a$$
$$\le (a - b)h_N + (u_N - a) \quad \text{since } b - u_{\tau_n} \le 0.$$

Thus Eq. (2.7.1) holds.

Combining (i), (ii), (iii) completes the proof. ∎

Let $\{U_k, \mathscr{F}_k, 1 \le k \le N\}$ be a submartingale. (The definition of a submartingale is modified in the obvious way to include the case of a finite indexing set as we have here.) Fix $\omega \in \Omega$ and let $u_k = U_k(\omega)$. Define $d_N$, $h_N$, and $\{i_k, 1 \leq k \leq N\}$ as we did above. Define random variables $D_N$, $H_N$, and $\{I_k, 1 \leq k \leq N\}$ by $D_N(\omega) = d_N$, $H_N(\omega) = h_N$, and $I_k(\omega) = i_k$ for $1 \leq k \leq N$. We now give the upcrossings inequality. The result for the martingale case is due to Doob [1953]. The result for the submartingale case as well as the proof given below is due to Snell.

**Theorem 2.7.1.**  (See Doob [1953, p. 316].)

$$E(H_N) \leq E[(U_N - a)^+]/(b - a). \tag{2.7.2}$$

**Proof.**  Let $U_0 = 0$ a.s. By Eq. (2.7.1),

$$\sum_{k=1}^{N} I_k(U_k - U_{k-1}) \leq (a - b)H_N + (U_N - a)^+. \tag{2.7.3}$$

$I_k$ is $\mathscr{B}(U_j, j \leq k - 1)$ measurable and hence $\mathscr{F}_{k-1}$ is measurable, a key observation.

$$E\left[\sum_{k=1}^{N} I_k(U_k - U_{k-1})\right] = E\left\{\sum_{k=1}^{N} E[I_k(U_k - U_{k-1}) \mid \mathscr{F}_{k-1}]\right\}$$

$$= E\left\{\sum_{k=1}^{N} I_k E[U_k - U_{k-1} \mid \mathscr{F}_{k-1}]\right\} \geq 0,$$

the inequality holding since $\tau_2 \geq 2$ and hence $I_1 = 0$ a.s., and since the submartingale assumption implies

$$E[U_k - U_{k-1} \mid \mathscr{F}_{k-1}] \geq 0 \qquad \text{a.s. for} \quad k \geq 2.$$

Taking expectations in Eq. (2.7.3) implies that

$$(a - b)EH_N + E[(U_N - a)^+] \geq 0$$

and hence that

$$E(H_N) \leq E[(U_N - a)^+]/(b - a). \quad \blacksquare$$

It is now an easy matter to derive the submartingale convergence theorem. This result is without doubt the most important martingale convergence result, as demonstrated by its wide applicability.

**Theorem 2.7.2.**  [Doob, 1953, p. 319]. If $\{S_n, n \geq 1\}$ is a submartingale with $\sup E \mid S_n \mid < \infty$, then there exists a random variable $S$ such that $S_n \to S$ a.s. and $E \mid S \mid \leq \sup E \mid S_n \mid$.

**Proof.**    It is first shown that $S_n$ cannot diverge in an oscillatory manner. Let

$$O_{a,b} = [\liminf S_n < a < b < \limsup S_n]$$

and

$$O = \bigcup O_{a,b}$$

where the union is over all pairs of rational numbers $a < b$ and, hence, is a countable union. We prove that $P(O) = 0$. $P(O) \leq \sum P(O_{a,b})$. Hence it suffices to show that $P(O_{a,b}) = 0$ for every pair $a < b$. Let $H_N$ be the number of upcrossings of $[a, b]$ by $[U_i, 1 \leq i \leq N]$ and $H = \lim H_N$. $O_{a,b} \subset [H = \infty]$. Hence it suffices to show that $P[H = \infty] = 0$. By the monotone convergence theorem, $\lim EH_N = EH$. Thus by Eq. (2.7.2) and hypothesis

$$EH \leq \sup_{k \geq 1} E[(S_k - a)^+]/(b - a) \leq \sup(E \mid S_k \mid + \mid a \mid)/(b - a) < \infty.$$

Thus $P[H = \infty] = 0$, establishing that $P(O) = 0$. $P(O) = 0$ implies

$$P[S_n \text{ converges}] + P[\mid S_n \mid \to \infty] = 1.$$

By the Fatou lemma

$$E[\liminf \mid S_n \mid] \leq \liminf E \mid S_n \mid < \infty.$$

Hence $P[\mid S_n \mid \to \infty] = 0$, $P[S_n \text{ converges}] = 1$, and $E[\lim \mid S_n \mid] \leq \sup E \mid S_n \mid < \infty$, establishing the theorem. ∎

**Exercise 2.7.1.**    Let $\{S_n, n \geq 1\}$ be a submartingale with $\sup E(S_n^+) < \infty$. Prove that $\sup E \mid S_n \mid < \infty$. (Thus the apparent weakening of the hypothesis of Theorem 2.7.2 achieved by replacing $\sup E \mid S_n \mid < \infty$ with $\sup E(S_n^+) < \infty$ is illusory.) We will use Exercise 2.7.1 occasionally. ∎

We mention here an interesting open question related to martingale convergence. A stochastic sequence $\{T_n, n \geq 1\}$ satisfying

$$E[T_n \mid T_k] = T_k \qquad \text{a.s. for each} \quad 1 \leq k < n < \infty$$

is called a weak martingale. It is at present an open question whether any interesting results about almost sure convergence for weak martingales can be proved. In his work [1970], P. Nelson studies the properties of weak martingales.

**Exercise 2.7.2.** Show that every martingale is a weak martingale. There exist examples (rather difficult) of weak martingales which are not martingales (see [Nelson, 1970]).

**Exercise 2.7.3.** Let $\{X_i, i \geq 1\}$ be defined by

$$X_1 = X_3 = \cdots = X_{2n+1} = \cdots \qquad \text{and} \qquad X_2 = X_4 = \cdots = X_{2n} = \cdots .$$

Suppose $E[X_1 \mid X_2] = X_2$ and $E[X_2 \mid X_1] = X_1$ a.s.

(i)   Assume $P[X_1 = X_2] < 1$ is possible. Show that $\{X_i, i \geq 1\}$ is a weak martingale but not a martingale. Note that $\sup E \mid X_n \mid < \infty$. Does $X_n$ converge almost surely?

(ii)   (i) would seem to contribute information to the open question discussed above. However $P[X_1 = X_2] < 1$ is impossible! Prove this under the assumption $EX_1^2 < \infty$ and $EX_2^2 < \infty$. Hint: Prove that $E[Y \mid X] \perp E[Y \mid X] - X$. ∎

## 2.8.   Applications of the Basic Submartingale Convergence Theorem

Throughout the remainder of the book we shall mean by $A$ implies $B$ that $P[A \cap B^c] = 0$. By $A = B$ we shall mean that $A$ implies $B$ and $B$ implies $A$. Let

$$A \triangle B = (A \cap B^c) \cup (A^c \cap B).$$

$A \triangle B$ is called the symmetric difference of $A$ and $B$. Note that $A = B$ simply means that $P[A \triangle B] = 0$.

The basic convergence theorem implies convergence in several of the examples of martingales given in Section 2.5. In Example 2.5.4 (the Polya urn scheme) the proportion of black balls, $T_n$, satisfies $\mid T_n \mid \leq 1$ a.s. for each $n \geq 1$. Thus $\sup_{n \geq 1} E \mid T_n \mid < \infty$, and hence $T_n$ converges almost surely to a random variable $T$ by the basic submartingale convergence theorem.

In Example 2.5.5, the likelihood ratio $T_n \geq 0$ a.s. for each $n \geq 1$. Thus $\sup_{n \geq 1} ET_n^- < \infty$ trivially. Using Exercise 2.7.1, $\sup_{n \geq 1} E \mid T_n \mid < \infty$ and hence $T_n$ converges almost surely. (See Doob [1953, p. 348] for more on the behavior of likelihood ratios.)

**Example 2.5.2. (cont.).** Since $f$ is of bounded variation, $\sup E \mid T_n \mid \leq$ total variation of $f < \infty$. Thus $T_n$ converges almost surely by the basic submartingale convergence theorem. This result can be used to help prove

Lebesgue's basic result that $f'$ exists and is finite almost everywhere. (See Doob [1953, p. 343] for more information.) ▮

The martingale convergence theorem has an interesting application to the asymptotic behavior of branching processes.

**Example 2.8.1.** Let

$$\{X_1^{(1)}, X_2^{(1)}, \ldots, X_1^{(2)}, X_2^{(2)}, \ldots, X_1^{(n)}, X_2^{(n)}, \ldots\}$$

be a sequence of independent identically distributed nonnegative integer valued random variables with $EX_1^{(1)} = \mu > 0$. Let

$$T_0 = 1 \quad \text{and} \quad T_n = \sum_{i=1}^{T_{n-1}} X_i^{(n)} \quad \left(\sum_{i=1}^{0}(\cdot) = 0 \text{ by convention}\right)$$

for $n \geq 1$. Note that $\{X_1^{(n)}, X_2^{(n)}, \ldots\}$ is independent of $T_1, T_2, \ldots, T_{n-1}$ for each $n \geq 1$. $\{T_n, n \geq 0\}$ is called a branching process. $T_n$ can be thought of as describing the size of a population at the $n$th generation, with each individual giving birth to a random number of offspring to determine the size of the population in the next generation. Let $U_n = T_n/\mu^n$ for each $n \geq 1$. Fix $n \geq 2$.

$$E[U_n \mid U_1, \ldots, U_{n-1}] = (U_{n-1}\mu^{n-1})EX_1^{(1)}/\mu^n = U_{n-1} \quad \text{a.s.}$$

Thus $\{U_n, n \geq 1\}$ is a martingale.

$$\sup E \mid U_n \mid = \sup EU_n = EU_1 = EX_1^{(1)}/\mu = 1.$$

Thus, by the basic convergence theorem $T_n/\mu^n$ converges almost surely to a random variable. From this follows one of the standard results of branching processes; namely, that $T_n \to 0$ a.s. for $\mu < 1$. (The material in this paragraph is due to T. E. Harris.)

The submartingale convergence theorem can be applied to a more interesting and more realistic version of Example 2.5.1.

**Example 2.5.1 (cont.).** Let $\{X_i, i \geq 1\}$ be a sequence of independent identically distributed random variables with

$$P[X_1 = 1] + P[X_1 = -1] = 1 \quad \text{and} \quad P[X_1 = 1] \leq \tfrac{1}{2}$$

(that is, coin tossing random variables with the coin possibly unfair). Let

$a_i$ be $\mathscr{B}(X_1, \ldots, X_{i-1})$ measurable for each $i \geq 2$ and $a_1$ constant. For each $i \geq 1$, $a_i X_i$ represents the gambler's winnings on trial $i$, given that the gambler bets $a_i$ on trial $i$. Let

$$T_n = \sum_{i=1}^{n} a_i X_i \qquad \text{for each} \quad n \geq 1 \quad \text{and} \quad T_0 = 1$$

(that is, the gambler's initial fortune is 1). Suppose that the gambler cannot bet more than he has at each trial, i.e., $0 \leq a_i \leq T_{i-1}$ a.s. for each $i \geq 1$. What is the asymptotic behavior of $T_n$ for various betting systems $\{a_i, i \geq 1\}$? First, note that $\{-T_n, n \geq 1\}$ is a submartingale since

$$E[-T_n \mid T_1, \ldots, T_{n-1}] = -T_{n-1} - a_n EX_n \geq -T_{n-1} \qquad \text{a.s.}$$

Since $0 \leq a_i \leq T_{i-1}$ a.s. for each $i \geq 1$, $-T_n \leq 0$ a.s. for all $n \geq 1$ follows. Thus $\sup E[(-T_n)^+] < \infty$ and by Exercise 2.7.1 $\sup E \mid -T_n \mid$ $< \infty$. Thus by the basic submartingale convergence theorem there exists a random variable $T$ such that $T_n \to T$ a.s. Therefore, asymptotically the gambler's accumulated winnings stop fluctuating and converge to a fixed amount. In one sense, fluctuations are the patient gambler's best friend. For, if the gambler's goal is to win an amount $C$, he would like to stay in the game until $T_n \geq C$, at which point he can quit ($a_m = 0$ for $m \geq n + 1$). But in our situation $T_n \to T$ a.s. whether the gambler quits at some time or plays forever. Thus, for a given betting system and $C$ sufficiently large, $P[T_n \geq C$ for some $n \geq 1]$ will be close to zero, thus likely foiling the gambler's hopes.

What can we say about the distribution of $S$ corresponding to various betting systems $\{a_i, i \geq 1\}$? First we note that $a_i \to 0$ a.s. as $i \to \infty$ holds for all systems since $\mid T_{i+1} - T_i \mid = \mid a_i \mid \to 0$.

It is quite easy for the gambler to lose his fortune rather quickly. For example, let $a_i = T_{i-1}$ for each $i \geq 1$. Since $P[X_i = -1$ for some $i \geq 1] = 1$, the gambler goes broke with probability one with this double or nothing system of bets.

Can the gambler avoid going broke; that is, do there exist systems such that $P[T_n \to 0] < 1$? Clearly such systems exist. For example, let $a_i = 0$ for all $i \geq 1$! An interesting way to view the gambler's options is to let $b_n$ be the proportion of $T_{n-1}$ bet on trial $n$ for each $n \geq 1$. That is, $a_n = b_n T_{n-1}$ with $b_n$ being $\mathscr{B}(X_1, X_2, \ldots, X_{n-1})$ measurable for each $n \geq 1$. From this viewpoint, one system that looks plausible is to let $b_n = \frac{1}{2}$ for each $n \geq 1$. Clearly the gambler does not lose his fortune in a finite amount of time and perhaps he does not lose all in the limit. But $T_n/T_{n-1} = 1 + X_n/2$

and since $P[X_n \to 0] = 0$ it follows that $T_n \to 0$ a.s. Indeed it is clear that the only way the gambler has any chance to avoid $T_n \to 0$ a.s. is to choose the $b_n$ such that $P[b_n \to 0$ as $n \to \infty] > 0$. Thus the gambler must either bet economically or quit at some finite time if he is to avoid financial ruin! ∎

Recall that $\{X_i, i \geq 1\}$ orthogonal with

$$\sum_{i=1}^{\infty} (\log i)^2 EX_i^2 < \infty$$

implies that $S_n$ converges almost surely. This result can be sharpened if in addition $\{S_n, n \geq 1\}$ is a martingale:

**Exercise 2.8.1.**   Let $\{S_n, n \geq 1\}$ be a submartingale with

$$\sum_{i=1}^{\infty} E \,|\, X_i \,|^p < \infty$$

for some $0 < p \leq 2$. Prove that $S_n$ converges almost surely. ∎

**Remark.**   This immediate corollary of the basic martingale convergence theorem will be used often. It is especially important in the case in which $p = 2$ and the $X_i$ are independent with mean zero.

**Theorem 2.8.1.**   [Doob, 1953, p. 320].   Let $\{X_i, i \geq 1\}$ be a martingale difference sequence with

$$E[\sup X_i] < \infty.$$

Then

$$\sup S_n < \infty \qquad \text{implies that} \quad S_n \quad \text{converges.}$$

**Remark.**   Note that $X_i \leq K$ a.s. for all $i \geq 1$ and some constant $K < \infty$ implies that $E[\sup X_i] < \infty$. Theorem 2.8.1 is often applied to random variables $X_i$ bounded above by a constant uniformly in $i \geq 1$.

**Proof.**   (This is the first use of what we will call the stopping rule method. It will be widely used throughout this chapter as well as in Chapter 3.) Fix $M > 0$. Let $t_M$ be the smallest integer $n \geq 1$ such that $S_n > M$ if such an $n$ exists; otherwise let $t_M = \infty$. Suppressing $M$, denote $t_M$ by $t$.

$t$ is clearly a stopping rule. Let

$$S_n^{(t)} = S_{\min(t,n)} \quad \text{for} \quad n \geq 1.$$

$\{S_n^{(t)}, n \geq 1)\}$ is a martingale by Theorem 2.6.2 (i). $S_n^{(t)} \leq M$ if $t > n$ and $S_n^{(t)} = S_{t-1} + X_t \leq M + \sup X_i$ if $t \leq n$. Hence

$$\sup E[(S_n^{(t)})^+] \leq E[\sup(S_n^{(t)})^+] < \infty.$$

By Exercise 2.7.1, $\sup E \mid S_n^{(t)} \mid < \infty$. Thus by the submartingale convergence theorem, $S_n^{(t)}$ converges almost surely. On $[t_M = \infty]$, $S_n^{(t)} = S_n$ for all $n \geq 1$; hence, $S_n$ converges almost surely on $\bigcup_{M=1}^{\infty} [t_M = \infty] = [\sup S_n < \infty]$. ∎

**Remark.** Since the above proof is uncluttered with computation, it clearly illuminates the stopping rule method. It should be remarked that the stopping rule method is parasitic in the sense that it depends on known convergence results in order to produce new ones. Without Doob's basic submartingale convergence theorem, the stopping rule method would not be nearly as useful.

The stopping rule method plays a role in the study of martingales very similar to the role that truncation plays in the study of independent random variables. Both methods produce a stochastic sequence with the same dependence structure as the original sequence but more regular in the sense of moments being better behaved.

Theorem 2.8.1 has interesting implications for the gambling situation. Suppose $\{Y_i, i \geq 1\}$ is a martingale difference sequence representing a gambler's winnings on successive trials of a fair game. Suppose the game is such that the gambler can never win or lose more than $C$ dollars on any one trial; that is, $\mid Y_i \mid \leq C$ a.s. for each $i \geq 1$. Is it possible that

$$P\left[ \sum_{i=1}^{n} Y_i \to \infty \right] > 0?$$

That is, can the gambler's accumulated winnings converge to infinity with positive probability? It is easy to see by Theorem 2.8.1 that the answer is no. For suppose that $P[\sum_{i=1}^{n} Y_i \to \infty] > 0$. Then $\sum_{i=1}^{n} Y_i \to \infty$ implies $\inf \sum_{i=1}^{n} Y_i > -\infty$. Applying Theorem 2.8.1 to $\{-Y_i, i \geq 1\}$, it follows that $\sum_{i=1}^{\infty} Y_i$ converges, producing a contradiction. Thus $P[\sum_{i=1}^{n} Y_i \to \infty] = 0$.

In general, is it possible to have a martingale difference sequence $\{Y_i, i \geq 1\}$ such that $\sum_{i=1}^{n} Y_i \to \infty$ a.s.? It is easy to construct an example showing that the answer is yes:

**Example 2.8.2.**   Let $\{Y_i, i \geq 1\}$ be independent with

$$P[Y_{i-1} = -i] = i^{-2}$$

and

$$P[Y_{i-1} = i/(i^2 - 1)] = (i^2 - 1)/i^2$$

for $i \geq 2$.

Since $EY_i = 0$ for $i \geq 1$, $\{Y_i, i \geq 1\}$ is a martingale difference sequence. $\sum_{i=2}^{\infty} P[Y_{i-1} = -i] < \infty$. Thus by the Borel–Cantelli lemma, $P[Y_{i-1} = -i$ i.o.$] = 0$ and hence $P[Y_{i-1} = i/(i^2 - 1)$ eventually$] = 1$. Thus $\sum_{i=1}^{\infty} Y_i \to \infty$ a.s.   ∎

**Exercise 2.8.2.**   Let $\{b_n, n \geq 1\}$ be an arbitrary sequence of real numbers. Construct a martingale $\{S_n, n \geq 1\}$ such that $P[S_n = b_n$ eventually$] = 1$. Is such a construction possible with $\{S_n - S_{n-1}, n \geq 2\}$ independent?   ∎

Let $\{Y_i, i \geq 1\}$ be a martingale difference sequence. It is the asymmetry of the $Y_i$ in Example 2.8.2 which enables us to use the Borel–Cantelli lemma and hence obtain $\sum_{i=1}^{n} Y_i \to \infty$ a.s. In this example, $\inf \sum_{i=1}^{n} Y_i > -\infty$ a.s. Thus (replacing $Y_i$ by $-Y_i$ for each $i \geq 1$), an alternative way to look at Example 2.8.2 is that $\sup_{n \geq 1} \sum_{i=1}^{n} Y_i < \infty$ a.s. does not in general imply $\sum_{i=1}^{\infty} Y_i$ converges almost surely. If certain regularity conditions are assumed, it follows that $\sup_{n \geq 1} \sum_{i=1}^{n} Y_i < \infty$ implies that $\sum_{i=1}^{\infty} Y_i$ converges. According to Theorem 2.8.1, $E \sup Y_i < \infty$ is such a condition. We will now study other regularity conditions which impose some degree of symmetry on the $Y_i$ and thereby yield convergence results.

**Definition 2.8.1.**   (i)   An adapted stochastic sequence $\{Y_i, \mathscr{F}_i, i \geq 1\}$ is said to be symmetric if

$$P[(Y_n, Y_{n+1}, \ldots, Y_{n+m}) \in A \mid \mathscr{F}_{n-1}]$$
$$= P[(-Y_n, -Y_{n+1}, \ldots, -Y_{n+m}) \in A \mid \mathscr{F}_{n-1}] \qquad \text{a.s.}$$

for all $(m + 1)$-dimensional Borel sets $A$, all $m \geq 0$, and all $n \geq 1$.

(ii)   An adapted stochastic sequence is said to be weakly symmetric if for some integer $M \geq 1$,

$$\sum_{i=1}^{\infty} E[D_i I(|D_i| \leq m) \mid \mathscr{F}_{i-1}] \qquad \text{converges a.s.}$$

for each $m \geq M$.   ∎

**Exercise 2.8.3.** According to Berman [1965], $\{X_i, i \geq 1\}$ is sign-invariant if $\{X_i, i \geq 1\}$ and $\{r_i X_i, i \geq 1\}$ have the same distribution for each choice of $r_i = \pm 1, i \geq 1$.

(i)  Prove that $\{X_i, i \geq 1\}$ is sign invariant if and only if $\{X_i, \mathscr{B}_i, i \geq 1\}$ is an adapted symmetric sequence $[\mathscr{B}_i = \mathscr{B}(X_1, \ldots, X_i)]$.

(ii)  Prove that $\{X_i, \mathscr{B}_i, i \geq 1\}$ an adapted symmetric sequence with $E|X_i| < \infty$ for each $i \geq 1$ implies that $\{X_i, i \geq 1\}$ is a martingale difference sequence. ∎

Note that symmetry implies weak symmetry and that $\{Y_i, \mathscr{F}_i, i \geq 1\}$ symmetric with $E|Y_i| < \infty$ for each $i \geq 1$ implies that $\{Y_i, \mathscr{F}_i, i \geq 1\}$ is a martingale difference sequence.

**Theorem 2.8.2.**  Let $\{Y_i, \mathscr{F}_i, i \geq 1\}$ be a weakly symmetric adapted stochastic sequence. Then

$$\sup_{n \geq 1} \sum_{i=1}^{n} Y_i < \infty \qquad \text{and} \qquad \sup |Y_i| < \infty$$

together imply that $\sum_{i=1}^{\infty} Y_i$ converges.

**Proof.**  Let $M$ be the integer specified in the definition of weak symmetry. Truncating and centering to form a martingale, we let

$$U_n = \sum_{i=1}^{n} Y_i I(|Y_i| \leq m) - E[Y_i I(|Y_i| \leq m) \mid \mathscr{F}_{i-1}]$$

for each $n \geq 1$ and a fixed $m \geq M$. $\{U_n, n \geq 1\}$ is a martingale with its differences uniformly bounded by $2m$. Thus $\sup U_n < \infty$ implies that $U_n$ converges by Theorem 2.8.1. Let $A_m = [|Y_i| \leq m$ for all $i \geq 1]$. On $A_m$,

$$U_n = \sum_{i=1}^{n} Y_i - \sum_{i=1}^{n} E[Y_i I(|Y_i| \leq m) \mid \mathscr{F}_{i-1}]$$

for each $n \geq 1$. Let

$$A = \left[\sup \sum_{i=1}^{n} Y_i < \infty, \qquad \sup |Y_i| < \infty\right].$$

$A \cap A_m$ implies that $\sup U_n < \infty$ and thus $A \cap A_m$ implies that $U_n$ converges. Thus $\bigcup_{m=M}^{\infty} (A \cap A_m)$ implies that $\sum_{i=1}^{\infty} Y_i$ converges. Since $A \subset [\sup |Y_i| < \infty] \subset \bigcup_{m=M}^{\infty} A_m$, $A = \bigcup_{m=M}^{\infty} (A \cap A_m)$, thus establishing the theorem. ∎

**Corollary 2.8.1.** Let $\{Y_i, \mathscr{F}_i, i \geq 1\}$ be a weakly symmetric adapted stochastic sequence. Then

$$\sup \left| \sum_{i=1}^n Y_i \right| < \infty \qquad \text{implies} \qquad \sum_{i=1}^n Y_i \quad \text{converges.}$$

**Proof.** $\sup | \sum_{i=1}^n Y_i | < \infty$ implies $\sup | Y_i | < \infty$. Hence the result is immediate from Theorem 2.8.2. ∎

$\sup | Y_i | < \infty$ cannot be dropped as a condition in the statement of Theorem 2.8.2:

**Example 2.8.3.** Let $\{Y_i, i \geq 1\}$ be independent with

$$P[Y_{i-2} = -\log i] = i^{-1}, \qquad P[Y_{i-2} = i] = (\log i)/i^2$$

and

$$P[Y_{i-2} = 0] = 1 - i^{-1} - (\log i)/i^2$$

for each $i \geq 3$. Let $\mathscr{F}_i = \mathscr{B}(Y_1, Y_2, \ldots, Y_i)$ for each $i \geq 1$. Using the Borel–Cantelli lemma, $P[Y_i \leq 0 \text{ eventually}] = 1$ and hence $\sup \sum_{i=1}^n Y_i < \infty$ a.s. $\sum_{i=1}^\infty EY_i I(| Y_i | \leq m)$ converges for all $m \geq 1$. But

$$\sum_{i=3}^\infty P[Y_{i-2} = -\log i] = \infty$$

and thus by the converse to the Borel–Cantelli lemma for independent events, $P[Y_{i-2} = -\log i \text{ i.o.}] = 1$. Thus $\sum_{i=1}^n Y_i \to -\infty$ a.s. ∎

Symmetry does however allow us to drop the assumption $\sup | Y_i | < \infty$ and make the weaker statement that $\sup_{n \geq 1} \sum_{i=1}^n Y_i < \infty$ implies $\sum_{i=1}^\infty Y_i$ converges.

**Lemma 2.8.1.** Let $\{Y_i, \mathscr{F}_i, i \geq 1\}$ be an adapted symmetric stochastic sequence. Then

$$\left[ \sup_{n \geq 1} \sum_{i=1}^n Y_i < \infty \right] = \left[ \inf_{n \geq 1} \sum_{i=1}^n Y_i > -\infty \right].$$

**Proof.** Let $U_n = \sum_{i=1}^n Y_i$ for each $n \geq 1$. Suppose $P[\sup_{n \geq 1} U_n < \infty] > 0$ and choose $K > 0$ such that $P[\sup_{n \geq 1} U_n < K] > 0$. By symmetry,

$$P[\sup_{n > N} U_n < K \mid \sup_{n \leq N} U_n < K] = P[\sup_{n > N} (U_n - U_N) \lessapprox K - U_N \mid \sup_{n \leq N} U_n < K]$$

$$= P[\inf_{n > N} U_n > 2U_N - K \mid \sup_{n \leq N} U_n < K]$$

for all integers $N \geq 1$. Since

$$P[\sup_{n>N} U_n < K \mid \sup_{n \leq N} U_n < K] = \frac{P[\sup_{n \geq 1} U_n < K]}{P[\sup_{n \leq N} U_n < K]} \to 1$$

as $N \to \infty$, it follows that

$$P[\inf_{n>N} U_n > 2U_N - K \mid \sup_{n \leq N} U_n < K] \to 1$$

as $N \to \infty$ and hence that

$$P[\inf_{n \geq 1} U_n > -\infty \mid \sup_{n \leq N} U_n < K] \to 1$$

as $N \to \infty$. Hence $[\sup_{n \geq 1} U_n < K] \subset [\inf_{n \geq 1} U_n > -\infty]$. Since

$$[\sup_{n \geq 1} U_n < \infty] = \bigcup_{K=1}^{\infty} [\sup_{n \geq 1} U_n < K],$$

it follows that

$$[\sup_{n \geq 1} U_n < \infty] \subset [\inf_{n \geq 1} U_n > -\infty]$$

when $P[\sup_{n \geq 1} U_n < \infty] > 0$. $[\sup_{n \geq 1} U_n < \infty] \subset [\inf_{n \geq 1} U_n > -\infty]$ holds trivially when $P[\sup_{n \geq 1} U_n < \infty] = 0$. By the symmetry of the above argument, the lemma then follows. ∎

**Theorem 2.8.3.** Let $\{Y_i, \mathscr{F}_i, i \geq 1\}$ be an adapted symmetric stochastic sequence. Then

$$\sup_{n \geq 1} \sum_{i=1}^{n} Y_i < \infty \qquad \text{implies} \qquad \sum_{i=1}^{\infty} Y_i \text{ converges.}$$

*Proof.* By Lemma 2.8.1,

$$\sup_{n \geq 1} \sum_{i=1}^{n} Y_i < \infty \qquad \text{implies} \qquad \sup_{n \geq 1} \left| \sum_{i=1}^{n} Y_i \right| < \infty$$

and hence that $\sup |Y_i| < \infty$. Thus Theorem 2.8.3 follows from Theorem 2.8.2. ∎

Theorem 2.8.2 has an interesting application to orthogonal series of real functions. Let $\Omega = [0, 1], \mathscr{F}$ be the Borel sets of $[0, 1]$ and $P$ be Lebesgue measure. We first introduce the orthogonal system of Rademacher functions

$\{R_i, i \geq 0\}$. Let

$$R_0(\omega) = \begin{cases} 1 & \text{if } 0 \leq \omega \leq \frac{1}{2} \\ -1 & \text{if } \frac{1}{2} < \omega \leq 1. \end{cases}$$

In order to define $R_i$ for each $i \geq 1$, extend $R_0$ periodically to $[0, \infty)$ and let

$$R_i(\omega) = R_0(2^i \omega)$$

for $0 \leq \omega \leq 1$ and each $i \geq 1$. It is instructive to graph the Rademacher functions (see Figure 2.8.1). As is easily shown, the $R_i$ are independent identically distributed with $P[R_i = 1] = P[R_i = -1] = \frac{1}{2}$. Thus $\{R_i, i \geq 0\}$ is merely one representation of a very familiar object, the "coin tossing" random sequence. It is interesting to note the connection between $\{R_i, i \geq 0\}$ and the binary expansion of numbers in the interval $[0, 1]$. Let $B_i = (1 - R_i)/2$ for $i \geq 0$. Then it is easily seen that for each $\omega \in [0, 1]$, $0.B_0(\omega)B_1(\omega)\cdots$ is the binary expansion of $\omega$. Thus results about $\{R_i, i \geq 0\}$ can be interpreted as results about binary expansions of real numbers.

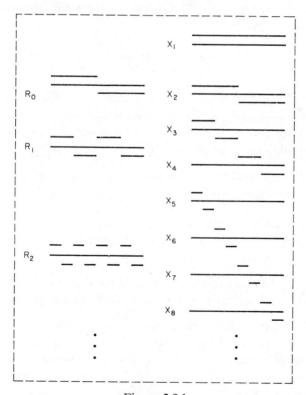

Figure 2.8.1.

Our purpose for introducing the Rademacher system is the construction of the associated Haar system $\{X_i, i \geq 1\}$. For each $\omega \in [0, 1]$, let

$$X_1(\omega) = 1,$$

$$X_2(\omega) = R_0(\omega),$$

and

$$X_i(\omega) = \begin{cases} 2^{j/2} R_j(\omega) & \text{if } (k-1)/2^j \leq \omega < k/2^j, \\ & \text{for } i = 2^j + k, 1 \leq k \leq 2^j, j \geq 1, \\ 0 & \text{otherwise} \end{cases}$$

for $i \geq 3$.

In order to make sense out of the definition of the Haar system, it is very instructive to graph the Haar functions (see Figure 2.8.1). The Haar system is of interest to analysts because it is a complete orthonormal system with a very simple structure. Moreover, given $f \in L_1[0, 1]$ the associated partial sums of the Fourier series (well defined since the Haar functions are bounded) converges to $f$ in $L_1$. For any choice of real constants $\{a_i, i \geq 1\}$, the $\sum_{i=1}^{n} a_i X_i$ are referred to as the partial sums of a Haar series.

**Lemma 2.8.2.** The Haar functions form a weakly symmetric adapted sequence.

**Proof.** Refer to Figure 2.8.1. It is clear from the geometry of the Haar functions that $\{X_n, n \geq 1\}$ is a weakly symmetric adapted sequence. It is left to the reader to provide a formal proof by computing the appropriate conditional expectations. ∎

Theorem 2.8.2 combines with Lemma 2.8.2 to yield a simple martingale proof of an interesting result in the theory of Haar series:

**Theorem 2.8.4.** [Talalyan and Arutyunyan, 1965]. A Haar series satisfying $\sup |a_i X_i| < \infty$ a.s. cannot converge to $+\infty$ (or $-\infty$) on a set of positive Lebesgue measure.

**Remark.** Talalyan and Arutyunyan prove this result without supposing $\sup |a_i X_i| < \infty$ a.s. See p. 91 for a martingale proof of this deeper result.

**Proof.** Suppose that a Haar series $\sum_{i=1}^{\infty} a_i X_i$ does converge to $+\infty$ on a set $A$ of positive Lebesgue measure. Thus $\inf \sum_{i=1}^{n} a_i X_i > -\infty$ on $A$. Since the Haar functions are weakly symmetric (Lemma 2.8.2), it follows by the symmetric analog of Theorem 2.8.2 that $\sum_{i=1}^{\infty} a_i X_i$ converges on $A$, a contradiction. ∎

**Exercise 2.8.4.** Prove that $\sum_{i=1}^{n} a_i X_i$ can be representend by $\sum_{i=1}^{n} v_i R_i + c$, where $v_i$ is a measurable function of $R_1, R_2, \ldots, R_{i-1}$, and $c$ is a constant. ∎

The basic submartingale convergence theorem allows an easy extension of the Borel–Cantelli lemma to sums of conditional probabilities.

**Theorem 2.8.5.** Let $\{B_i, i \geq 1\}$ be a sequence of events and $\{\mathscr{F}_i, i \geq 1\}$ an increasing sequence of $\sigma$ fields such that $B_i \in \mathscr{F}_i$ for each $i \geq 1$. Then

$$[B_i \text{ i.o.}] = \left[\sum_{i=1}^{\infty} P(B_i \mid \mathscr{F}_{i-1}) = \infty\right],$$

that is, $\sum_{i=1}^{\infty} P[B_i \mid \mathscr{F}_{i-1}] < \infty$ implies the $B_i$ occur at most finitely often and $\sum_{i=1}^{\infty} P[B_i \mid \mathscr{F}_{i-1}] = \infty$ implies the $B_i$ occur infinitely often.

**Proof.** Let $U_n = \sum_{i=1}^{n} \{I(B_i) - E[I(B_i) \mid \mathscr{F}_{i-1}]\}$ for $n \geq 1$. $\{U_n, n \geq 1\}$ is a martingale with its martingale differences satisfying

$$| I(B_i) - E[I(B_i) \mid \mathscr{F}_{i-1}]| \leq 1 \qquad \text{a.s.}$$

Clearly,

$$[B_i \text{ at most finitely often}] = \left[\sum_{i=1}^{\infty} I(B_i) < \infty\right].$$

$[\sum_{i=1}^{\infty} I(B_i) < \infty]$ implies $[\sup U_n < \infty]$ by the definition of the $U_n$. Hence, applying Theorem 2.8.1, $[B_i$ at most finitely often] implies that $U_n$ converges and hence that

$$\sum_{i=1}^{\infty} E[I(B_i) \mid \mathscr{F}_{i-1}] = \sum_{i=1}^{\infty} P[B_i \mid \mathscr{F}_{i-1}] < \infty.$$

$\sum_{i=1}^{\infty} P[B_i \mid \mathscr{F}_{i-1}] < \infty$ implies $\inf U_n > -\infty$. Hence, by the symmetric analog of Theorem 2.8.1,

$$\sum_{i=1}^{\infty} P[B_i \mid \mathscr{F}_{i-1}] < \infty \qquad \text{implies} \qquad \sum_{i=1}^{\infty} I(B_i) < \infty,$$

that is, $\sum_{i=1}^{\infty} P[B_i \mid \mathscr{F}_{i-1}] < \infty$ implies that $B_i$ occurs at most finitely often. ∎

**Example 2.8.4.** Consider a gambling casino with infinitely many different games to choose from. On each trial a gambler chooses a particular game to play. If he chooses game number $i$ on a particular trial the probability that he loses on that trial is $1/i$. The gambler is constrained for each

$i \geq 1$ to play game $i$ at least once before playing game $i + 1$. Beyond this the gambler is free to employ any strategy he wishes in choosing which game to play on each trial. Is there any way the gambler can keep from losing infinitely often?

In order to answer this question, for each $i \geq 1$ let $A_i$ be the event that the gambler loses on trial $i$ and $\mathscr{F}_i$ be the $\sigma$ field generated by the events $A_1, A_2, \ldots, A_i$. Because of the constraint on his strategy, on trial $i$ the gambler is playing one of the games 1 through $i$. Thus $P[A_i \mid \mathscr{F}_{i-1}] \geq i^{-1}$ for each $i \geq 1$. Hence $\sum_{i=1}^{\infty} P[A_i \mid \mathscr{F}_{i-1}] = \infty$. Therefore, by Theorem 2.8.5 $P[A_i \text{ i.o.}] = 1$, answering the question. ∎

We mention one other simple application of Theorem 2.8.5. Let $\{A_i, i \geq 1\}$ be a sequence of events such that

$$\sum_{n=2}^{\infty} P\left[A_n \,\middle|\, \bigcap_{i=1}^{n-1} A_i^c\right] = \infty.$$

Let $\mathscr{F}_i$ be the $\sigma$ field generated by $A_1, A_2, \ldots, A_i$ for each $i \geq 1$. Note that $\bigcap_{i=1}^{\infty} A_i^c$ implies

$$\sum_{n=2}^{\infty} P[A_n \mid \mathscr{F}_{n-1}] = \sum_{n=2}^{\infty} P\left[A_n \,\middle|\, \bigcap_{i=1}^{n-1} A_i^c\right] = \infty \qquad \text{a.s.}$$

Thus $\bigcap_{n=1}^{\infty} A_n^c$ implies $A_n$ i.o. But the only way this is possible is that $P[\bigcap_{n=1}^{\infty} A_n^c] = 0$. That is, $\sum_{n=2}^{\infty} P[A_n \mid \bigcap_{i=1}^{n-1} A_i^c] = \infty$ implies that $P[A_n$ occurs for at least one $n \geq 1] = 1$.

Many of the results of Section 2.8 have been "local" results. Rather than saying that someting (for example, convergence) happens with probability one or zero, local results state conditions which imply that something happens on part of the probability space.

**Exercise 2.8.5.** Construct a martingale $\{U_n, n \geq 1\}$ such that $0 < P[U_n \text{ converges}] < 1$. Construct a martingale $\{U_n, n \geq 1\}$ such that $\sup |U_n| < \infty$ a.s. and $U_n$ diverges a.s. ∎

**Corollary 2.8.2.** (The ordinary Borel–Cantelli lemma).

$$\sum_{i=2}^{\infty} P(B_i) < \infty$$

implies that

$$P[B_i \text{ i.o.}] = 0.$$

**Proof.** For each $i \geq 1$ let $\mathscr{F}_i$ be the $\sigma$ field generated by $B_1$, $B_2$, ..., $B_i$.

$$\infty > \sum_{i=1}^{\infty} P(B_i) = E\left\{ \sum_{i=1}^{\infty} E[I(B_i) \mid \mathscr{F}_{i-1}] \right\}.$$

Hence

$$\sum_{i=1}^{\infty} P[B_i \mid \mathscr{F}_{i-1}] < \infty \qquad \text{a.s.}$$

Thus, the desired result follows from Theorem 2.8.5. ∎

**Corollary 2.8.3.** (Partial converse to the ordinary Borel–Cantelli lemma.) If the $B_i$ are independent events, then

$$\sum_{i=1}^{\infty} P(B_i) = \infty \qquad \text{implies} \quad P[B_i \text{ i.o.}] = 1.$$

**Proof.** Essentially the same as the proof of Corollary 2.8.2. ∎

**Corollary 2.8.4.** Let $\{Y_i, \mathscr{F}_i, i \geq 1\}$ be an adapted stochastic sequence and $\{b_i, i \geq 1\}$ a sequence of positive constants. Then on the event $[\sum_{i=2}^{\infty} P(|Y_i| > b_i \mid \mathscr{F}_{i-1}) < \infty]$,

$$\left[ \sum_{i=1}^{\infty} Y_i \text{ converges} \right] = \left[ \sum_{i=1}^{\infty} Y_i I(|Y_i| \leq b_i) \text{ converges} \right].$$

(To say that on $A$, $B = C$ means that $P[B \triangle C \mid A] = 0$.)

**Proof.** The proof is left as an exercise. ∎

Let $\mathscr{B}_n = \mathscr{B}(X_i, 1 \leq i \leq n)$ for each $n \geq 1$, $\mathscr{B}_{\infty} = \mathscr{B}(X_i, i \geq 1)$, and $Z$ be an integrable random variable. Let

$$T_n = E[Z \mid \mathscr{B}_n]$$

for $1 \leq n \leq \infty$. We have seen by Example 2.5.3 that $\{T_n, \mathscr{B}_n, 1 \leq n < \infty\}$ is a martingale. Note that this martingale has a "last" member $T_\infty$ in the sense that $E[T_\infty \mid \mathscr{B}_n] = T_n$ a.s.

Let $\mathscr{B}_{-n} = \mathscr{B}(X_i, i \geq n)$ for each $n \geq 1$, and $\mathscr{B}_{-\infty} = \bigcap_{n=1}^{\infty} \mathscr{B}(X_i, i \geq n)$. Let

$$T_n = E[Z \mid \mathscr{B}_n]$$

for each $-\infty \leq n \leq -1$. By Example 2.5.3 we see that $\{T_n, \mathscr{B}_n, -\infty \leq n \leq -1\}$ is a martingale with a first member $T_{-\infty}$. What are the convergence properties of these two martingales? For $n \geq 1$, interpreting $E[Z \mid \mathscr{B}_n]$ as the expected value of $Z$, given the behavior of $X_1, X_2, \ldots, X_n$, it seems plausible that

$$E[Z \mid \mathscr{B}_n] \to E[Z \mid \mathscr{B}_\infty]$$

in some sense as $n \to \infty$. Likewise it seems plausible that

$$E[Z \mid \mathscr{B}_n] \to E[Z \mid \mathscr{B}_{-\infty}]$$

as $n \to -\infty$.

**Theorem 2.8.6.** (i) $E[Z \mid \mathscr{B}_n] \to E[Z \mid \mathscr{B}_\infty]$ a.s. and in $L_1$ as $n \to \infty$.

(ii) $E[Z \mid \mathscr{B}_n] \to E[Z \mid \mathscr{B}_{-\infty}]$ a.s. and in $L_1$ as $n \to -\infty$.

**Proof.** (i) Fix $n \geq 1$. $E \mid T_n \mid = E \mid E[Z \mid \mathscr{B}_n] \mid \leq E\{E[\mid Z \mid \mid \mathscr{B}_n]\}$ $= E \mid Z \mid$. Hence $\sup_{n \geq 1} E \mid T_n \mid < \infty$. Thus, $T_n$ converges almost surely to a random variable $T$ as $n \to \infty$ by the basic convergence theorem. We must establish that $T = E[Z \mid \mathscr{B}_\infty]$ a.s. Let

$$\phi(A) = \int_A T \, dP \quad \text{and} \quad \psi(A) = \int_A E[Z \mid \mathscr{B}_\infty] \, dP$$

define $\phi$ and $\psi$ for all $A \in \mathscr{F}$. Recall from basic measure theory that two finite measures equal on a field are equal on the $\sigma$ field generated by the field.

Thus, noting that $T$ and $E[Z \mid \mathscr{B}_\infty]$ are $\mathscr{B}_\infty$ measurable (without loss of generality modifying $T$ on a set of probability zero to make $T$ $\mathscr{B}_\infty$ measurable) it suffices to show that $\phi(A) = \psi(A)$ for all $A \in \bigcup_{i=1}^\infty \mathscr{B}_i$ in order to prove $T = E[Z \mid \mathscr{B}_\infty]$ a.s. Hence it suffices to show that $\phi(A) = \psi(A)$ for all $A \in \mathscr{B}_n$ for arbitrary fixed $n \geq 1$. Fix $n \geq 1$. Recalling that $\{T_n, \mathscr{B}_n, 1 \leq n \leq \infty\}$ is a martingale, by Corollary 2.6.1, $\{\mid T_n \mid, \mathscr{B}_n, 1 \leq n \leq \infty\}$ is a submartingale. Hence

$$\int_{\mid T_n \mid > N} \mid T_n \mid \, dP \leq \int_{\mid T_n \mid > N} \mid T_\infty \mid \, dP$$

$$\leq \int_{\sup_{n \geq 1} \mid T_n \mid > N} \mid T_\infty \mid \, dP \to 0$$

as $N \to \infty$ since $T_\infty$ is integrable and $\sup_{n \geq 1} \mid T_n \mid < \infty$ a.s. Thus $\{T_n, n \geq 1\}$ is uniformly integrable. Choose $A \in \mathscr{F}_n$. Since $T_m \to T$ a.s. and the $T_m$ are uniformly integrable, $T_m \to T$ in $L_1$ according to a basic result of measure theory. Hence,

$$\lim_{m \to \infty} \int_A T_m \, dP = \int_A T \, dP.$$

By the martingale property

$$\int_A T_m \, dP = \int_A E[Z \mid \mathscr{B}_\infty] \, dP \qquad \text{for all} \quad m \geq n.$$

Hence

$$\int_A E[Z \mid \mathscr{B}_\infty] \, dP = \int_A T \, dP$$

for all $A \in \mathscr{F}_n$; that is $\phi(A) = \psi(A)$ for all $A \in \mathscr{F}_n$ as desired, establishing (i).

(ii)   To prove this result, we cannot appeal to the basic submartingale convergence theorem. Rather, one must go back and derive an upcrossings inequality for $\{T_n, \mathscr{B}_n, n \leq -1\}$. Refer to Lemma 2.7.1 and Theorem 2.7.1. Let $H_N$ be the number of upcrossings of $[a, b]$ by $T_{-N}, T_{-N+1}, \ldots, T_{-1}$. The result of Theorem 2.7.1 then becomes

$$E(H_N) \leq E[(T_{-1} - a)^+]/(b - a).$$

It then follows exactly as in the proof of the basic convergence theorem that

$$P[\limsup_{n \to -\infty} T_n = \liminf_{n \to -\infty} T_n] = 1,$$

that is

$$P[(T_n \to \infty) \cup (T_n \to -\infty) \cup (T_n \text{ converges}) \text{ as } n \to -\infty] = 1.$$

By the Fatou lemma

$$E[\liminf_{n \to -\infty} | T_n |] \leq \liminf_{n \to -\infty} E \mid T_n \mid \leq E \mid T_{-1} \mid < \infty.$$

Thus there exists a random variable $T$ such that $T_n \to T$ a.s.

Since $\{T_n, \mathscr{B}_n, -\infty \leq n \leq -1\}$ is a martingale, $\{| T_n |, \mathscr{B}_n, -\infty \leq n \leq -1\}$ is a submartingale. Fix $n \leq -1$ and $N \geq 1$. Hence

$$\int_{|T_n|>N} | T_n | \, dP \leq \int_{|T_n|>N} | T_1 | \, dP$$

$$\leq \int_{\sup_{n \geq 1}|T_n|>N} | T_1 | \, dP \to 0 \qquad \text{as} \quad N \to \infty.$$

Thus $\{T_n, n \le -1\}$ is uniformly integrable. Choose $A \in \mathscr{B}_{-\infty}$. Since $T_n \to T$ a.s. as $n \to -\infty$ and the $T_n$ are uniformly integrable, $T_n \to T$ in $L_1$ as $n \to -\infty$. It then follows as in (i) using the martingale property that

$$\int_A E[Z \mid \mathscr{B}_{-\infty}]\, dP = \int_A T_n\, dP \to \int_A T\, dP.$$

Hence

$$T = E[Z \mid \mathscr{B}_{-\infty}] \quad \text{a.s.}$$

as desired. ∎

**Exercise 2.8.6.** (Doob). In addition to the hypotheses of Theorem 2.8.6, suppose $Z$ is $\mathscr{B}_\infty$ measurable. Letting $T_n = E[Z \mid \mathscr{B}_n]$ for $n \ge 1$, prove that $E[|\,T_n - Z\,|\,|\,\mathscr{B}_n] \to 0$ a.s. and in $L_1$. Hint: If you discover the right trick, the proof is very short. ∎

In Chapter 3, Theorem 2.8.6 (ii) will be an essential tool in the presentation of a martingale proof of the famous strong law of large numbers for independent identically distributed random variables.

The following example gives an interesting application of Theorem 2.8.6 to periodic functions.

**Example 2.8.5.** Let $\Omega = [0, 1)$, $\mathscr{F}$ be the Borel sets and $P$ be Lebesgue measure. Let $f$ be a measurable integrable real-valued function defined on $\Omega$. Extend $f$ periodically to $[0, 2)$. Let

$$f_n(x) = \sum_{i=1}^{2^{-n}} 2^n f(x + i2^n)$$

for each $n \le -1$ and $0 \le x \le 1$ define a sequence of functions on $\Omega$. Note that $f$ continuous implies that

$$f_n(x) \to \int_0^1 f(y)\, dy \quad \text{for each} \quad 0 \le x < 1$$

since the $f_n$ are successive Riemann sums of $\int_0^1 f(y)\, dy$. We use Theorem 2.8.6 to drop this assumption of continuity. For each $n \le -1$, let $\mathscr{A}_n$ be the Borel sets of $[0, 1)$ obtained by taking the Borel sets of $[0, 2^n)$ and extending them periodically to $[0, 1)$. For example, for $n = -2$, $[0, 1/5]$ $\subset [0, 1/4)$ extended becomes

$$[0, 1/5] \cup [1/4, 9/20] \cup [1/2, 7/10] \cup [3/4, 19/20] \subset [0, 1].$$

Clearly $\mathscr{A}_{n-1} \subset \mathscr{A}_n$ and $f_n$ is $\mathscr{A}_n$ measurable for each $n \leq -1$. Moreover, that

$$E[f \mid \mathscr{A}_n] = f_n \quad \text{a.s. for each} \quad n \leq -1$$

can be shown. To see this, fix $n \leq -1$ and $A \in \mathscr{A}_n$. Let $A_n$ be the Borel set of $[0, 2^n)$ which extends periodically to form $A$.

$$\int_A f_n(x)\, dx = 2^{-n} \int_{A_n} f_n(x)\, dx = \sum_{i=1}^{2^{-n}} \int_{A_n} f(x + i2^n)\, dx$$
$$= \int_A f(x)\, dx,$$

thus establishing that $E[f \mid \mathscr{A}_n] = f_n$ a.s. Hence $\{f_n, \mathscr{A}_n, n \leq -1\}$ is a martingale. By Theorem 2.8.6,

$$f_n \to E[f \mid \mathscr{A}_\infty] \quad \text{a.s. as} \quad n \to -\infty,$$

where $\mathscr{A}_\infty = \bigcap_{n \leq -1} \mathscr{A}_n$. We already know that $f$ continuous implies that $E[f \mid \mathscr{A}_\infty] = \int_0^1 f(x)\, dx$ a.s. It follows easily that $\mathscr{A}_\infty = \{\varnothing, \Omega\}$, except possibly for null sets. Thus for all measurable integrable $f$,

$$f_n \to \int_0^1 f(x)\, dx \quad \text{a.s.} \tag{2.8.1}$$

This result has an interesting application to classical real analysis. Suppose $f$ is periodic with period $2^{-n}$ for each $n \geq 1$. Then using Eq. (2.8.1) and the fact that $f = f_n$ for each $n \leq -1$, it follows that $f$ is constant almost surely. ∎

Theorem 2.8.6 (ii) has an easy generalization:

**Exercise 2.8.7.** Let $\{T_n, \mathscr{F}_n, n \leq -1\}$ be a martingale. Prove that $T_n$ converges almost surely as $n \to -\infty$. ∎

Theorem 2.8.6 has application to the study of differentiation of integrals as the following example shows.

**Example 2.8.6.** Let $0 = x_0^{(n)} < x_1^{(n)} < \cdots < x_n^{(n)} = 1$ define a set of increasingly finer partitions becoming dense in $[0, 1]$. Let $f$, defined on $[0, 1]$, be measurable and Lebesgue integrable. Let $A_1^{(n)} = [0, x_1^{(n)}]$ and $A_i^{(n)} = (x_{i-1}^{(n)}, x_i^{(n)}]$ for $2 \leq i \leq n$ and each $n \geq 1$. Let

$$f_n(x) = \int_{A_i^{(n)}} f(y)\, dy / (x_i^{(n)} - x_{i-1}^{(n)})$$

when $x \in A_i^{(n)}$ for each $x \in [0, 1]$ define a sequence of functions $\{f_n, n \geq 1\}$. For each $n \geq 1$, let $\mathscr{F}_n$ be the $\sigma$ field generated by unions of the $A_i^{(n)}$. Let $\Omega = [0, 1]$, $\mathscr{F}$ be the Borel sets of $[0, 1]$, and $P$ be Lebesgue measure. Choose $A \in \mathscr{F}_n$ for fixed $n \geq 1$.

$$\int_A f(x)\, dx = \int_A f_n(x)\, dx$$

is easily verified. Thus

$$E[f \mid \mathscr{F}_n] = f_n \qquad \text{a.s.}$$

for each $n \geq 1$.

Let $\mathscr{F}_\infty$ be the $\sigma$ field generated by $\bigcup_{n=1}^\infty \mathscr{F}_n$. By Theorem 2.8.6 $f_n \to E[f \mid \mathscr{F}_\infty]$ a.s. as $n \to \infty$. But since the $x_i^{(n)}$ become dense in $[0, 1]$, $\mathscr{F}_\infty$ is just the Borel sets and $E[f \mid \mathscr{F}_\infty] = f$ a.s. Thus $f_n \to f$ a.s. $\text{Lim}_{n \to \infty} f_n$ can be viewed as the "derivative" of $\int f$ with respect to the given partition. Thus we get that the "derivative" of $\int f$ is $f$. To get the Lebesgue result that the derivative of the integral is the function a.s. (independent of the choice of partition) requires a deeper analysis. ∎

A simple but clever application of Theorem 2.8.6 is given by Doob [1953, p. 342–343] in proving a result of Jessen [1934] about integration in infinitely many dimensions:

***Example 2.8.7.*** Let $\Omega$ be $[0, 1] \times [0, 1] \times \cdots = [0, 1]^\infty$ and $P$ be infinite dimensional Lebesgue measure. Let $Y_i(\omega)$ be the $i$th coordinate of $\omega$ for each $\omega \in \Omega$ define a sequence of random variables $\{Y_i, i \geq 1\}$. The $Y_i$ are easily shown to be independent identically distributed random variables, each uniformly distributed on $[0, 1]$. Let $T$ be an integrable random variable. For each $\omega = (y_1, y_2, \ldots) \in \Omega$, let

$$T_{-n}(\omega) = \int_0^1 \cdots \int_0^1 T(x_1, x_2, \ldots, x_n, y_{n+1}, y_{n+2}, \ldots)\, dx_1\, dx_2 \cdots dx_n$$

for each $n \geq 1$ and

$$T_n(\omega) = \int_0^1 \cdots \int_0^1 \cdots T(y_1, y_2, \ldots, y_n, x_{n+1}, x_{n+2}, \ldots)\, dx_{n+1}\, dx_{n+2} \cdots$$

for each $n \geq 1$ define random variable sequences $\{T_n, n \leq -1\}$ and $\{T_n, n \geq 1\}$. It seems intuitive that $T_n \to T$ and $T_{-n} \to ET$ as $n \to \infty$. Let $\mathscr{G}_n = \mathscr{B}(Y_1, Y_2, \ldots, Y_n)$ and $\mathscr{G}_{-n} = \mathscr{B}(Y_n, Y_{n+1}, \ldots)$ for each $n \geq 1$.

Let $A$ be a Borel set of $R_n$, $n$-dimensional Euclidean space. Fix $n \geq 1$.

$$\int_{A \times [0,1]^\infty} T_n \, dP$$

$$= \int_0^1 \cdots \int_0^1 \cdots \left[ \int \cdots \int_A T(y_1, y_2, \ldots, y_n, \ldots) \, dy_1 \, dy_2 \cdots dy_n \right]$$

$$dy_{n+1} \, dy_{n+2} \cdots = \int_{A \times [0,1]^\infty} T \, dP.$$

Thus

$$E[T \mid \mathscr{G}_n] = T_n \qquad \text{a.s. for each} \quad n \geq 1. \tag{2.8.2}$$

Let $A$ be a Borel set of $R_\infty$, an infinite dimensional Euclidean space. Fix $n \geq 1$.

$$\int_{[0,1]^n \times A} T_{-n} \, dP$$

$$= \int_A \cdots \int \left[ \int_0^1 \cdots \int_0^1 T(y_1, y_2, \ldots, y_n, \ldots) \, dy_1 \, dy_2 \cdots dy_n \right]$$

$$dy_{n+1} \, dy_{n+2} \cdots$$

$$= \int_{[0,1]^n \times A} T \, dP.$$

Thus

$$E[T \mid \mathscr{G}_{-n}] = T_{-n} \qquad \text{a.s.} \tag{2.8.3}$$

Let $\mathscr{G}_\infty = B(Y_i, i \geq 1)$ (the Borel sets of $R_\infty$) and $\mathscr{G}_{-\infty} = \bigcap_{n=1}^\infty \mathscr{G}_{-n}$. Applying Theorem 2.8.6 to Eqs. (2.8.2) and (2.8.3) we obtain that $T_n \to E[T \mid \mathscr{G}_\infty]$ a.s. and $T_{-n} \to E[T \mid \mathscr{G}_{-\infty}]$ a.s. as $n \to \infty$. Since $T$ is $\mathscr{G}_\infty$ measurable, $E[T \mid \mathscr{G}_\infty] = T$ a.s. $E[T \mid \mathscr{G}_{-\infty}]$ is, of course, $\mathscr{G}_{-\infty}$ measurable. It is a basic result of the theory of independence that all events of $\mathscr{G}_{-\infty}$ have $P$ measure 0 or 1 (see Theorem 2.12.3). Thus $E[T \mid \mathscr{G}_{-\infty}]$ is constant almost surely. Hence, $E[T \mid \mathscr{G}_{-\infty}] = ET$ a.s. We thus have $T_n \to T$ a.s. and $T_{-n} \to ET$ a.s. as $n \to \infty$, as desired. ∎

It was previously mentioned that the Haar functions form a basis for $L_1[0, 1]$ in the strong sense that the Fourier series of an integrable function converges to that function in $L_1$. The Haar functions also form a basis for $L_1[0, 1]$ in this same strong sense with respect to almost sure convergence. These important facts are easy consequences of martingale theory:

***Example 2.8.8.*** Let $\{X_i, i \geq 1\}$ be the Haar functions. Let $\mathscr{B}_i = \mathscr{B}(X_1, X_2, \ldots, X_i)$ for each $i \geq 1$. Each $\mathscr{B}_i$ is generated by a partition of $[0, 1]$ into subintervals. Given $\mathscr{B}_i$ for fixed $i \geq 1$, the partition generating $\mathscr{B}_{i+1}$ is gotten by splitting one particular subinterval of the partition generating $\mathscr{B}_i$ in half. $\mathscr{B}_1 = \{\varnothing, [0, 1]\}$. Thus for each $i \geq 1$, $\mathscr{B}_i$ is generated by a partition of $[0, 1]$ into exactly $i$ subintervals of $[0, 1]$. Fix $i \geq 1$. Consider the linear space $M_i$ of $\mathscr{B}_i$ measurable random variables. Each element of $M_i$ is constant on each of the $i$ subintervals generating $\mathscr{B}_i$. Thus each element of $M_i$ is a linear combination of the same $i$ indicator random variables; that is, $M_i$ is $i$-dimensional and $M_i \subset L_2[0, 1]$ also. $X_1$, $X_2, \ldots, X_i$ are orthonormal and $\mathscr{B}_i$ measurable. Thus $\{X_1, X_2, \ldots, X_i\}$ is an orthonormal basis for $M_i$. Let $T \in L_2[0, 1]$. $M_i$ is a closed subspace of $L_2[0, 1]$ since it is a finite-dimensional subspace. Let $T_i$ be the orthogonal projection of $T$ on $M_i$. Thus

$$E(T - T_i)Y = 0$$

for each $Y \in M_i$. Taking $Y = I(B)$ for $B \in \mathscr{B}_i$, it follows that

$$\int_B T \, dP = \int_B T_i \, dP$$

for each $B \in \mathscr{B}_i$. But this says that $T_i = E[T \mid \mathscr{B}_i]$ a.s. According to the theory of Fourier series

$$T_i = \sum_{j=1}^{i} E(TX_j)X_j.$$

Thus $E[T \mid \mathscr{B}_i]$ is the $i$th partial sum of the Fourier series of $T$ with respect to the Haar functions, a most useful fact.

Now suppose merely that $T \in L_1[0, 1]$. Is $E[T \mid \mathscr{B}_i]$ still the $i$th partial sum of the Fourier series of $T$? We know that

$$\int_B U \, dP = \int_B \sum_{j=1}^{i} E(UX_j)X_j \, dP$$

for all $U \in L_2[0, 1]$ and all $B \in \mathscr{B}_i$. In particular, this result holds for all indicator functions $U$. But, using a standard measure theoretic argument, this result holds for all simple random variables $U$, all nonnegative integrable random variables $U$, and, hence, for all $U \in L_1[0, 1]$. Thus

$$E[T \mid \mathscr{B}_i] = \sum_{j=1}^{i} E(TX_j)X_j \qquad \text{a.s.}$$

for all $T \in L_1[0, 1]$.

But, using Theorem 2.8.6 (i), we get that

$$\sum_{j=1}^{i} E(TX_j)X_j = E[T \mid \mathcal{B}_i] \to E[T \mid \mathcal{B}_\infty] \qquad \text{a.s.}$$

and in $L_1$, where $\mathcal{B}_\infty$ is the $\sigma$ field generated by $\bigcup_{i=1}^{\infty} \mathcal{B}_i$. $\bigcup_{i=1}^{\infty} \mathcal{B}_i$ is easily seen to generate the Borel sets. Thus $E[T \mid \mathcal{B}_\infty] = T$ a.s. $\{X_i, i \geq 1\}$ is a basis for $L_1[0, 1]$ with respect to both $L_1$ convergence and almost sure convergence and in fact the partial sums of the Fourier series of each function $T$ of $L_1[0, 1]$ converge almost surely and in $L_1$ to $T$. It is worth emphasizing the ease with which martingale theory has enabled us to obtain these results. For related material the reader is referred to the work of Gundy [1966]. For example, it is shown that *every* complete orthonormal system of martingale differences must consist of random variables with at most two nonzero values and with $M_i$ $i$-dimensional. ∎

We have already seen by Exercise 2.8.1 that $\{X_i, \mathcal{F}_i, i \geq 1\}$ a martingale difference sequence with $\sum_{i=1}^{\infty} E(X_i^2) < \infty$ implies that $S_n$ converges almost surely. There is a local version of this result based on the behavior of $\sum_{i=1}^{\infty} E[X_i^2 \mid \mathcal{F}_{i-1}]$.

**Theorem 2.8.7.**  [Doob, 1953, p. 320].   Let $\{X_i, \mathcal{F}_i, i \geq 1\}$ be a martingale difference sequence. Then

$$\sum_{i=1}^{\infty} E[X_i^2 \mid \mathcal{F}_{i-1}] < \infty \qquad \text{implies} \quad S_n \quad \text{converges.}$$

**Proof.**  (Stopping rule method again).  Suppose (replacing $X_1$ with 0, if necessary) without loss of generality that $X_1 = 0$. Fix $K > 0$. Let $t$ be the smallest integer $n \geq 1$ such that $\sum_{i=1}^{n+1} E[X_i^2 \mid \mathcal{F}_{i-1}] > K$ if such an $n$ exists, otherwise let $t = \infty$. $t$ is a stopping rule. Fix $n \geq 1$. Let $S_n^{(t)} = S_{\min(t,n)}$. $\{S_n^{(t)}, n \geq 1\}$ is a martingale with $S_n^{(t)} = \sum_{i=1}^{n} I(t \geq i)X_i$. By orthogonality,

$$E(S_n^{(t)})^2 = E\sum_{i=1}^{n} X_i^2 I(t \geq i) = E\left\{ \sum_{i=1}^{n} E[I(t \geq i)X_i^2 \mid \mathcal{F}_{i-1}] \right\}$$

$$= E\left\{ \sum_{i=1}^{n} I(t \geq i)E[X_i^2 \mid \mathcal{F}_{i-1}] \right\}$$

$$= E\left\{ \sum_{i=1}^{\min(t,n)} E[X_i^2 \mid \mathcal{F}_{i-1}] \right\} \leq K.$$

Thus using the fact that $|a| \leq a^2 + 1$,

$$E |S_n^{(t)}| \leq E(S_n{}^t)^2 + 1 \leq K + 1.$$

Thus $S_n^{(t)}$ converges almost surely by the basic convergence theorem. Therefore, $S_n$ converges almost surely on $t = \infty$ for each $K > 0$ and hence on $[\sum_{i=1}^{\infty} E(X_i{}^2 \mid \mathscr{F}_{i-1}) < \infty]$. ∎

Theorem 2.8.7 can be applied to obtain a conditional three series theorem, a result which will fully flower when the $X_i$ are independent.

**Theorem 2.8.8.** Let $\{X_i, \mathscr{F}_i, i \geq 1\}$ be an adapted sequence of random variables. Let $C$ be a positive constant. Then $S_n$ converges almost surely on the event where

$$\sum_{i=1}^{\infty} [P \mid X_i \mid \geq C \mid \mathscr{F}_{i-1}] < \infty, \tag{2.8.4}$$

$$\sum_{i=1}^{\infty} E[X_i I(\mid X_i \mid \leq C) \mid \mathscr{F}_{i-1}] \qquad \text{converges,} \tag{2.8.5}$$

and

$$\sum_{i=1}^{\infty} \{E[X_i{}^2 I(\mid X_i \mid \leq C) \mid \mathscr{F}_{i-1}] - E^2[X_i I(\mid X_i \mid \leq C) \mid \mathscr{F}_{i-1}]\} \qquad \text{converges.} \tag{2.8.6}$$

**Proof.** Let $A$ be the event where Eqs. (2.8.4), (2.8.5), and (2.8.6) hold. Since Eq. (2.8.4) holds on $A$, it follows from Corollary 2.8.4 that

$$\left[ \sum_{i=1}^{\infty} X_i \text{ converges} \right] = \left[ \sum_{i=1}^{\infty} X_i I(\mid X_i \mid \leq C) \text{ converges} \right] \qquad \text{on} \quad A.$$

Thus, since Eq. (2.8.5) holds on $A$,

$$\left[ \sum_{i=1}^{\infty} X_i \text{ converges} \right] = \left\{ \sum_{i=1}^{\infty} X_i I(\mid X_i \mid \leq C) - E[X_i I(\mid X_i \mid \leq C) \mid \mathscr{F}_{i-1}] \right\}$$

$$\text{converges on} \quad A.$$

Let

$$Y_i = X_i I(\mid X_i \mid \leq C) - E[X_i I(\mid X_i \mid \leq C) \mid \mathscr{F}_{i-1}]$$

for $i \geq 1$. $\{Y_i, i \geq 1\}$ is a martingale difference sequence with

$$E[Y_i{}^2 \mid \mathscr{F}_{i-1}] = E[X_i{}^2 I(\mid X_i \mid \leq C) \mid \mathscr{F}_{i-1}] - E^2[X_i I(\mid X_i \mid \leq C) \mid \mathscr{F}_{i-1}].$$

By Theorem 2.8.7, since Eq. (2.8.6) holds on $A$, $\sum_{i=1}^{\infty} Y_i$ converges on $A$. Since

$$\left[\sum_{i=1}^{\infty} Y_i \text{ converges}\right] = \left[\sum_{i=1}^{\infty} X_i \text{ converges}\right] \quad \text{on} \quad A,$$

the theorem is proved. ∎

**Corollary 2.8.5.** [Chow, 1965]. Let $(X_i, \mathscr{F}_i, i \geq 1)$ be a martingale difference sequence. Then

$$\sum_{i=1}^{\infty} E[|X_i|^p \mid \mathscr{F}_{i-1}] < \infty \quad \text{for some} \quad 1 \leq p \leq 2 \qquad (2.8.7)$$

implies that $S_n$ converges.

**Proof.** We show that Eq. (2.8.7) implies Eq. (2.8.4), (2.8.5), and (2.8.6) after which the result follows by Theorem 2.8.8. Fix $C > 0$. For $i \geq 1$, noting that $p \geq 1$,

$$P[|X_i| \geq C \mid \mathscr{F}_{i-1}] = E[I(|X_i| \geq C) \mid \mathscr{F}_{i-1}]$$
$$\leq E[|X_i|^p I(|X_i| \geq C) \mid \mathscr{F}_{i-1}]/C^p$$
$$\leq E[|X_i|^p \mid \mathscr{F}_{i-1}]/C^p \quad \text{a.s.}$$

(Note we have really proved a conditional Chebyshev inequality.) Hence Eq. (2.8.7) implies Eq. (2.8.4).

Since $(X_i, \mathscr{F}_i, i \geq 1)$ is a martingale difference sequence and noting that $p \geq 1$

$$\sum_{i=2}^{\infty} |E[X_i I(|X_i| \leq C) \mid \mathscr{F}_{i-1}]|/C = \sum_{i=2}^{\infty} |E[X_i I(|X_i| > C) \mid \mathscr{F}_{i-1}]|/C$$
$$\leq \sum_{i=2}^{\infty} E[(|X_i|/C)I(|X_i| > C) \mid \mathscr{F}_{i-1}]$$
$$\leq \sum_{i=2}^{\infty} E[|X_i|^p \mid \mathscr{F}_{i-1}]/C^p < \infty \quad \text{a.s.},$$

that is, Eq. (2.8.7) implies Eq. (2.8.5). Since $p \leq 2$,

$$\sum_{i=1}^{\infty} E[X_i^2 I(|X_i| \leq C) \mid \mathscr{F}_{i-1}]/C^2 \leq \sum_{i=1}^{\infty} E[|X_i|^p \mid \mathscr{F}_{i-1}]/C^p \quad \text{a.s.}$$

that is, Eq. (2.8.7) implies Eq. (2.8.4). ∎

**Exercise 2.8.8.** Let $\{X_i, \mathscr{F}_i, i \geq 1\}$ be an adapted sequence of random variables. Prove that $\sum_{i=1}^{\infty} E[|X_i|^p \mid \mathscr{F}_{i-1}] < \infty$ for some $0 < p < 1$ implies that $S_n$ converges. ∎

Corollary 2.8.5 and Exercise 2.8.8 have shown that $\{X_i, \mathscr{F}_i, i \geq 1\}$ a martingale difference sequence with $\sum_{i=1}^{\infty} E[|X_i|^p \mid \mathscr{F}_{i-1}] < \infty$ for some $0 < p \leq 2$ implies that $S_n$ converges. What can be said when $p > 2$? It seems plausible that $\sum_{i=1}^{\infty} E[|X_i|^p \mid \mathscr{F}_{i-1}] < \infty$ for some $p > 2$ implies that $S_n$ converges. This turns out to be untrue:

**Example 2.8.9.** Let $\{X_i, i \geq 1\}$ be independent with $P[X_i = i^{-1/2}] = P[X_i = -i^{-1/2}] = \frac{1}{2}$ for each $i \geq 1$.

$$\sum_{i=1}^{\infty} E |X_i|^p = \sum_{i=1}^{\infty} i^{-p/2} < \infty \qquad \text{for all} \quad p > 2.$$

$$\sum_{i=1}^{\infty} EX_i^2 = \sum_{i=1}^{\infty} i^{-1} = \infty.$$

Since $E \sup X_i^2 < \infty$, it follows by Theorem 2.11.2 of Section 2.11 (see the proof of (i) implying (ii)) that $S_n$ diverges almost surely. ∎

However, a convergence result concerning the conditional $p$th moments for $p > 2$ can be stated.

**Theorem 2.8.9.** [Chow, 1965]. Let $(X_i, \mathscr{F}_i, i \geq 1)$ be a martingale difference sequence and $\{b_i, i \geq 1\}$ be a sequence of positive constants such that $\sum_{i=1}^{\infty} b_i < \infty$. Then

$$\sum_{i=1}^{\infty} b_i^{1-p/2} E[|X_i|^p \mid \mathscr{F}_{i-1}] < \infty \qquad \text{for some} \quad p \geq 2 \qquad (2.8.8)$$

implies that $S_n$ converges.

**Proof.** Suppose $p > 2$ since the result is known for $p = 2$. We show that $\sum_{i=1}^{\infty} b_i^{1-p/2} E[|X_i|^p \mid \mathscr{F}_{i-1}] < \infty$ implies that $\sum_{i=1}^{\infty} E[X_i^2 \mid \mathscr{F}_{i-1}] < \infty$, thus establishing the result by Theorem 2.8.7. Fix $i \geq 1$. Let $Y_i = E^{2/p}[|X_i|^p \mid \mathscr{F}_{i-1}]$. $Y_i > b_i$ if and only if $E^{2/p-1}[|X_i|^p \mid \mathscr{F}_{i-1}] < b_i^{1-p/2}$. Thus

$$Y_i = Y_i[I(Y_i \leq b_i) + I(Y_i > b_i)]$$
$$\leq b_i + E^{2/p-1}[|X_i|^p \mid \mathscr{F}_{i-1}]E[|X_i|^p \mid \mathscr{F}_{i-1}]I(Y_i > b_i)$$
$$\leq b_i + b_i^{1-p/2}E[|X_i|^p \mid \mathscr{F}_{i-1}].$$

$\sum_{i=1}^{\infty} b_i < \infty$ by hypothesis. So (2.8.8) implies $\sum_{i=1}^{\infty} E^{2/p}[|X_i|^p \mid \mathscr{F}_{i-1}] < \infty$.

By the conditional convexity inequality $(E[\Phi(X) \mid \mathscr{G}] \geq \Phi(E[X \mid \mathscr{G}])$ a.s. when $\Phi$ is convex and $E \mid \Phi(X) \mid < \infty$, $E \mid X \mid < \infty$), $E[X_i^2 \mid \mathscr{F}_{i-1}] \leq E^{2/p}[\mid X_i \mid^p \mid \mathscr{F}_{i-1}]$ a.s. Thus Eq. (2.8.8) also implies $\sum_{i=1}^{\infty} E[X_i^2 \mid \mathscr{F}_{i-1}] < \infty$. ∎

**Corollary 2.8.6.** Let $(X_i, \mathscr{F}_i, i \geq 1)$ be a martingale difference sequence. Then

$$\sum_{i=1}^{\infty} E[\mid X_i \mid^p \mid \mathscr{F}_{i-1}]\{i(\log i)^{1+\varepsilon}\}^{p/2-1} < \infty \qquad (2.8.9)$$

for some $\varepsilon > 0$ and $p \geq 2$ implies $S_n$ converges.

**Proof.**   Immediate from Theorem 2.8.9. ∎

Corollary 2.8.6 is sharp in the sense that Eq. (2.8.9) can hold almost surely for $\varepsilon = 0$ and all $p > 2$ and yet the $S_n$ diverge almost surely:

**Example 2.8.10.**   Let $\{X_i, i \geq 1\}$ be independent with

$$P[X_i = (i \log i \log \log i)^{-1/2}] = P[X_i = -(i \log i \log \log i)^{-1/2}] = \tfrac{1}{2}$$

for each $i \geq 3$. Then $\sum_{i=1}^{\infty} EX_i^2 = \infty$ and hence as in Example 2.8.9 $S_n$ diverges almost surely. Yet Eq. (2.8.9) clearly holds for $\varepsilon = 0$ and all $p > 2$. ∎

Corollary 2.8.5, Exercise 2.8.8, and Theorem 2.8.9 make it seem plausible that

$$\sum_{i=1}^{\infty} b_i E[\mid X_i \mid^p \mid \mathscr{F}_{i-1}] < \infty \qquad \text{for some}\ \ 0 < p < 2$$

and constants $b_i \downarrow 0$ sufficiently slowly implies that $S_n$ converges. But this is also untrue.

**Example 2.8.11.**   Let $b_i \downarrow 0$. Choose $0 \leq p_i \leq \tfrac{1}{2}$ for all $i \geq 1$ such that $\sum_{i=1}^{\infty} p_i = \infty$ and $\sum_{i=1}^{\infty} b_i p_i < \infty$. (Convince yourself that such a choice is possible.) Let $\{X_i, i \geq 1\}$ be independent with $P[X_i = 1] = P[X_i = -1] = p_i$ and $P[X_i = 0] = 1 - 2p_i$ for all $i \geq 1$. Then

$$\sum_{i=1}^{\infty} EX_i^2 = 2 \sum_{i=1}^{\infty} p_i = \infty,$$

implying, as in Example 2.8.9, that $S_n$ diverges almost surely. Yet

$$\sum_{i=1}^{\infty} b_i E \mid X_i \mid^p = 2 \sum_{i=1}^{\infty} b_i p_i < \infty$$

for all $0 < p < 2$. ∎

The preceding results and examples should be carefully compared to understand fully the role played by the conditional absolute moments of the $X_i$ in implying the convergence of the martingale $\{S_n, n \geq 1\}$.

## 2.9. Sums of Squares and Martingale Transforms

Throughout Section 2.9, $(X_i, \mathscr{F}_i, i \geq 1)$ will be assumed to be a martingale difference sequence. Some of the results of Section 2.9 can be proved in several ways. Here proofs using stopping rule techniques will be given whenever possible. A rough outline of Section 2.9 is as follows: A decomposition result is established for martingales $\{Q_n, n \geq 1\}$ satisfying $\sup E \mid Q_n \mid < \infty$. This decomposition is then used to establish that $\sup E \mid S_n \mid < \infty$ implies $\sum_{i=1}^{\infty} X_i^2 < \infty$ a.s. This result is then used to study the convergence of martingale transforms.

The following lemma is often useful.

**Lemma 2.9.1.**  Let $t$ be a stopping rule and $\{Q_n, n \geq 1\}$ be a martingale. Let $Q_n^{(t)} = Q_{\min(t,n)}$ for each $n \geq 1$. Then

$$E \mid Q_t I(t < \infty) \mid \leq \liminf_{n \to \infty} E \mid Q_n^{(t)} \mid \leq \sup_n \mid EQ_n^{(t)} \mid$$

$$\leq \sup_n E \mid Q_n \mid.$$

**Proof.**  Clearly $Q_n^{(t)} \to Q_t$ a.s. on $t < \infty$. Thus

$$E \mid Q_t I(t < \infty) \mid = E[\liminf \mid Q_n^{(t)} \mid I(t < \infty)]$$
$$\leq \liminf E[\mid Q_n^{(t)} \mid I(t < \infty)]$$
$$\leq \liminf E \mid Q_n^{(t)} \mid.$$

Fix $n \geq 1$.

$$E \mid Q_n^{(t)} \mid = \int \mid Q_n^{(t)} \mid dP = \int_{t \geq n} \mid Q_n \mid dP + \int_{t < n} \mid Q_t \mid dP.$$

$$\int_{t < n} \mid Q_t \mid dP = \sum_{i=1}^{n-1} \int_{t=i} \mid Q_i \mid dP$$
$$\leq \sum_{i=1}^{n-1} \int_{t=i} \mid Q_n \mid dP = \int_{t < n} \mid Q_n \mid dP$$

by the fact that $\{\mid Q_n \mid, n \geq 1\}$ is a submartingale. Thus

$$E \mid Q_n^{(t)} \mid \leq \int_{t \geq n} \mid Q_n \mid dP + \int_{t < n} \mid Q_n \mid dP = E \mid Q_n \mid.$$

Hence

$$\sup E \mid Q_n^{(t)} \mid \;\leq \sup E \mid Q_n \mid$$

as desired.  ∎

If in addition $Q_n$ converges a.s. to a limiting random variable $Q_\infty$, then $I(t < \infty)$ can be replaced by 1 in the above lemma. The proof of this is essentially the same as above.

A basic tool in establishing convergence results in this section is a martingale decomposition theorem due to Gundy.

**Theorem 2.9.1.** [Gundy, 1968]. Let $(Q_n, \mathscr{F}_n, n \geq 1)$ be a martingale satisfying $\sup E \mid Q_n \mid < \infty$. Then for every $K > 0$, $Q_n$ may be decomposed: $Q_n = T_n + U_n + V_n$ for $n \geq 1$ where $\{T_n, n \geq 1\}$, $\{U_n, n \geq 1\}$, and $\{V_n, n \geq 1\}$ are martingales such that

$$\sup E \mid T_n \mid \;< \infty \qquad \text{and} \qquad P[\sup \mid T_n \mid \;> 0] \leq \sup E \mid Q_n \mid /K, \qquad (2.9.1)$$

$$E \sum_{n=1}^{\infty} \mid U_n - U_{n-1} \mid \;\leq 4 \sup E \mid Q_n \mid \qquad \text{(where } U_0 = 0\text{)}, \qquad (2.9.2)$$

and

$$\sup E(V_n)^2 \leq 2K \sup E \mid Q_n \mid \;< \infty. \qquad (2.9.3)$$

We give a proof of Theorem 2.9.1 due to Burkholder [1973]. Central to Burkholder's approach is the following inequality.

**Lemma 2.9.2.** Let $\{Q_n, \mathscr{F}_n, n \geq 1\}$ be a martingale satisfying $\sup E \mid Q_n \mid < \infty$. Let $t$ be the first $n \geq 1$ such that $\mid Q_n \mid \;\geq K$, where $K > 0$. Let $Y_i = Q_i - Q_{i-1}$ for $i \geq 1$, where $Q_0 = 0$. Then

$$E \sum_{i=1}^{t-1} Y_i^2 + E Q_{t-1}^2 = 2 E Q_t Q_{t-1} \leq 2K \sup E \mid Q_n \mid.$$

**Proof.**   Note that $Q_n \to Q_\infty$ a.s. Hence $Q_t$ and $Q_{t-1}$ are well defined. $\mid Q_{t-1} \mid \;\leq K$ and hence

$$E Q_t Q_{t-1} \leq K E \mid Q_t \mid \;\leq K \sup E \mid Q_n \mid$$

by Lemma 2.9.1 and the comment following Lemma 2.9.1. Fix $n \geq 1$.

$$Q_{n-1}^2 = \sum_{i=1}^{n-1} Y_i^2 + 2 \sum_{i=1}^{n-1} Q_{i-1} Y_i.$$

Thus

$$\sum_{i=1}^{n-1} Y_i^2 + Q_{n-1}^2 = 2Q_{n-1}^2 + 2Y_n Q_{n-1} - 2\sum_{i=1}^{n} Q_{i-1}Y_i.$$

That is,

$$\sum_{i=1}^{n-1} Y_i^2 + Q_{n-1}^2 = 2Q_n Q_{n-1} - 2\sum_{i=1}^{n} Q_{i-1}Y_i.$$

Let $v = \min(t, n)$. Clearly the above equality holds with $n$ replaced by $v$.

$$E \sum_{i=1}^{v} Q_{i-1}Y_i = \sum_{i=1}^{n} E[I(t \geq i)Q_{i-1}E(Y_i \mid \mathscr{F}_{i-1})] = 0$$

(recall Exercise 2.6.2). Here we have used the fact that $I(t \geq i)Q_{i-1}$ is bounded. Thus

$$E \sum_{i=1}^{v-1} Y_i^2 + EQ_{v-1}^2 = 2EQ_v Q_{v-1}. \tag{2.9.4}$$

On $t < \infty$, $|Q_v Q_{v-1}| \leq K |Q_v| \leq K |Q_t|$. On $t = \infty$, $|Q_v Q_{v-1}| \leq K^2$. Thus $\sup_{n \geq 1} |Q_v Q_{v-1}| \leq K^2 + K |Q_t|$ is integrable. $\sup_{n \geq 1} Q_{v-1}^2 \leq K^2$ is also integrable. Thus, letting $n \to \infty$ on both sides of Eq. (2.9.4) yields the desired result. ∎

**Proof of Theorem 2.9.1.** Fix $K > 0$. Let $t$ be the first $n \geq 1$ such that $|Q_n| \geq K$. Decompose, letting $Y_i = Q_i - Q_{i-1}$,

$$Y_i = \{Y_i I(t > i) - E[Y_i I(t > i) \mid \mathscr{F}_{i-1}]\}$$
$$+ \{Y_i I(t = i) - E[Y_i I(t = i) \mid \mathscr{F}_{i-1}]\} + \{Y_i I(t < i)\}.$$

Thus $\{Y_i, i \geq 1\}$ is decomposed into the sum of three martingale differences. Let $T_n = \sum_{i=1}^{n} Y_i I(t < i)$ for $n \geq 1$.

$$P[\sup |T_n| > 0] \leq P[t < \infty]$$
$$= P[|Q_t| I(t < \infty) \geq K]$$
$$\leq E |Q_t| I(t < \infty)/K$$
$$\leq \sup E |Q_n|/K$$

by Lemma 2.9.1, as desired. For $t < n$, $|T_n| = |Q_n - Q_t| \leq |Q_n| + |Q_t|$. Thus, $\sup E|T_n| < \infty$.

Let $U_n = \sum_{i=1}^{n} \{Y_i I(t = i) - E[Y_i I(t = i) \mid \mathscr{F}_{i-1}]\}$ for each $n \geq 1$.

$$E \sum_{i=1}^{\infty} \mid U_i - U_{i-1} \mid = E \sum_{i=1}^{\infty} \mid Y_i I(t = i) - E[Y_i I(t = i) \mid \mathscr{F}_{i-1}] \mid$$

$$\leq 2 \sum_{i=1}^{\infty} E \mid Y_i \mid I(t = i)$$

$$= 2E \mid Y_t \mid I(t < \infty)$$

$$\leq 4E \mid Q_t \mid I(t < \infty)$$

$$\leq 4 \sup E \mid Q_n \mid,$$

using Lemma 2.9.1, as desired.

Let $V_n = \sum_{i=1}^{n} \{Y_i I(t > i) - E[Y_i I(t > i) \mid \mathscr{F}_{i-1}]\}$ for each $n \geq 1$.

$$EV_n^2 = \sum_{i=1}^{n} E\{Y_i I(t > i) - E[Y_i I(t > i) \mid \mathscr{F}_{i-1}]\}^2$$

$$\leq \sum_{i=1}^{n} EY_i^2 I(t > i)$$

$$= E \sum_{i=1}^{t-1} Y_i^2$$

$$\leq 2K \sup E \mid Q_n \mid$$

by Lemma 2.9.2. ∎

It is now a simple matter to obtain the basic convergence result for sums of squares of martingale differences.

**Theorem 2.9.2.**   [Austin, 1966].   Suppose $\sup E \mid S_n \mid < \infty$ a.s. Then $\sum_{i=1}^{\infty} X_i^2 < \infty$ a.s.

**Proof.**   Decompose $S_n = T_n + U_n + V_n$ according to the statement of Theorem 2.9.1 for an integer $K \geq 1$. Let $A_k = [\sup \mid T_n \mid = 0]$. Let $Y_n = U_n - U_{n-1}$ and $Z_n = V_n - V_{n-1}$ for $n \geq 1$ where $U_0 = 0$ and $V_0 = 0$. On $A_K$,

$$\sum_{i=1}^{\infty} X_i^2 \leq 2\left(\sum_{i=1}^{\infty} Y_i^2 + \sum_{i=1}^{\infty} Z_i^2\right).$$

By Eq. (2.9.2), $E \sum_{i=1}^{\infty} \mid Y_i \mid < \infty$. Thus $\sum_{i=1}^{\infty} \mid Y_i \mid < \infty$ a.s. which implies that $\sum_{i=1}^{\infty} Y_i^2 < \infty$ a.s. By Eq. (2.9.3), $\sup_{n \geq 1} E(V_n^2) < \infty$. $\sum_{i=1}^{n} EZ_i^2 = EV_n^2$. Hence

$$\sup_{n \geq 1} \sum_{i=1}^{n} EZ_i^2 = E\left(\sum_{i=1}^{\infty} Z_i^2\right) < \infty.$$

This implies that $\sum_{i=1}^{\infty} Z_i^2 < \infty$ a.s. Hence $\sum_{i=1}^{\infty} X_i^2 < \infty$ a.s. on $A_K$ for each integer $K \geq 1$. But by Eq. (2.9.1),

$$P[\sup_{n \geq 1} |T_n| > 0] \leq 2 \sup_{n \geq 1} E |S_n|/K$$

and hence $\bigcup_{K=1}^{\infty} A_K = \Omega$. Thus $\sum_{i=1}^{\infty} X_i^2 < \infty$ a.s. ∎

Gundy's decomposition theorem was presented because of its usefulness in establishing Theorem 2.9.2. It should be emphasized that one can give a short direct proof of Austin's result (as Austin did [1966]). Theorem 2.9.2 has other interesting applications as well. It provides elementary proofs for certain martingale inequalities of Burkholder [1966]. For example, it can be used to show that there exists $C > 0$ such that

$$P\left[\sum_{i=1}^{\infty} X_i^2 \geq K^2\right] \leq C \sup E |S_n|/K$$

and

$$P[\sup_{n \geq 1} |S_n| \geq K] \leq CE\left(\sum_{i=1}^{\infty} X_i^2\right)^{1/2}\bigg/K$$

for every $K > 0$ (see Theorem 3.3.5).

**Exercise 2.9.1.** (An inequality of Doob's [1953, p. 314], related to the Burkholder inequalities mentioned above). Let $\{Q_n, n \geq 1\}$ be a sub-martingale. Fix $K > 0$, $p \geq 1$, and $N \geq 1$.

  (i)  Prove that

$$P[\sup_{n \leq N} Q_n \geq K] \leq EQ_N I(\max_{n \leq N} Q_n \geq K)/K \leq E |Q_N|/K.$$

  (ii)  If, moreover, $\{Q_n, n \geq 1\}$ is a martingale, prove that

$$P[\sup_{n \leq N} |Q_n|^p > K] \leq E |Q_N|^p/K \quad \text{for} \quad p \geq 1. \tag{2.9.5}$$

(When $p = 2$ and the $Q_n - Q_{n-1}$ are independent with mean 0 this is Kolmogorov's famous maximal inequality. It can be made fundamental to the study of the almost sure convergence of partial sums of independent random variables.) ∎

We now establish another martingale decomposition theorem closely related to Gundy's decomposition and then apply it to the study of convergence of martingale transforms.

**Theorem 2.9.3.**   [Chow, 1968].   Let $\{R_i, i \geq 1\}$ be a martingale difference sequence satisfying $E \sup | R_i | < \infty$. Fix $p \geq 1$ and let $\sigma = (\sum_{i=1}^{\infty} | R_i |^p)^{1/p}$. Then for every $K > 0$, $R_i$ may be decomposed $R_i = W_i + Y_i + Z_i$ for $i \geq 1$ where $\{W_i, i \geq 1\}$, $\{Y_i, i \geq 1\}$, and $\{Z_i, i \geq 1\}$ are martingale differences such that

$$P[\sup | W_n | > 0] \leq E\sigma/K$$

and

$$\sigma \leq K \quad \text{implies} \quad \sup | W_i | = 0, \tag{2.9.6}$$

$$E \sum_{n=1}^{\infty} | Y_n | \leq 2E(\sup | R_n |), \tag{2.9.7}$$

and

$$E\left( \sum_{n=1}^{\infty} | Z_n |^p \right) < \infty. \tag{2.9.8}$$

**Proof.**   (One should note the similarity of this proof to the proof of Gundy's martingale decomposition theorem — Theorem 2.9.1.) Fix $K > 0$. Let $t$ be the smallest integer $n \geq 1$ such that $\sigma_n \equiv (\sum_{i=1}^{n} | R_i |^p)^{1/p} > K$ if such an $n$ exists, otherwise let $t = \infty$. Let $\sigma_0 = 0$. For $i \geq 1$,

$$\begin{aligned} R_i &= \{R_i I(t < i)\} + \{R_i I(t = i) - E[R_i I(t = i) \mid \mathscr{F}_{i-1}]\} \\ &\quad + \{R_i I(t > i) - E[R_i I(t > i) \mid \mathscr{F}_{i-1}]\} \\ &\equiv W_i + Y_i + Z_i. \end{aligned} \tag{2.9.9}$$

This is a decomposition into a sum of three martingale differences. $\sigma \leq K$ implies $\sigma_n \leq K$ for all $n \geq 1$ implies $t = \infty$ implies $\sup | W_i | = 0$. Also

$$P[\sup | W_i | > 0] \leq P[t < \infty] = P[\sigma > K] \leq E(\sigma)/K.$$

$$E \sum_{i=1}^{\infty} | Y_i | \leq E \sum_{i=1}^{\infty} | R_i I(t = i) | + E \sum_{i=1}^{\infty} E[| R_i | I(t = i) \mid \mathscr{F}_{i-1}]$$

$$\leq 2 \sum_{i=1}^{\infty} E[| R_i | I(t = i)] = 2E[| R_t | I(t < \infty)] \leq 2E \sup | R_i |.$$

For each $i \geq 1$,

$$\begin{aligned} E | Z_i |^p &= E[| R_i I(t > i) - E[R_i I(t > i) \mid \mathscr{F}_{i-1}] |^p] \\ &\leq 2^p E[| R_i |^p I(t > i)], \end{aligned} \tag{2.9.10}$$

using the elementary inequality

$$| a + b |^p \leq 2^{p-1}(| a |^p + | b |^p).$$

According to an elementary inequality (see Hardy *et al.* [1964, p. 39]), for any $a > 0$, $b > 0$, and $p \geq 1$,

$$a^p - b^p \leq pa^{p-1}(a - b).$$

Applying this,

$$| R_i |^p = \sigma_i^p - \sigma_{i-1}^p \leq p\sigma_i^{p-1}(\sigma_i - \sigma_{i-1}).$$

Thus, using Eq. (2.9.10),

$$\sum_{i=1}^{\infty} E | Z_i |^p \leq 2^p p \sum_{i=1}^{\infty} E[\sigma_i^{p-1}(\sigma_i - \sigma_{i-1})I(t > i)].$$

$t > i$ implies $\sigma_i^{p-1} \leq K^{p-1}$. Hence

$$2^p p \sum_{i=1}^{\infty} E[\sigma_i^{p-1}(\sigma_i - \sigma_{i-1})I(t > i)] \leq 2^p p K^{p-1} E\left[\sum_{i=1}^{\infty} (\sigma_i - \sigma_{i-1})I(t > i)\right]$$

$$\leq 2^p p K^p,$$

completing the proof. ∎

**Definition 2.9.1.** Let $(Y_i, \mathscr{F}_i, i \geq 1)$ be a martingale difference sequence with $v_i$ $\mathscr{F}_{i-1}$ measurable for each $i \geq 1$. Then $(T_n, n \geq 1)$, defined by $T_n = \sum_{i=1}^{n} v_i Y_i$, is called a martingale transform and $(v_i, i \geq 1)$ is called the transforming sequence. ∎

If $E | v_i Y_i | < \infty$ for each $i \geq 1$, the martingale transform $\{T_n, n \geq 1\}$ is clearly a martingale. There is a nice gambling interpretation of a martingale transform. Let $Y_i$ be the gambler's winnings at trial $i$ given that he bets one dollar at each trial. Let $\mathscr{G}_i = \mathscr{B}(Y_1, Y_2, \ldots, Y_i)$ for each $i \geq 1$ and $\mathscr{G}_0 = \{\varnothing, \Omega\}$. Suppose the game is fair in the sense that $E[Y_i | \mathscr{G}_{i-1}] = 0$ a.s. for each $i \geq 1$. In order to make the gambler's role more interesting we can allow him to vary his bet from trial to trial. Let $v_i Y_i$ be the gambler's winnings at trial $i$ given that he bets $v_i$ dollars at trial $i$. Suppose the gambler may choose the amount he bets at each trial on the basis of his past luck with the game. That is, the gambler's betting system $\{v_i, i \geq 1\}$ is a stochastic sequence with each $v_i$ being $\mathscr{G}_{i-1}$ measurable. The gambler's accumulated winnings at trial $n$ are given by $\sum_{i=1}^{n} v_i Y_i$. All of the above assumptions imply that $\{T_n, n \geq 1\}$ is a martingale transform. We had a previous example of such a martingale transform in the continuation of Example 2.5.1 given in Section 2.8 (p. 45).

Suppose in the above gambling example that $\sup_{n \geq 1} E \mid \sum_{i=1}^{n} Y_i \mid < \infty$. If the gambler bets one dollar in each trial, then $\sum_{i=1}^{\infty} Y_i$ converges almost surely. Thus although initially $\{\sum_{i=1}^{n} Y_i, n \geq 1\}$ may fluctuate, $\sum_{i=1}^{n} Y_i$ eventually gets close to a fixed amount and stays close. Can the gambler change this asymptotic inactivity of the accumulated winnings by introducing a system $\{v_i, i \geq 1\}$ of bets? The answer is no, provided the gambler is limited in the amount he bets in the sense that $\sup \mid v_i \mid < \infty$ a.s.

**Theorem 2.9.4.**    (Burkholder's basic convergence result for martingale transforms [1966]). Let the martingale difference sequence $(X_i, \mathscr{F}_i, i \geq 1)$ have a transforming sequence $(v_i, i \geq 1)$. Assume

$$\sup E \mid S_n \mid < \infty \quad \text{and} \quad \sup \mid v_i \mid < \infty \quad \text{a.s.}$$

Then the martingale transform $T_n = \sum_{i=1}^{n} v_i X_i$ converges almost surely.

**Proof.**    (Due to Chow [1968]).    Fix a constant $L > 0$. Let $t$ be the smallest $n \geq 1$ such that $\mid S_n \mid \geq L$ if such an $n$ exists, otherwise let $t = \infty$. Let

$$R_i = v_i X_i I(t \geq i) I(\mid v_i \mid \leq L)$$

for $i \geq 1$. $\{R_i, i \geq 1\}$ is a martingale difference sequence since $v_i I(\mid v_i \mid \leq L) I(t \geq i)$ is $\mathscr{F}_{i-1}$ measurable. The main step of the proof is that $\sum_{i=1}^{\infty} R_i$ converges a.s. We shall use Chow's martingale decomposition theorem (Theorem 2.9.3). In order to apply this decomposition theorem $E \sup \mid R_i \mid < \infty$ must be shown. For $i \geq 1$,

$$\mid R_i \mid \leq L \mid X_i \mid I(t \geq i) = L I(t = i) \mid X_i \mid + L I(t > i) \mid X_i \mid$$
$$\leq L \mid X_t \mid I(t < \infty) + L I(t > i)(\mid S_i \mid + \mid S_{i-1} \mid)$$
$$\leq L(\mid S_t \mid + \mid S_{t-1} \mid) I(t < \infty) + 2L^2 \leq L \mid S_t \mid I(t < \infty) + 3L^2.$$

By Lemma 2.9.1, $E \mid S_t \mid I(t < \infty) < \infty$ and hence $E \sup \mid R_i \mid < \infty$. We apply Chow's decomposition theorem to $R_i$ for some $K > 0$ and $p = 2$. Thus

$$R_i \equiv W_i + Y_i + Z_i \quad \text{with} \quad \sum_{i=1}^{\infty} R_i^2 \leq K^2$$

implying

$$\sup \mid W_i \mid = 0, \quad \text{with} \quad E \sum_{i=1}^{\infty} \mid Y_i \mid < \infty \quad \text{a.s., and} \quad E \sum_{i=1}^{\infty} Z_i^2 < \infty.$$

Let $A_K = [\sum_{i=1}^{\infty} R_i^2 \le K^2]$. On $A_K$, $\sup |W_i| = 0$ as mentioned above; hence, $\sum_{i=1}^{\infty} W_i$ converges a.s. on $A_K$. Trivially $\sum_{i=1}^{\infty} Y_i$ converges a.s. since $\sum_{i=1}^{\infty} E|Y_i| < \infty$. $E\sum_{i=1}^{\infty} Z_i^2 < \infty$ implies that $\sup E|\sum_{i=1}^{\infty} Z_i| < \infty$ and hence that $\sum_{i=1}^{\infty} Z_i$ converges a.s. Thus, on $A_K$, $\sum_{i=1}^{\infty} R_i$ converges a.s. But

$$\sum_{i=1}^{\infty} R_i^2 \le L^2 \sum_{i=1}^{\infty} X_i^2 < \infty \qquad \text{a.s.}$$

by the definition of the $R_i$ and Austin's theorem (Theorem 2.9.2) since $\sup E|S_n| < \infty$. Thus $\Omega = \bigcup_{K=1}^{\infty} A_K$ and hence $\sum_{i=1}^{\infty} R_i$ converges a.s. Since

$$\sup |v_i| < \infty \quad \text{a.s. and} \qquad \sup |S_n| < \infty \quad \text{a.s.,}$$

it follows by the definition of the $R_i$ and $t$ that $T_n = \sum_{i=1}^{n} v_i X_i$ converges a.s., using the fact that $\Omega = \bigcup_{L=1}^{\infty} \{[\sup |v_n| \le L] \cap [\sup |S_n| \le L]\}$. ∎

**Exercise 2.9.2.** Convince yourself that a Haar series may be represented as a martingale transform of the Rademacher functions plus a constant. ∎

In the gambling example discussed prior to Theorem 2.9.4, there is (as Burkholder [1966] points out) one way in which the introduction of a gambling system $\{v_i, i \ge 1\}$ could make the gambler's fate more interesting. Without a gambling system, $\sum_{i=1}^{n} Y_i \to T$ a.s. with

$$E|T| \le \lim \inf \left| \sum_{i=1}^{n} Y_i \right| < \infty.$$

But, although $\sup |v_i| < \infty$ a.s. implies $\sum_{i=1}^{n} v_i Y_i \to U$ a.s., it is not necessarily true that $E|U| < \infty$. Thus, a judicious choice of a system $\{v_i, i \ge 1\}$ can increase the variability of the gambler's ultimate winnings. Example 2.9.1 illustrates this.

**Example 2.9.1.** Let $\Omega$ be the positive integers with $P(\{k\}) = k^{-1} -(k+1)^{-1}$ for $k \ge 1$. Let

$$T_n(k) = \begin{cases} n & \text{if } 0 \le n < k \\ -1 & \text{if } n \ge k. \end{cases}$$

It follows that $\{T_n, n \ge 1\}$ is a martingale satisfying $\sup E|T_n| < \infty$. (Verify this.) Clearly $T_n \to -1$ a.s. Let $v_i = (-1)^{i+1}$ for $i \ge 1$. Then $\sum_{i=1}^{n} v_i(k)(T_i(k) - T_{i-1}(k)) = 0$ if $n < k$ and $n$ even, or 1 if $n < k$ and

$n$ odd. When $n \geq k$, $\sum_{i=1}^{n} v_i(k)(T_i(k) - T_{i-1}(k)) = k + 1$ if $k$ is even, or $-k$ if $k$ is odd. Let $U(k) = \lim \sum_{i=1}^{n} v_i(k)(T_i(k) - T_{i-1}(k))$. $U$ exists as an almost sure limit. $P[| U | \geq k] \geq k^{-1}$ by the above computations. Hence $E | U | = \infty$.  ∎

**Exercise 2.9.3.**   Let $E \sup | X_i |^2 < \infty$. Prove that $\sum_{i=1}^{\infty} X_i^2 < \infty$ implies $S_n$ converges. Hint: Let $t$ be the smallest $n \geq 1$ such that $\sum_{i=1}^{n} X_i^2 \geq K$, etc.  ∎

Recall that $E(\sum_{i=1}^{\infty} X_i^2) < \infty$ implies that $\sum_{i=1}^{\infty} X_i$ converges a.s. Recently Davis [1970] has shown that

$$E\left( \sum_{i=1}^{\infty} X_i^2 \right)^{1/2} < \infty \qquad \text{if and only if} \quad E \sup | S_n | < \infty.$$

Since $E \sup | S_n | < \infty$ implies that $\sup E | S_n | < \infty$ which in turn implies that $S_n$ converges a.s., we have that

$$E\left( \sum_{i=1}^{\infty} X_i^2 \right)^{1/2} < \infty \qquad \text{implies} \quad S_n \text{ converges a.s.,}$$

a result due to Burkholder [1966]. Recall that

$$E \sum_{i=1}^{\infty} | X_i |^p a_i^{1-p/2} < \infty \qquad \text{and} \qquad \sum_{i=1}^{\infty} a_i < \infty \qquad (2.9.11)$$

for constants $a_i > 0$ and some $p \geq 2$ implies that $S_n$ converges a.s. The preceeding two remarks suggest

**Theorem 2.9.5.**   [Stout, 1967].   If

$$E\left( \sum_{i=1}^{\infty} | X_i |^p a_i^{1-p/2} \right)^{1/p} < \infty \qquad \text{and} \qquad \sum_{i=1}^{\infty} a_i < \infty$$

for constants $a_i > 0$ and some $p \geq 2$, then $S_n$ converges a.s.

**Proof.**   [Chow, 1968].   Let $R_n = X_n a_n^{1/p-1/2}$ for $n \geq 1$ and use Chow's martingale decomposition theorem and Eq.(2.9.11).  ∎

**Exercise 2.9.4.**   Finish the proof of Theorem 2.9.5.  ∎

There is an interesting result by Davis [1968] that deserves mention, although we will not prove it:

**Theorem 2.9.6.** Let $\{X_i, i \geq 1\}$ and $\{Y_i, i \geq 1\}$ each be martingale difference sequences with respect to the *same* $\sigma$ fields $\{\mathscr{F}_i, i \geq 1\}$. Assume $\sup E \mid \sum_{i=1}^{n} X_i \mid < \infty$. Then

$$\sum_{i=1}^{\infty} Y_i^2 \leq \sum_{i=1}^{\infty} X_i^2 \quad \text{a.s. implies} \quad \sum_{i=1}^{\infty} Y_i \text{ converges a.s.} \quad (2.9.12)$$

and

$$\sup_{i \geq 1} \mid Y_i \mid \leq \sup_{n \geq 1} \left| \sum_{i=1}^{n} X_i \right| \quad \text{a.s.} \quad (2.9.13)$$

implies that $[\sum_{i=1}^{\infty} Y_i \text{ converges}] = [\sum_{i=1}^{\infty} Y_i^2 < \infty] = [\sup_{n \geq 1} \sum_{i=1}^{n} Y_i < \infty]$. Eq. (2.9.12) sharpens an earlier result of Burkholder's by replacing the hypothesis $\sum_{i=1}^{n} Y_i^2 \leq \sum_{i=1}^{n} X_i^2$ a.s. for each $n \geq 1$ by $\sum_{i=1}^{\infty} Y_i^2 \leq \sum_{i=1}^{\infty} X_i^2$ a.s.

**Exercise 2.9.5.** According to a remark made above, $E(\sum_{i=1}^{\infty} X_i^2)^{1/2} < \infty$ implies $S_n$ converges almost surely. Use this result to sharpen Exercise 2.9.3 by replacing $E \sup \mid X_i \mid^2 < \infty$ by $E \sup \mid X_i \mid < \infty$. ▌

## 2.10. Convergence of Martingale Transforms Satisfying a Local Regularity Condition

In Section 2.10 we introduce a regularity condition for martingale differences which had its origins with Marcinkiewicz and Zygmund [1938] and was formulated in the martingale case by Gundy [1967]. This condition is rather general and leads to some very nice characterizations of local convergence of martingale transforms.

**Definition 2.10.1.** A martingale difference sequence $\{X_i, \mathscr{F}_i, i \geq 1\}$ is said to be regular in the sense of Marcinkiewicz and Zygmund (regular MZ) if

$$\infty > E[\mid X_i \mid \mid \mathscr{F}_{i-1}] \geq \delta E^{1/2}[X_i^2 \mid \mathscr{F}_{i-1}] \quad \text{a.s.} \quad (2.10.1)$$

for all $i \geq 1$ and some $\delta > 0$. ▌

**Definition 2.10.2.** Given a stochastic sequence $(U_i, i \geq 1)$ defined on the probability space $(\Omega, \mathscr{F}, P)$, a stochastic sequence $(V_i, i \geq 1)$ defined on a possibly different probability space is said to be a representation of $(U_i, i \geq 1)$ if the two stochastic sequences have the same joint distributions. ▌

Our purpose is to study the convergence of martingale transforms $T_n = \sum_{i=1}^{n} v_i X_i$, when the $X_i$ are regular MZ. Often, in applications, $E[X_i^2 \mid \mathcal{F}_{i-1}] = 1$ a.s. for each $i \geq 1$ will also be assumed. One could think of a martingale difference sequence $\{X_i, i \geq 1\}$ satisfying $E[X_i^2 \mid \mathcal{F}_{i-1}] = 1$ a.s. for each $i \geq 1$ as being "conditionally" orthonormal. We refer to a regular MZ martingale difference sequence satisfying $E[X_i^2 \mid \mathcal{F}_{i-1}] = 1$ a.s. for each $i \geq 1$ as being *normed* regular MZ. A martingale transform $\{T_n, n \geq 1\}$ with the $X_i$ regular MZ may always be represented without loss of generality as a transform of a normed regular MZ martingale difference sequence as the following theorem implies.

**Theorem 2.10.1.** Let $(X_i, \mathcal{F}_i, i \geq 1)$ be regular MZ. Then there exists a martingale difference sequence $(Y_i, \mathcal{G}_i, i \geq 1)$ with transforming sequence $\{E^{1/2}[X_i^2 \mid \mathcal{F}_{i-1}], i \geq 1\}$ such that $(Y_i, \mathcal{G}_i, i \geq 1)$ is normed regular MZ and $(E^{1/2}[X_i^2 \mid \mathcal{F}_{i-1}] Y_i, i \geq 1)$ is a representation of $(X_i, i \geq 1)$.

*Proof.* Since $(X_i, \mathcal{F}_i, i \geq 1)$ is regular MZ, there exists a constant $\delta > 0$ for which

$$\infty > E[\mid X_i \mid \mid \mathcal{F}_{i-1}] \geq \delta E^{1/2}[X_i^2 \mid \mathcal{F}_{i-1}] \qquad \text{a.s. for each } i \geq 1.$$

Thus $E[X_i^2 \mid \mathcal{F}_{i-1}] < \infty$ a.s. for each $i \geq 1$. Enlarging the probability space, if necessary, let $(R_i, i \geq 1)$ be a stochastic sequence independent of $(X_i, i \geq 1)$ with the $R_i$ mutually independent and $P[R_i = 1] = P[R_i = -1] = \frac{1}{2}$. For $i \geq 1$ let

$$\mathcal{G}_i = \mathcal{B}(X_1, X_2, \ldots, X_i, R_1, R_2, \ldots, R_i), \qquad \mathcal{G}_0 = (\varnothing, \Omega),$$

and

$$Y_i = \begin{cases} \dfrac{X_i}{E^{1/2}[X_i^2 \mid \mathcal{F}_{i-1}]} & \text{if } E[X_i^2 \mid \mathcal{F}_{i-1}] > 0, \\ R_i & \text{if } E[X_i^2 \mid \mathcal{F}_{i-1}] = 0. \end{cases}$$

$E[X_i^2 \mid \mathcal{F}_{i-1}] = 0$ implies $X_i = 0$, by the definition of conditional expectation. Thus $E^{1/2}[X_i^2 \mid \mathcal{F}_{i-1}] Y_i = X_i$ a.s.; hence $\{E^{1/2}[X_i^2 \mid \mathcal{F}_{i-1}] Y_i, i \geq 1\}$ is a representation of $\{X_i, i \geq 1\}$.

$$E[Y_i^2 \mid \mathcal{G}_{i-1}] = E\left[\frac{X_i^2}{E[X_i^2 \mid \mathcal{F}_{i-1}]} I(E[X_i^2 \mid \mathcal{F}_{i-1}] > 0) \mid \mathcal{G}_{i-1}\right]$$

$$+ E[R_i^2 I(E[X_i^2 \mid \mathcal{F}_{i-1}] = 0) \mid \mathcal{G}_{i-1}]$$

$$= I(E[X_i^2 \mid \mathcal{F}_{i-1}] > 0) \cdot 1 + I(E[X_i^2 \mid \mathcal{F}_{i-1}] = 0) \cdot 1 = 1 \quad \text{a.s.}$$

Similarly $E[Y_i \mid \mathscr{G}_{i-1}] = 0$ a.s. $E^{1/2}[X_i^2 \mid \mathscr{F}_{i-1}]$ is $\mathscr{G}_{i-1}$ measurable and hence is a transforming sequence. ∎

Thus, if $T_n = \sum_{i=1}^{n} v_i X_i$ forms a martingale transform with respect to $\sigma$ fields $\mathscr{F}_n$ and regular $X_i$ there exists a representation $T_n' = \sum_{i=1}^{n} (v_i E^{1/2} [X_i^2 \mid \mathscr{F}_{i-1}]) Y_i$ such that $T_n'$ forms a martingale transform with respect to $\sigma$-fields $\mathscr{G}_n$ and the martingale differences $Y_i$ are normed regular MZ.

The following simple lemma clarifies the probabilistic meaning of normed MZ regularity.

**Lemma 2.10.1.** (Gundy [1967]). Let the martingale difference sequence $(X_i, \mathscr{F}_i, i \geq 1)$ be normed in the sense that $E[X_i^2 \mid \mathscr{F}_{i-1}] = 1$ a.s. for each $i \geq 1$. Then $\{X_i, i \geq 1\}$ is MZ regular if and only if there exists $\alpha > 0$ and $\beta > 0$ such that

$$P[|X_i| \geq \alpha \mid \mathscr{F}_{i-1}] \geq \beta \qquad \text{a.s.}$$

for all $i \geq 1$.

**Proof.** Assume $\{X_i, i \geq 1\}$ MZ regular. Thus there exists a constant $\delta > 0$ such that $E[|X_i| \mid \mathscr{F}_{i-1}] \geq \delta > 0$ a.s. for all $i \geq 1$. Choose $0 < \alpha < \delta$. Fix $i \geq 1$.

$$\delta \leq E[|X_i| I(|X_i| < \alpha) \mid \mathscr{F}_{i-1}] + E[|X_i| I(|X_i| \geq \alpha) \mid \mathscr{F}_{i-1}]$$
$$\leq \alpha + E[|X_i| I(|X_i| \geq \alpha) \mid \mathscr{F}_{i-1}] \qquad \text{a.s.}$$

Hence, using the conditional Cauchy–Schwarz inequality

$$(\delta - \alpha)^2 \leq E^2\{|X_i| I(|X_i| \geq \alpha) \mid \mathscr{F}_{i-1}\}$$
$$\leq E[|X_i|^2 \mid \mathscr{F}_{i-1}] P[|X_i| \geq \alpha \mid \mathscr{F}_{i-1}]$$
$$= 1 \cdot P[|X_i| \geq \alpha \mid \mathscr{F}_{i-1}] \qquad \text{a.s.}$$

Take $\beta = (\delta - \alpha)^2$, establishing the result one way.

Assume there exists constants $\alpha > 0$ and $\beta > 0$ such that $P[|X_i| \geq \alpha \mid \mathscr{F}_{i-1}] \geq \beta$ a.s. for all $i \geq 1$.

$$E[|X_i| \mid \mathscr{F}_{i-1}] \geq E[|X_i| I(|X_i| \geq \alpha) \mid \mathscr{F}_{i-1}] \geq \alpha P[|X_i| \geq \alpha \mid \mathscr{F}_{i-1}]$$
$$\geq \alpha\beta > 0 \qquad \text{a.s.}$$

for all $i \geq 1$. ∎

The following result is important in that it indicates the wide variety of martingale differences that are normed regular MZ. Let $\mathscr{G}_i = \mathscr{B}(X_1, X_2, \ldots, X_i)$ throughout Section 2.10.

**Theorem 2.10.2.**  (i)  Let $(X_i, \mathscr{F}_i, i \geq 1)$ be a martingale difference sequence satisfying $|X_i| \leq K < \infty$. a.s. and $E[X_i^2 \mid \mathscr{F}_{i-1}] = 1$ a.s. for all $i \geq 1$. Then $(X_i, \mathscr{F}_i, i \geq 1)$ is normed regular MZ.

(ii)  Let the $X_i$ be independent with $EX_i = 0$ and $EX_i^2 = 1$ for all $i \geq 1$ and the $X_i^2$ uniformly integrable. (That is, $E[X_i^2 I(|X_i| \geq N)] \to 0$ as $N \to \infty$ uniformly in $i$.) Then $(X_i, \mathscr{G}_i, i \geq 1)$ is normed regular MZ.

(iii)  Let $X_i$ be independent and symmetric with $EX_i^2 < \infty$ for all $i \geq 1$. Then $(X_i, i \geq 1)$ can be represented by a martingale difference sequence $(w_i Y_i, i \geq 1)$ with $w_i \mathscr{H}_{i-1}$ measurable for each $i \geq 1$ and $(Y_i, \mathscr{H}_i, i \geq 1)$ normed regular MZ.

**Proof.**  (i)  $E[|X_i| \mid \mathscr{F}_{i-1}] \geq E[X_i^2 \mid \mathscr{F}_{i-1}]/K = 1/K$ a.s. for $i \geq 1$.

(ii)  $E[X_i^2 \mid \mathscr{G}_{i-1}] = EX_i^2 = 1$      a.s.

$$
\begin{aligned}
E[|X_i| \mid \mathscr{G}_{i-1}] &= E|X_i| \geq E[|X_i| I(|X_i| \leq N)] \\
&\geq E[X_i^2 I(|X_i| \leq N)]/N \\
&= (1 - E[X_i^2 I(|X_i| > N)])/N \qquad \text{a.s.}
\end{aligned}
$$

Since the $X_i^2$ are uniformly integrable, it follows that there exists a constant $\delta > 0$ such that $E[|X_i| \mid \mathscr{G}_{i-1}] \geq \delta$ a.s. for all $i \geq 1$.

(iii)  Enlarging the probability space, if necessary, let $\{R_i, i \geq 1\}$ be a stochastic sequence independent of $\{X_i, i \geq 1\}$ with the $R_i$ mutually independent and $P[R_i = 1] = P[R_i = -1] = \frac{1}{2}$. Fix $i \geq 1$. Let

$$
\mathscr{H}_{i-1} = \mathscr{B}(X_1, X_2, \ldots, X_{i-1}, |X_i|, R_1, R_2, \ldots, R_{i-1}),
$$

$$
Y_i = \begin{cases} \dfrac{X_i}{|X_i|} & \text{if } X_i \neq 0 \\ R_i & \text{if } X_i = 0, \end{cases}
$$

and $w_i = |X_i|$. $(w_i Y_i, i \geq 1)$ is trivially a representation of $(X_i, i \geq 1)$. $w_i$ is $\mathscr{H}_{i-1}$ measurable by definition of $w_i$ and $\mathscr{H}_{i-1}$.

$$
\begin{aligned}
E[Y_i^2 \mid \mathscr{H}_{i-1}] &= E[(X_i/|X_i|)^2 I(|X_i| \neq 0) \mid \mathscr{H}_{i-1}] \\
&\quad + E[R_i^2 I(|X_i| = 0) \mid \mathscr{H}_{i-1}] = 1 \qquad \text{a.s.}
\end{aligned}
$$

$$
\begin{aligned}
E[w_i Y_i \mid \mathscr{H}_{i-1}] &= E[X_i I(|X_i| \neq 0) \mid \mathscr{H}_{i-1}] \\
&= I(|X_i| \neq 0) E[X_i \mid \mathscr{H}_{i-1}] = 0 \qquad \text{a.s.}
\end{aligned}
$$

by independence and symmetry. $|Y_i| = 1$ a.s. implies that $\{Y_i, \mathscr{H}_i, i \geq 1\}$ is MZ regular. Hence $(w_i Y_i, \mathscr{H}_i, i \geq 1)$ is a martingale difference sequence with $(Y_i, \mathscr{H}_i, i \geq 1)$ normed regular MZ. ∎

Recalling the definition of a symmetric stochastic sequence (Definition 2.8.1 (i)), it would seem that Theorem 2.10.2 (iii) could be extended to symmetric stochastic sequences:

**Exercise 2.10.1.** Let $\{X_i, \mathscr{F}_i, i \geq 1\}$ be a symmetric adapted stochastic sequence. Prove that $\{X_i, i \geq 1\}$ can be represented by a martingale difference sequence $\{w_i Y_i, i \geq 1\}$ with $w_i$ $\mathscr{H}_{i-1}$ measurable for each $i \geq 1$ and $\{Y_i, \mathscr{H}_i, i \geq 1\}$ normed regular MZ. Hint: Proceed as in the proof of Theorem 2.10.2 (iii). The conclusion of Exercise 2.8.3 can be helpful. ∎

Theorem 2.10.2 and Exercise 2.10.1 show that certain conditions imposed on the martingale difference sequence $\{X_i, i \geq 1\}$ imply MZ regularity. Another approach is to impose conditions on the $\sigma$ fields of the martingale difference sequence.

**Definition 2.10.3.** An event $A$ contained in a $\sigma$ field $\mathscr{G}$ is an atom of $\mathscr{G}$ if $B \subset A$, $B \in \mathscr{G}$ implies $P(B) = P(A)$ or $P(B) = 0$. An atom of a $\sigma$ field $\mathscr{G}$ is thus an event which cannot be split into two $\mathscr{G}$ events, both of positive probability. ∎

**Definition 2.10.4.** A $\sigma$ field $\mathscr{G}$ is said to be atomic if it is generated by a countable set of disjoint atoms. That is, there exists disjoint atoms $\{A_i, i \geq 1\}$ of $\mathscr{G}$ such that $A = \bigcup_j A_{i_j}$ for each $A \in \mathscr{G}$. ∎

**Definition 2.10.5.** An increasing sequence of atomic $\sigma$ fields is said to be regular if there exists a constant $\delta > 0$ such that

$$P(E_{k+1})/P(E_k) \geq \delta$$

for all $k \geq 1$, and all atoms $E_k \in \mathscr{F}_k$, atoms $E_{k+1} \in \mathscr{F}_{k+1}$ such that $P(E_k) > 0$, $P(E_{k+1}) > 0$, and $E_{k+1} \subset E_k$. ∎

**Theorem 2.10.3.** Let $(X_i, \mathscr{F}_i, i \geq 1)$ be a martingale difference sequence with atomic $\sigma$ fields $\mathscr{F}_i$ being regular. Then $(X_i, \mathscr{F}_i, i \geq 1)$ is regular MZ.

**Proof.**  Fix $i \geq 1$. Letting

$$E[X_i^2 \mid \mathscr{F}_{i-1}] = E[X_i^2 \mid E_{i-1}]I(E_{i-1})$$

on each positive probability atom $E_{i-1}$ of $\mathscr{F}_{i-1}$ yields one version of the conditional expectation of $X_i^2$ given $\mathscr{F}_{i-1}$. ($EX_i^2 = \infty$ is possible, but $E[X_i^2 \mid \mathscr{F}_{i-1}]$ behaves like a conditional expectation.) Let $E$ be a positive probability atom of $\mathscr{F}_{i-1}$ and let $\{E_j^i, j = 1, \ldots, j_i\}$ be a collection of positive probability atoms of $\mathscr{F}_i$ such that $\bigcup_j E_j^i = E$.

$$E[X_i^2 \mid E] = \sum_{j=1}^{j_i} E[X_i^2 \mid E_j^i]P[E_j^i \mid E]$$

$$\geq \sum_{j=1}^{j_i} E[X_i^2 \mid E_j^i]\delta$$

for some $1 > \delta > 0$, using the assumption that the $\mathscr{F}_i$ are regular. Since $X_i^2 = E[X_i^2 \mid E_j^i]$ a.s. on $E_j^i$ it follows that

$$| X_i | \leq \delta^{-1/2}E^{1/2}[X_i^2 \mid E] \qquad \text{a.s.}$$

on $E$. Hence

$$| X_i | \leq \delta^{-1/2}E^{1/2}[X_i^2 \mid \mathscr{F}_{i-1}] \qquad \text{a.s.}$$

Enlarging the probability space if necessary, let $\{R_i, i \geq 1\}$ be a stochastic sequence independent of $\{X_i, i \geq 1\}$ with the $R_i$ independent and $P[R_i = 1] = P[R_i = -1] = \frac{1}{2}$ for $i \geq 1$. Fix $i \geq 1$. Let

$$Y_i = \begin{cases} X_i/E^{1/2}[X_i^2 \mid \mathscr{F}_{i-1}] & \text{if } E^{1/2}[X_i^2 \mid \mathscr{F}_{i-1}] > 0 \\ R_i & \text{if } E^{1/2}[X_i^2 \mid \mathscr{F}_{i-1}] = 0. \end{cases}$$

$| Y_i | \leq \delta^{-1/2}$ a.s. $\{Y_i, i \geq 1\}$ is a martingale difference sequence. By Theorem 2.10.3 (i), $\{Y_i, i \geq 1\}$ is normed regular MZ. It follows that $\{X_i, i \geq 1\}$ is regular MZ. ∎

It is instructive to look at some of the examples of martingale difference sequences and see whether the associated $\sigma$ fields are regular (thus implying regularity of the martingale difference sequences).

Recall in Example 2.8.8 that the partial sums of a Haar series form a martingale. Here the $\sigma$ fields are clearly atomic and regular since each $\sigma$ field is generated from the preceding one by splitting one of the atoms of the preceding $\sigma$ field into two sets of equal Lebesgue measure.

However, in Example 2.8.2 the asymmetry of the independent random variables $Y_i$ implies that the $\sigma$ fields $\mathscr{B}_i = \mathscr{B}(Y_1, Y_2, \ldots, Y_i)$ are not regular. Let $E_i = (Y_1 = -2, Y_2 = -3, \ldots, Y_i = -(i + 1))$ and $E_{i+1} = (Y_1 = -2, Y_2 = -3, \ldots, Y_{i+1} = -(i + 2))$ for each $i \geq 1$. $E_i \in \mathscr{B}_i$ and $E_{i+1} \in \mathscr{B}_{i+1}$ for each $i \geq 1$. $P(E_i) > 0$ and $E_i$ is an atom of $\mathscr{B}_i$ for each $i \geq 1$.

$$P(E_{i+1})/P(E_i) = (i + 2)^{-2} \to 0$$

as $i \to \infty$. Thus the $\sigma$ fields are not regular. We shall see that this non-regularity is necessary in order that $\sum_{i=1}^{n} Y_i \to \infty$ a.s.

We now come to the main result of Section 2.10. Let $T_n = \sum_{i=1}^{n} v_i X_i$ be a martingale transform sequence with $\{X_i, \mathscr{F}_i, i \geq 1\}$ normed regular MZ. Using Theorem 2.8.7, $\sum_{i=1}^{\infty} v_i^2 < \infty$ implies $T_n$ converges. This, of course, has nothing to do with MZ regularity. However the necessity of $\sum_{i=1}^{\infty} v_i^2 < \infty$ for the convergence of $T_n$ as well as certain other charac-terizations of the convergence of $T_n$ do follow when $\{X_i, \mathscr{F}_i, i \geq 1\}$ is normed MZ regular. These results are combined in the next theorem. Recall that $A$ if and only if $B$ means that $P[A \triangle B] = 0$.

**Theorem 2.10.4.** Let $(X_i, \mathscr{F}_i, i \geq 1)$ be a normed regular MZ martingale difference sequence and $(v_i, i \geq 1)$ a transforming sequence. Then

(i)   $\sum_{i=1}^{\infty} v_i X_i$ converges        if and only if

(ii)  $\sup \sum_{i=1}^{n} v_i X_i < \infty$        if and only if

(iii) $\sup \left| \sum_{i=1}^{n} v_i X_i \right| < \infty$        if and only if

(iv)  $\sum_{i=1}^{\infty} v_i^2 < \infty$        if and only if

(v)   $\sum_{i=1}^{\infty} v_i^2 X_i^2 < \infty.$

**Proof.**   As remarked above, (iv) implying (i) is known to us. (i) implying (ii) and (i) implying (iii) are trivially true. We shall present the rather involved argument which shows that (iii) implies (iv) and (v). (v) implies (i) is proved in a somewhat similar manner and is omitted. Thus, except for the omission, the equivalence of (i), (iii), (iv), and (v) is then proved. The rather involved proof that (ii) implies (iii) is also omitted.

Assume (iii). Fix $M > v_1$ and $K > 0$ to be chosen later and let $t$ be the smallest integer $n \geq 1$ such that $|\sum_{i=1}^{n} v_i X_i| > K$ or $|v_{n+1}| > M$ if such an $n$ exists. Otherwise, let $t = \infty$. We first show that $t = \infty$ implies that

$$\sum_{i=1}^{\infty} v_i^2 X_i^2 < \infty \qquad \text{and} \qquad \sum_{i=1}^{\infty} v_i^2 < \infty.$$

Then we show that

$$P\left[t = \infty \,\bigg|\, \sup |\sum_{i=1}^{n} v_i X_i| < \infty\right] \to 1$$

as $K \to \infty$ and then $M \to \infty$. Once these facts are proved they imply that (iii) implies (iv) and (iii) implies (v).

If we can produce a finite upper bound for

$$\int_{\Omega} \sum_{i=1}^{n} I(t \geq i) v_i^2 \, dP = \int_{\Omega} \sum_{i=1}^{n} I(t \geq i) v_i^2 X_i^2 \, dP$$

independent of $n$, this will prove that $t = \infty$ implies that $\sum_{i=1}^{\infty} v_i^2 < \infty$ and $\sum_{i=1}^{\infty} v_i^2 X_i^2 < \infty$ as desired. This unfortunately takes a large amount of computation. Fix $i \geq 1$. We first show that there exists a constant $\delta > 0$ such that

$$E[v_i^2 X_i^2 I(|v_i X_i| < K^2) \,|\, \mathscr{F}_{i-1}] \geq \delta^2 v_i^2/9 \qquad \text{on} \quad t \geq i. \qquad (2.10.2)$$

By hypotheses, there exists $\delta > 0$ such that $E[|X_i| \,|\, \mathscr{F}_{i-1}] \geq \delta$ a.s. Choose $K$ large enough so that $K^2 \geq 3M/(2\delta)$. $t \geq i$ implies

$$\delta - K^{-2}|v_i| \geq \delta - K^{-2}M \geq \delta/3.$$

Thus, to establish Eq. (2.10.2), it suffices to show that

$$E[v_i^2 X_i^2 I(|v_i X_i| < K^2) \,|\, \mathscr{F}_{i-1}] \geq \delta^2 v_i^2/9 \qquad (2.10.3)$$

on the event $[\delta - K^{-2}|v_i| \geq \delta/3]$.

$$E[|v_i X_i| \, I(|v_i X_i| \geq K^2) \,|\, \mathscr{F}_{i-1}] \leq K^{-2} E[v_i^2 X_i^2 I(|v_i X_i| \geq K^2) \,|\, \mathscr{F}_{i-1}]$$
$$\leq K^{-2} v_i^2 \qquad \text{a.s.}$$

Thus

$$E[|v_i X_i| \, I(|v_i X_i| < K^2) \,|\, \mathscr{F}_{i-1}]$$
$$= E[|v_i X_i| \,|\, \mathscr{F}_{i-1}] - E[|v_i X_i| \, I(|v_i X_i| \geq K^2) \,|\, \mathscr{F}_{i-1}]$$
$$\geq E[|v_i X_i| \,|\, \mathscr{F}_{i-1}] - K^{-2} v_i^2 \qquad \text{a.s.}$$

Using the convexity inequality for conditional expectations,

$$E^{1/2}[v_i^2 X_i^2 I(|v_i X_i| < K^2) | \mathscr{F}_{i-1}] \geq E[|v_i X_i| I(|v_i X_i| < K^2) | \mathscr{F}_{i-1}]$$
$$\geq E[|v_i X_i| | \mathscr{F}_{i-1}] - K^{-2} v_i^2$$
$$\geq \delta |v_i| - K^{-2} v_i^2 \qquad \text{a.s.}$$

The last inequality is by the MZ regularity.

$$\delta |v_i| - K^{-2} v_i^2 = |v_i| (\delta - K^{-2} |v_i|) \geq |v_i| \delta/3$$

on $[\delta - K^{-2} |v_i| \geq \delta/3]$. Thus, Eq. (2.10.3) and hence Eq. (2.10.2) are established.

$$E[v_i^2 X_i^2 I(t \geq i)] = E\{I(t \geq i) E[v_i^2 X_i^2 | \mathscr{F}_{i-1}]\}$$
$$= E[v_i^2 I(t \geq i)] \leq M^2.$$

Using the fact that $t$ is a stopping rule, it follows that $\{v_i X_i I(t \geq i), i \geq 1\}$ is a square integrable martingale difference sequence and hence is orthogonal. Thus, adopting the convention $\sum_{i=1}^{0} (\cdot) = 0$,

$$E\left[\sum_{i=1}^{n} v_i^2 X_i^2 I(t \geq i)\right]$$

$$= E\left\{\left[\sum_{i=1}^{n} v_i X_i I(t \geq i)\right]^2\right\}$$

$$= E\left[\left(\sum_{i=1}^{\min(t,n)} v_i X_i\right)^2\right]$$

$$= E\left[\left(\sum_{i=1}^{n} v_i X_i\right)^2 I(t > n)\right] + E\left[\left(\sum_{i=1}^{t} v_i X_i\right)^2 I(t \leq n)\right]$$

$$\leq K^2 + E\left\{\left[\left(\sum_{i=1}^{t-1} v_i X_i\right)^2 + 2\left(\sum_{i=1}^{t-1} v_i X_i\right) v_t X_t + v_t^2 X_t^2\right] I(t \leq n)\right\}$$

$$\leq 2K^2 + \int_{t \leq n} \left[2\left(\sum_{i=1}^{t-1} v_i X_i\right) v_t X_t + v_t^2 X_t^2\right] dP$$

$$\leq (K^4 + 2K^3 + 2K^2) + \int_{t \leq n, |v_t X_t| \geq K^2} \left[2\left(\sum_{i=1}^{t-1} v_i X_i\right) v_t X_t + v_t^2 X_t^2\right] dP$$

$$= (K^4 + 2K^3 + 2K^2) + \sum_{j=1}^{n} \int_{t=j, |v_j X_j| \geq K^2} \left[2\left(\sum_{i=1}^{t-1} v_i X_i\right) v_j X_j + v_j^2 X_j^2\right] dP$$

$$\leq (K^4 + 2K^3 + 2K^2) + \sum_{j=1}^{n} \int_{t \geq j, |v_j X_j| \geq K^2} (2K^{-1} + 1) v_j^2 X_j^2 \, dP.$$

Hence

$$(K^4 + 2K^3 + 2K^2)$$

$$\geq E\left[\sum_{i=1}^{n} v_i^2 X_i^2 I(t \geq i)\right] - \sum_{i=1}^{n} \int_{t \geq i, |v_i X_i| \geq K^2} (2K^{-1} + 1)v_i^2 X_i^2 \, dP$$

$$= \sum_{i=1}^{n} \int_{t \geq i} v_i^2 X_i^2 \{I(|v_i X_i| < K^2) - 2K^{-1}I(|v_i X_i| \geq K^2)\} \, dP.$$

Since $(t \geq i) \in \mathcal{F}_{i-1}$,

$$(K^4 + 2K^3 + 2K^2) \geq \sum_{i=1}^{n} \int_{t \geq i} E[(v_i^2 X_i^2)\{I(|v_i X_i| < K^2)$$

$$-2K^{-1}I(|v_i X_i| \geq K^2)\} \mid \mathcal{F}_{i-1}] \, dP. \qquad (2.10.4)$$

Using Eqs. (2.10.2) and (2.10.4),

$$(K^4 + 2K^3 + 2K^2) \geq \sum_{i=1}^{n} \int_{t \geq i} (\delta^2/9 - 2K^{-1})v_i^2 \, dP.$$

Hence

$$\infty > (K^4 + 2K^3 + 2K^2)(\delta^2/9 - 2K^{-1})^{-1} \geq \int \sum_{i=1}^{n} I(t \geq i)v_i^2 \, dP$$

for all $K$ sufficiently large for all $n \geq 1$ thus establishing that $t = \infty$ implies that $\sum_{i=1}^{\infty} v_i^2 < \infty$ and $\sum_{i=1}^{\infty} v_i^2 X_i^2 < \infty$.

By definition of $t$, in order to show that

$$\lim_{M \to \infty} \lim_{K \to \infty} P\left[t = \infty \mid \sup |\sum_{i=1}^{n} v_i X_i| < \infty\right] \to 1,$$

it suffices to show that

$$\sup |\sum_{i=1}^{n} v_i X_i| < \infty \qquad \text{implies that} \quad \sup |v_i| < \infty.$$

$\sup |\sum_{i=1}^{n} v_i X_i| < \infty$ implies that $\sup |v_i X_i| < \infty$. Let $A_L \equiv A = [\sup |v_i X_i| \leq L]$ for a constant $L > 0$. By the conditional Borel–Cantelli lemma (Theorem 2.8.5),

$$\sum_{i=1}^{\infty} P[|v_i X_i| > L \mid \mathcal{F}_{i-1}] < \infty \qquad \text{on} \quad A. \qquad (2.10.5)$$

By Lemma 2.10.1, using the hypothesis of normed MZ regularity,

$$P[|v_iX_i| \geq \alpha |v_i|\,|\,\mathscr{F}_{i-1}] \geq \beta \qquad \text{a.s.}$$

for constants $\alpha > 0$ and $\beta > 0$.

$$E[I(|v_iX_i| \geq L)\,|\,\mathscr{F}_{i-1}]$$
$$\geq I(\alpha |v_i| \geq L)E[I(|v_iX_i| \geq L)\,|\,\mathscr{F}_{i-1}]$$
$$\geq I(\alpha |v_i| \geq L)E[I(|v_iX_i| \geq \alpha |v_i|)\,|\,\mathscr{F}_{i-1}]$$
$$\geq \beta I(\alpha |v_i| \geq L).$$

Using Eq. (2.10.5),

$$\sum_{i=1}^{\infty} I(|v_i| \geq L/\alpha) < \infty \qquad \text{on} \quad A.$$

Hence $\sup |v_i| < \infty$ on $A$.
  Since

$$\bigcup_{L=1}^{\infty} A_L \supset \left[\sup \left|\sum_{i=1}^{n} v_iX_i\right| < \infty\right],$$

it follows that $\sup |\sum_{i=1}^{n} v_iX_i| < \infty$ implies $\sup |v_i| < \infty$ as desired. ∎

**Exercise 2.10.2.** (Rather tricky). Prove under the hypotheses of Theorem 2.10.5 that $\sum_{i=1}^{\infty} v_i^2X_i^2 < \infty$ implies that $\sum_{i=1}^{\infty} v_iX_i$ converges; that is, that (v) implies (i). Hint: For fixed $M > v_1$ and $K > 0$, let $t$ be the smallest integer $n \geq 1$ such that $\sum_{i=1}^{n} v_i^2X_i^2 > K$ or $|v_{n+1}| > M$ if such an $n$ exists, otherwise let $t = \infty$. Show for all $n \geq 1$ that $E\{[\sum_{i=1}^{n} I(t \geq i)v_iX_i]^2\} \leq L$ for some constant $L < \infty$. This implies by Doob's martingale convergence theorem that $\sum_{i=1}^{n} I(t \geq i)v_iX_i$ converges a.s. Thus on $t = \infty$, $\sum_{i=1}^{n} v_iX_i$ converges. To complete the proof, show that

$$\lim_{M\to\infty} \lim_{K\to\infty} P\left[t = \infty \,\bigg|\, \sum_{i=1}^{\infty} v_i^2X_i^2 < \infty\right] \to 1. \quad ∎$$

**Remark.** The equivalence of the conditions $\sum_{i=1}^{\infty} v_iX_i$ converges, $\sum_{i=1}^{\infty} v_i^2X_i^2 < \infty$, and $\sum_{i=1}^{\infty} v_i^2 < \infty$ is due to Gundy [1967]. Chow [1969] established the equivalence of $\sum_{i=1}^{\infty} v_iX_i$ converging and $\sup |\sum_{i=1}^{\infty} v_iX_i| < \infty$. Davis [1969] improved Chow's result by establishing the equivalence of $\sum_{i=1}^{\infty} v_iX_i$ converging and $\sup \sum_{i=1}^{\infty} v_iX_i < \infty$. Chow [1963] had already established this one sided result in the special case where the $\mathscr{F}_i$ are atomic. The proof that (iii) implies (iv) and (v) given above is due to Chow [1968]. Another proof can be found in the work of Burkholder and Gundy [1973].

It has already been shown, assuming $\sup | a_i X_i | < \infty$ a.s., that a Haar series cannot converge to $+\infty$ on a set of positive Lebesgue measure (Theorem 2.8.4). The Haar functions are regular MZ because of the regularity of their $\sigma$ fields. Thus Theorem 2.10.4 yields a proof that a Haar series cannot converge to $+\infty$ on a set of positive Lebesgue measure, without assuming $\sup | a_i X_i | < \infty$. For a treatment of the convergence of orthogonal series using the concept of regular $\sigma$ fields, see Gundy [1966].

It is (once again!) interesting to consider a gambling interpretation:

**Example 2.5.1 (cont.).** (See Sections 2.5 and 2.8 (p. 45) for earlier segments of Example 2.5.1.) Let $\{X_i, i \geq 1\}$ be a sequence of independent identically distributed random variables with $P[X_1 = 1] = P[X_1 = -1] = \frac{1}{2}$. Let $v_i$ be $\mathscr{G}_{i-1}$ measurable for each $i \geq 1$. For each $i \geq 1$, $v_i X_i$ represents the gambler's winnings on trial $i$, given that the gambler bets $v_i$ dollars on trial $i$. Let $T_n = \sum_{i=1}^{n} v_i X_i$ for each $n \geq 1$. $T_n$ is for each $n \geq 1$ the gambler's accumulated winnings at trial $n$. As before, our interest is in the asymptotic behavior of $T_n$ for various betting systems $\{v_i, i \geq 1\}$. Theorem 2.10.4 is very informative in this respect. ($\{X_i, \mathscr{G}_i, i \geq 1\}$ is trivially a normed regular MZ martingale difference sequence.) For example any system such that $\sum_{i=1}^{\infty} v_i^2 = \infty$ a.s. guarantees that $\sup T_n = \infty$ a.s., thus enabling the gambler to win as much as he wants, if he is patient enough to wait long enough and knows when to quit. However, $\sum_{i=1}^{\infty} v_i^2 < \infty$ a.s. prevents this possibility since $T_n \to T$ a.s. Thus one crucial aspect of the gambler's betting system is the finiteness of $\sum_{i=1}^{\infty} v_i^2$. ∎

One of the main applications of the material of Section 2.10 is to the convergence of partial sums of independent random variables. This will be presented in Section 2.12.

## 2.11.  Global Regularity

MZ regularity is a local condition. There are certain global regularity conditions which yield similar results to those implied by MZ regularity.

**Theorem 2.11.1.**  Let $\{X_i, i \geq 1\}$ be a martingale difference sequence satisfying the global regularity condition $E \sup | X_i | < \infty$. Then

(i)   $S_n$ converges          if and only if

(ii)  $\sup S_n < \infty$      if and only if

(iii) $\sum\limits_{i=1}^{\infty} X_i^2 < \infty.$

**Proof.** That (ii) implies (i) follows from Theorem 2.8.1, and (i) implies (ii) trivially. Thus (i) and (ii) are equivalent.

That (iii) implies (i) was the content of Exercise 2.9.5. Fix $K > 0$. Let $t$ be the first $n \geq 1$ such that $S_n \geq K$ if such an integer $n$ exists; otherwise let $t = \infty$. Let $S_n^{(t)} = S_{\min(t,n)}$ for each $n \geq 1$. Standard computation shows that the martingale $\{S_n^{(t)}, n \geq 1\}$ satisfies $E \sup_n | S_n^{(t)} | < \infty$. By Austin's result, $\sum_{i=1}^{\infty} X_i^2 I(t \geq i) < \infty$ a.s. Thus, using what is by now a standard argument, (i) implies (iii). ∎

**Theorem 2.11.2.** Let $(X_i, \mathscr{F}_i, i \geq 1)$ be a martingale difference sequence satisfying the global regularity condition $E(\sup | X_i |^2) < \infty$. Then

(i)   $S_n$ converges                  if and only if

(ii)  $\sum\limits_{i=1}^{\infty} E[X_i^2 \,|\, \mathscr{F}_{i-1}] < \infty$      if and only if

(iii) $\sum\limits_{i=1}^{\infty} X_i^2 < \infty.$

**Proof.** (ii) implies (i) by Theorem 2.8.7; this assertion has nothing to do with regularity.

(iii) if and only if (i) follows from Theorem 2.11.1 since

$$E(\sup | X_i |^2) < \infty \qquad \text{implies} \quad E(\sup | X_i |) < \infty.$$

It remains to show that (i) implies (ii). Fix a constant $K > 1$. Let $t$ be the smallest integer $n \geq 1$ such that $| \sum_{i=1}^{n} X_i | > K$ if such an $n$ exists, otherwise, let $t = \infty$. Let $S_n^t = \sum_{i=1}^{n} X_i I(t \geq i)$.

$$(S_n^t)^2 \leq K^2 \qquad \text{if} \quad t > n$$

and

$$(S_n^t)^2 \leq 2(K^2 + \sup X_i^2) \qquad \text{if} \quad t \leq n.$$

Hence

$$\infty > \sup_n E(S_n^t)^2 = \sum_{i=1}^{\infty} EX_i^2 (t \geq i)$$

since $\{X_i I(t \geq i), i \geq 1\}$ is a martingale difference sequence and $E \sup X_i^2 < \infty$.

$$\sum_{i=1}^{\infty} E[X_i^2 I(t \geq i)] = E\left\{ \sum_{i=1}^{\infty} I(t \geq i) E[X_i^2 \mid \mathcal{F}_{i-1}]\right\}.$$

Thus

$$\sum_{i=1}^{\infty} I(t \geq i) E[X_i^2 \mid \mathcal{F}_{i-1}] < \infty \qquad \text{a.s.}$$

Hence $t = \infty$ implies $\sum_{i=1}^{\infty} E[X_i^2 \mid \mathcal{F}_{i-1}] < \infty$ a.s. ∎

**Remark.** The equivalence of (i) and (ii) above is due to Doob [1953, p. 319].

Chow [1968] establishes an interesting global result analogous to Theorem 2.10.5:

**Theorem 2.11.3.** Let $(X_i, i \geq 1)$ be a martingale difference sequence with $EX_i^2 = 1$ and $E \mid X_i \mid \geq \delta$ for some $\delta > 0$ and all $i \geq 1$. Then, given a sequence of numbers $\{a_i, i \geq 1\}$

(i)   $\displaystyle\sum_{i=1}^{\infty} a_i X_i$ converges      if and only if

(ii)   $\displaystyle\sup \left| \sum_{i=1}^{n} a_i X_i \right| < \infty$      if and only if

(iii)   $\displaystyle\sum_{i=1}^{\infty} a_i^2 < \infty$.

**Proof.** Omitted. See Chow [1968]. ∎

## 2.12. Convergence of Partial Sums of Independent Random Variables

Throughout Section 2.12, $\{X_i, i \geq 1\}$ will denote a sequence of independent random variables and $\mathcal{B}_i = B(X_1, X_2, \ldots, X_i)$ for each $i \geq 1$. In studying the convergence of $S_n$, three lines of attack will be used. First, the martingale theory of previous sections will be quite helpful. Second, Kolmogorov's 0–1 law will establish the impossibility of local convergence. Third, centering at medians and symmetrization will produce characterizations of the convergence of $S_n$ not possible in the more general case of martingale differences.

Many martingale results of the previous sections translate immediately into important results about convergence in the case of independence. The reason is that a sequence of independent random variables with mean zero forms a martingale difference sequence:

**Theorem 2.12.1.** Suppose $EX_i = 0$ for $i \geq 2$. Then $\{X_i, \mathscr{B}_i, i \geq 2\}$ is a martingale difference sequence.

**Proof.** For each $i \geq 2$, $E[X_i \mid \mathscr{B}_{i-1}] = EX_i = 0$ a.s. ∎

For convenient reference we list some of the important convergence results for partial sums of independent random variables which follow immediately from Theorem 2.12.1 and already established martingale convergence results:

**Theorem 2.12.2.** (i) $\sup E \mid S_n \mid < \infty$ and $EX_i = 0$ if $i \geq 2$ imply that $S_n$ converges almost surely to a random variable $S$ for which $E \mid S \mid < \infty$.

(ii) $\sum_{i=1}^{\infty} E \mid X_i \mid^p < \infty$ for $1 < p \leq 2$ and $EX_i = 0$ for $i \geq 2$ imply $S_n$ converges almost surely. (Trivially for any sequence of random variables $Y_i$, $\sum_{i=1}^{\infty} E \mid Y_i \mid^p < \infty$ for $0 < p \leq 1$ implies $\sum_{i=1}^{\infty} Y_i$ converges a.s.).

(iii) $\sum_{i=1}^{\infty} P[\mid X_i \mid \geq C] < \infty$, $\sum_{i=1}^{\infty} E[X_i I(X_i \leq C)]$ converges and $\sum_{i=1}^{\infty} \text{var}[X_i I(\mid X_i \mid \leq C)] < \infty$ for some constant $0 < C < \infty$ imply that $S_n$ converges almost surely.

(iv) Let $EX_i = 0$ and $EX_i^2 = 1$ for $i \geq 1$ with either the $X_i^2$ uniformly integrable or the $X_i$ symmetric. Then, given a sequence of constants $\{c_i, i \geq 1\}$,

$$\sum_{i=1}^{\infty} c_i X_i \qquad \text{converges a.s.}$$

if and only if

$$\sum_{i=1}^{\infty} c_i^2 X_i^2 < \infty \qquad \text{a.s.}$$

if and only if

$$\sup \sum_{i=1}^{n} c_i X_i < \infty \qquad \text{a.s.}$$

(v) Suppose $EX_i = 0$ for $i \geq 2$ and $E \sup_{i \geq 1} X_i^2 < \infty$. Then $\sum_{i=1}^{\infty} X_i$ converges a.s. if and only if $\sum_{i=1}^{\infty} EX_i^2 < \infty$. $\sum_{i=1}^{\infty} X_i$ diverges a.s. if and only if $\sum_{i=1}^{\infty} EX_i^2 = \infty$.

**Exercise 2.12.1.** Which martingale results imply (i)–(v)? ∎

Recall that from Exercise 2.8.5 a martingale $\{T_n, n \geq 1\}$ can converge on part of the space and diverge on part of the space in the sense that $0 < P(T_n \text{ converges}) < 1$. This local convergence is not possible for partial sums of independent random variables:

**Theorem 2.12.3.**  (Kolmogorov's 0–1 law).  Let $\mathscr{G}_n = \mathscr{B}(X_m, m \geq n)$ for each $n \geq 1$ and $\mathscr{G} = \bigcap_{n=1}^{\infty} \mathscr{G}_n$. Then $A \in \mathscr{G}$ implies $P(A) = 0$ or $P(A) = 1$.

**Remarks.**  $\mathscr{G}$ is called the tail $\sigma$ field for obvious reasons. Its events are called tail events. Theorem 2.12.3 is sometimes stated by saying that the tail $\sigma$ field in the case of independence is (almost surely) trivial. Although not emphasized at the time, the Borel–Cantelli lemma together with its converse for independent events provides another example of a 0–1 law. Theorem 2.12.2 (v) and Theorem 2.12.4 yield still other examples.  ∎

In order to prove Theorem 2.12.3, we need a measure-theoretic lemma that is often useful.

**Lemma 2.12.1.**  Let $\mu$ be a finite measure on a field $\mathscr{F}_0$ and let $\mathscr{F}$ be the $\sigma$ field generated by $\mathscr{F}_0$. Then, given $A \in \mathscr{F}$ and $\varepsilon > 0$, there exists $A_0 \in \mathscr{F}_0$ such that $\mu(A \bigtriangleup A_0) \leq \varepsilon$.

**Proof of Lemma 2.12.1.**  Omitted. See Halmos [1950, p. 56].  ∎

**Proof of Theorem 2.12.3.**  Choose $A \in \mathscr{G}$. $\bigcup_{n=1}^{\infty} \mathscr{B}_n$ is a field generating $\mathscr{B}_{\infty} \equiv \mathscr{B}(X_i, i \geq 1)$. $\mathscr{G} \subset \mathscr{B}_{\infty}$ and thus $A \in \mathscr{B}_{\infty}$. Note that $B \in \bigcup_{n=1}^{\infty} \mathscr{B}_n$ implies $B \in \mathscr{B}_n$ for some $n$. Hence, applying Lemma 2.12.1, there exists $A_n \in \mathscr{B}_n$ such that $P[A_n \bigtriangleup A] \to 0$. Hence

$$P(A_n) \to P(A) \qquad (2.12.1)$$

as $n \to \infty$. $\mathscr{B}_n$ and $\mathscr{G}$ are independent families of events for each $n \geq 1$. Hence $A_n$ and $A$ are independent; that is, $P(A_n \cap A) = P(A_n)P(A)$. Letting $n \to \infty$, and applying Eq. (2.12.1) we obtain $P(A) = P^2(A)$. Thus $P(A) = 0$ or $P(A) = 1$.  ∎

Besides the above classical proof, following Doob [1953, p. 334] one can give a short and cute martingale proof of the 0–1 law: Let $A \in \mathscr{G}$. Fix $i \geq 1$. $E[I(A) \mid \mathscr{B}_i] = PA$ a.s. by the independence of $\mathscr{G}$ and $\mathscr{B}_i$. By Theorem 2.8.6 (i),

$$E[I(A) \mid \mathscr{B}_i] \to E[I(A) \mid \mathscr{B}_{\infty}] = I(A) \qquad \text{a.s.,}$$

using the fact that $I(A)$ is $\mathcal{B}_\infty$ measurable. Thus $P(A) = I(A)$ a.s. which is *only* possible if $P(A) = 0$ or $P(A) = 1$. ∎

**Corollary 2.12.1.** If $\{X_i, i \geq 1\}$ are independent, then either $P[\sum_{i=1}^\infty X_i \text{ converges}] = 0$ or $P[\sum_{i=1}^\infty X_i \text{ converges}] = 1$.

***Proof.*** Fix an integer $N \geq 1$ and assume $n \geq N + 1$.

$$\sum_{i=1}^n X_i = \sum_{i=1}^N X_i + \sum_{i=N+1}^n X_i.$$

Thus $\sum_{i=1}^n X_i$ converges as $n \to \infty$ if and only if $\sum_{i=N+1}^n X_i$ converges as $n \to \infty$. Thus $[\sum_{i=1}^\infty X_i \text{ converges}] \in \mathcal{G}_N$ for all $N \geq 1$ implying that $[\sum_{i=1}^\infty X_i \text{ converges}]$ is a tail event. Apply Theorem 2.12.3. ∎

Although our immediate purpose in giving Theorem 2.12.3 was the establishment of Corollary 2.12.1, Theorem 2.12.3 will prove to be extremely useful. This is because so many of the strong events of interest to us are tail events. For example,

$$(S_n/n \text{ converges}),$$

$$\{\sum_{i=1}^n a_i X_i / \sum_{i=1}^n a_i \text{ converges}\}$$

(assuming $a_i > 0$ for all $i$ and $\sum_{i=1}^n a_i \to \infty$),

$$(\lim \sup S_n/(2n \log_2 n)^{1/2} = 1)$$

are all easily shown (do as an exercise) to be tail events. These events relate, respectively, to major topics of Chapters 3, 4, and 5.

There is a companion 0–1 law to the Kolmogorov 0–1 law, namely the Hewitt–Savage 0–1 law [1955]. The Hewitt–Savage 0–1 law is more powerful than the Kolmogorov 0–1 law in the special case when it applies; that is, when random variables are independent identically distributed.

**Definition 2.12.1.** Let $\{Y_i, i \geq 1\}$ be a stochastic sequence. $B \in \mathcal{B}\{Y_i, i \geq 1\}$ is said to be symmetric if there is a Borel set $C_\infty$ of $R_\infty$ such that for each $n \geq 1$ and each permutation $\{i_1, i_2, \ldots, i_n\}$ of $\{1, 2, \ldots, n\}$,

$$B = [(Y_1, Y_2, \ldots, Y_n, \ldots) \in C_\infty]$$
$$= [(Y_{i_1}, Y_{i_2}, \ldots, Y_{i_n}, Y_{n+1}, \ldots) \in C_\infty]. ∎$$

**Exercise 2.12.2.** Show that each tail event is a symmetric event. Construct some symmetric events which are not tail events.

**Theorem 2.12.4.** (Hewitt–Savage 0–1 law). Let the $X_i$ be identically distributed. Each symmetric event of $\mathscr{B}(X_i, i \geq 1)$ has probability zero or one.

**Proof.** Choose $B$ symmetric. By Lemma 2.12.1 there exists $B_n \in \mathscr{B}_n$ such that

$$P[B_n \triangle B] \to 0. \qquad (2.12.2)$$

Choose $C_n$, a Borel set of $R_n$, and $C_\infty$, a Borel set of $R_\infty$ such that

$$B_n = [(X_1, X_2, \ldots, X_n) \in C_n] \quad \text{and} \quad B = [(X_1, X_2, \ldots, X_n, \ldots) \in C_\infty].$$

Let

$$B_n{}' = [(X_{n+1}, X_{n+2}, \ldots, X_{2n}) \in C_n]$$

and

$$B' = [(X_{n+1}, X_{n+2}, \ldots, X_{2n}, X_1, X_2, \ldots, X_n, X_{2n+1}, X_{2n+2}, \ldots) \in C_\infty].$$

$$P(B_n \cap B_n{}') = P(B_n)P(B_n{}') \to P^2(B) \qquad (2.12.3)$$

by Eq. (2.12.2) since $B_n$ and $B_n{}'$ are independent with $P(B_n) = P(B_n{}')$ for all $n \geq 1$. Since the $X_i$ are independent identically distributed,

$$P(B_n{}' \triangle B') = P(B_n \triangle B) \quad \text{for all} \quad n \geq 1. \qquad (2.12.4)$$

Since $B$ is symmetric,

$$P(B_n{}' \triangle B) = P(B_n{}' \triangle B') \quad \text{for all} \quad n \geq 1.$$

Thus using Eqs. (2.12.2) and (2.12.4), $P(B_n{}' \triangle B) \to 0$ as well as $P(B_n \triangle B) \to 0$. Thus

$$P(B_n \cap B_n{}') \to P(B).$$

Using Eq. (2.12.3), $P^2(B) = P(B)$. Thus $P(B) = 1$ or $P(B) = 0$. ∎

Suppose that $\{X_i, i \geq 1\}$ is a sequence of discrete (as well as independent) random variables for which $\sum_{i=1}^{\infty} X_i \to X$ a.s. It is a startling consequence of the Kolmogorov 0–1 law that $X$ is of "pure type"; that is, $X$ is either discrete, $X$ is absolutely continuous, or $X$ is singular. Of course, in general it is quite easy to construct random variables that are mixtures of the three basic types. The point is that $X$ cannot be such a mixture.

**Theorem 2.12.5.** Let $\{X_i, i \geq 1\}$ be a sequence of independent and discrete random variables such that

$$\sum_{i=1}^{\infty} X_i \to X \qquad \text{a.s.}$$

Then $X$ is of pure type.

**Proof.** Let the $D_k$ be countable sets such that $P[X_k \in D_k] = 1$ for $k \geq 1$. Let $D = \bigcup_{k=1}^{\infty} D_k$. Let $G$ be the smallest group in $R_1$ containing $D$. $G$ consists of all real numbers of the form $\sum_{i=1}^{n} m_i x_i$ for all $n$, all integers $\{m_i, 1 \leq i \leq n\}$, and all numbers $\{x_i, 1 \leq i \leq n\} \subset D$. Since $D$ is countable, it follows that $G$ is countable. For any Borel set $B \in R_1$, let

$$G \oplus B = [x \in R_1 \mid x = x_1 + x_2, x_1 \in G, x_2 \in B].$$

Let

$$A = [X \in G \oplus B] \cap \left[ \sum_{i=1}^{m} X_i \text{ converges as } m \to \infty \right].$$

The key step is to use the group structure of $G$ to show that $A$ is a tail event. On $[\sum_{i=1}^{\infty} X_i \text{ converges}]$, $X - \sum_{i=n}^{\infty} X_i \in G$ for each $n \geq 1$. Hence on $[\sum_{i=1}^{\infty} X_i \text{ converges}]$, since $G$ is a group, $X \in G \oplus B$ if and only if $\sum_{i=n}^{\infty} X_i \in G \oplus B$. Thus

$$A = \left[ \sum_{i=n}^{\infty} X_i \in G \oplus B \right] \cap \left[ \sum_{i=1}^{m} X_i \text{ converges as } m \to \infty \right]$$

for each $n \geq 1$. $[\sum_{i=1}^{m} X_i \text{ converges as } m \to \infty]$ is a tail event. Hence $A$ by the above representation is a tail event.

Thus either (i) $P[X \in G \oplus B] = 0$ for all countable sets of the form $G \oplus B$ or (ii) there exists a countable set of the form $G \oplus B$ such that $P[X \in G \oplus B] = 1$. If (ii), $X$ is discrete and hence of pure type. Assume (i). Either (iii) $P[X \in G \oplus B] = 0$ for all sets $G \oplus B$ of Lebesgue measure 0 or (iv) there exists a set $G \oplus B$ of Lebesgue measure 0 such that $P[X \in G \oplus B] = 1$. If (iv), recalling (i), then $X$ is singular and hence of pure type. Assume (iii), the only remaining case. Let $B$ have Lebesgue measure zero. $G \oplus B$ has Lebesgue measure zero since $G$ is countable. $[X \in G \oplus B] \supset [X \in B]$. Hence $P[X \in B] = 0$, that is, $P_X$ defined by $P_X(B) = P[X \in B]$

is absolutely continuous. Thus $X$ has density $dP_X/d\mu$ (the Radon–Nikodym derivative of $P_X$ with respect to Lebesgue measure $\mu$) and is hence a random variable of continuous type. Thus $X$ is of pure type. ∎

The above result is due to Jessen and Wintner [1935]. Breiman [1968, pp. 49–50] discusses some unanswered questions about the characterization of the limiting random variable $X$. Even in the simple case of $\sum_{i=1}^{n} a_i X_i$ with $\sum_{i=1}^{\infty} a_i^2 < \infty$ and $X_i$ identically distributed with $P[X_1 = 1] = P[X_1 = 0] = \frac{1}{2}$, the type of the random variable $\sum_{i=1}^{\infty} a_i X_i$ is unknown except for very special choices of the $a_i$. The following example looks at some of the possibilities.

**Example 2.12.1.**   Let $\{X_i, i \geq 1\}$ be identically distributed with $P[X_1 = 1] = P[X_1 = 0] = \frac{1}{2}$. Let $a_i = 2^{-i}$ for $i \geq 1$. In order to find the distribution of $\sum_{i=1}^{\infty} 2^{-i} X_i$, we take a particular representation of the $X_i$: Let $\Omega = [0, 1]$, $\mathscr{F}$ be the Borel sets of $[0, 1]$, and $P$ be Lebesgue measure. Let $\{R_i, i \geq 0\}$ be the Rademacher functions. (Refer to the remarks following Theorem 2.8.3 for the definition of the Rademacher functions.) Let

$$X_i = (1 - R_{i-1})/2 \qquad (2.12.5)$$

define a particular representation of the $X_i$. (It is easy to see that the $X_i$ as defined by Eq. (2.12.5) are independent identically distributed with $P[X_1 = 1] = P[X_1 = 0] = \frac{1}{2}$.) Then $\sum_{i=1}^{\infty} 2^{-i} X_i(\omega)$ is merely the binary expansion of $\omega$ for each $\omega \in [0, 1]$. Thus $\sum_{i=1}^{\infty} 2^{-i} X_i(\omega) = \omega$ for each $\omega \in [0, 1]$. Thus

$$\sum_{i=1}^{n} 2^{-i} X_i \to X \qquad \text{a.s.}$$

where $X$ is uniformly distributed on $[0, 1]$.

In order to obtain a limiting singular distribution, let $a_i = 2 \cdot 3^{-i}$ for $i \geq 1$. Let the $X_i$ be represented as above. $a_i X_i$ equals $0$ or $2 \cdot 3^{-i}$ for each $i \geq 1$. It is easy to show that except for a set of Lebesgue measure zero the Cantor set of $[0, 1]$ consists of exactly those numbers with non-terminating ternary expansions consisting entirely of $0$'s and $2$'s. Thus the Cantor set $C$ satisfies $P[\sum_{i=1}^{\infty} a_i X_i \in C] = 1$. The mapping $\omega \to \sum_{i=1}^{\infty} a_i X_i(\omega)$ is 1:1 from $[0, 1]$ to the Cantor set. Thus $P[\sum_{i=1}^{\infty} a_i X_i = c] = 0$ for each $c \in [0, 1]$. The Cantor set has Lebesgue measure zero. Thus $\sum_{i=1}^{\infty} a_i X_i$ has a singular distribution.

A limiting discrete distribution can be obtained trivially by letting $a_i = 0$ for all $i$ sufficiently large. Somewhat more interestingly, let $\{Y_i, i \geq 1\}$ be a sequence of independent discrete random variables satisfying

$$\sum_{i=1}^{\infty} P[Y_i \neq b_i] < \infty$$

for some sequence of constants $\{b_i, i \geq 1\}$ such that $\sum_{i=1}^{\infty} b_i$ converges. Clearly $\sum_{i=1}^{\infty} Y_i$ converges a.s. It is easy to see that the range of $\sum_{i=1}^{\infty} Y_i$ is discrete: Let $t$ be the largest $i \geq 1$ such that $Y_i \neq b_i$. Since $\sum_{i=1}^{\infty} P[Y_i \neq b_i]$ $< \infty$, $P[t < \infty] = 1$.

$$\sum_{i=1}^{\infty} Y_i = \sum_{i=1}^{t} Y_i + \sum_{i=t+1}^{\infty} b_i$$

$$= \sum_{j=1}^{\infty} I(t=j) \left\{ \sum_{i=1}^{j} Y_i + \sum_{i=j+1}^{\infty} b_i \right\}.$$

$\sum_{i=1}^{j} Y_i + \sum_{i=j+1}^{\infty} b_i$ is a discrete random variable for each $j \geq 1$, thus establishing that the limiting random variable $\sum_{i=1}^{\infty} Y_i$ is discrete. According to Levy [1931], the above conditions on the $Y_i$ are necessary (as well as sufficient) for the limiting random variable to be discrete. That is, $\sum_{i=1}^{\infty} Y_i$ discrete implies that there exists $\{b_i, i \geq 1\}$ such that $\sum_{i=1}^{\infty} P[Y_i \neq b_i] < \infty$ and $\sum_{i=1}^{\infty} b_i$ converges.  ∎

## 2.13.  Centering at Medians, the Levy-Type Inequality, and Symmetrization

The Levy maximal inequalities and inequalities of similar character are basic to the study of the almost sure behavior of sums of independent random variables. In particular, they provide the way for showing that $\{X_i, i \geq 1\}$ independent and $\sum_{i=1}^{n} X_i \to X$ in probability together imply that $\sum_{i=1}^{n} X_i \to X$ a.s.

Recall that given a random variable $X$, any constant $\mu(X)$ is a median of $X$ if

$$P[X \geq \mu(X)] \geq \tfrac{1}{2} \quad \text{and} \quad P[X \leq \mu(X)] \geq \tfrac{1}{2}.$$

Medians (unlike means) always exist. Throughout, $\mu(X)$ will denote a median of $X$.

***Exercise 2.13.1.*** (A weakness of working with medians). Show that a random variable need not have a unique median. Show that there exists $X$ and $Y$ independent with the same distribution and with unique medians such that $\mu(X + Y) \neq \mu(X) + \mu(Y)$. Given a constant $a$ and $X$, show that there exists medians $\mu(X)$ and $\mu(aX)$ such that $\mu(aX) = a\mu(X)$.   ∎

**Theorem 2.13.1.**   (Levy maximal inequalities).   Let $\{X_i, 1 \leq i \leq n\}$ be independent for some $n \geq 1$. Then

(i)   $P\{\max_{k \leq n}[S_k + \mu(S_n - S_k)] \geq \varepsilon\} \leq 2P[S_n \geq \varepsilon]$

and

(ii)   $P[\max_{k \leq n} | S_k + \mu(S_n - S_k) | \geq \varepsilon] \leq 2P[| S_n | \geq \varepsilon]$   for all  $\varepsilon > 0$.

**Proof.**   We establish (i) first. Fix $\varepsilon > 0$. Let $t$ be the smallest integer $k \leq n$ such that $S_k + \mu(S_n - S_k) \geq \varepsilon$ if such an integer exists; otherwise let $t = n + 1$.

$$[S_n \geq \varepsilon] = [S_n \geq \varepsilon, t \leq n] \supset \bigcup_{k=1}^{n} [t = k, S_n - S_k - \mu(S_n - S_k) \geq 0].$$

Thus

$$\sum_{k=1}^{n} P[t = k, S_n - S_k - \mu(S_n - S_k) \geq 0] \leq P[S_n \geq \varepsilon]. \quad (2.13.1)$$

Fix $k \leq n$.

$$P[t = k, S_n - S_k - \mu(S_n - S_k) \geq 0]$$
$$= P[t = k]P[S_n - S_k - \mu(S_n - S_k) \geq 0]$$

by independence.

$$P[S_n - S_k - \mu(S_n - S_k) \geq 0] \geq \tfrac{1}{2}$$

since $\mu(S_n - S_k)$ is a median of $S_n - S_k$. Thus

$$P[t = k, S_n - S_k - \mu(S_n - S_k) \geq 0] \geq \tfrac{1}{2}P[t = k].$$

Combining with Eq. (2.13.1),

$$P[S_n \geq \varepsilon] \geq \tfrac{1}{2} \sum_{k=1}^{n} P[t = k] = \tfrac{1}{2}P\{\max_{k \leq n}[S_k + \mu(S_n - S_k)] \geq \varepsilon\},$$

establishing (i).

Replacing $X_i$ by $-X_i$ for $i \leq n$ yields

$$P[S_n \leq -\varepsilon] \geq \tfrac{1}{2}P\{\min_{k \leq n}[S_k + \mu(S_n - S_k)] \leq -\varepsilon]\}.$$

Combining,

$$
\begin{aligned}
P[|\,S_n\,| \geq \varepsilon] &= P[S_n \geq \varepsilon] + P[S_n \leq -\varepsilon] \\
&\geq \tfrac{1}{2}P\{\max_{k \leq n}[S_k + \mu(S_n - S_k)] \geq \varepsilon\} \\
&\quad + \tfrac{1}{2}P\{\min_{k \leq n}[S_k + \mu(S_n - S_k)] \leq -\varepsilon\} \\
&\geq \tfrac{1}{2}P[\max_{k \leq n}|\,S_k + \mu(S_n - S_k)\,| \geq \varepsilon],
\end{aligned}
$$

establishing (ii). ∎

**Remark.** For future reference, Theorem 2.13.1 (i) will be referred to as the first Levy maximal inequality and Theorem 2.13.1 (ii) as the second Levy maximal inequality.

It is instructive to note the simplicity of the inequalities (i) and (ii) in Theorem 2.13.1 if in addition the $X_i$ are assumed to be symmetric.

**Exercise 2.13.2.** [Skorokhod, 1957; see Breiman [1968], p. 45]. Let $\{X_i, 1 \leq i \leq n\}$ be independent for some $n \geq 2$. Prove that

$$P[\sup_{k \leq n}|\,S_k\,| \geq 2\varepsilon] \leq P[|\,S_n\,| \geq \varepsilon]/\inf_{k \leq n} P[|\,S_n - S_k\,| < \varepsilon].$$

(This inequality can take the place of Levy's inequalities in some situations. It avoids consideration of medians.) ∎

**Theorem 2.13.2.** Let $\{X_i, i \geq 1\}$ be independent with $S_n \to S$ in probability. Then $S_n \to S$ a.s.

**Proof.** We proceed essentially by the method of subsequences (Lemma 2.3.1), using the second Levy maximal inequality to analyze the maximal term. $S_n \to S$ in probability implies there exists an increasing positive integer sequence $\{n_k, k \geq 1\}$ such that

$$\sum_{k=1}^{\infty} P[|\,S_{n_k} - S\,| > 2^{-k-1}] < \infty.$$

It follows that $S_{n_k} \to S$ a.s. and

$$\sum_{k=2}^{\infty} P[|\,S_{n_k} - S_{n_{k-1}}\,| > 2^{-k}] < \infty, \qquad (2.13.2)$$

since

$$[|\, S_{n_k} - S_{n_{k-1}}\,| > 2^{-k}] \subset [|\, S_{n_k} - S\,| > 2^{-k-1}] \cup [|\, S_{n_{k-1}} - S\,| > 2^{-k-1}]$$

for each $k \geq 1$. By the second Levy maximal inequality applied to $X_{n_{k-1}+1}$, $X_{n_{k-1}+2}, \ldots, X_{n_k}$,

$$P[\max_{n_{k-1} < n \leq n_k} |\, S_n - S_{n_{k-1}} + \mu(S_{n_k} - S_n)\,| \geq \varepsilon] \leq 2P[|\, S_{n_k} - S_{n_{k-1}}\,| \geq \varepsilon]$$

for all $\varepsilon > 0$ and $k \geq 1$. Thus, by the Borel–Cantelli lemma and Eq. (2.13.2),

$$\max_{n_{k-1} < n \leq n_k} |\, S_n - S_{n_{k-1}} + \mu(S_{n_k} - S_n)\,| \to 0 \qquad \text{a.s.}$$

as $k \to \infty$. Consider integers $n$ and $k$ such that $n_{k-1} < n \leq n_k$.

$$\begin{aligned}
|\, S_n &- S + \mu(S_{n_k} - S_n)\,| \\
&\leq |\, S_n - S_{n_{k-1}} + \mu(S_{n_k} - S_n)\,| + |\, S_{n_{k-1}} - S\,| \\
&\leq \max_{n_{k-1} < n \leq n_k} |\, S_n - S_{n_{k-1}} + \mu(S_{n_k} - S_n)\,| + |\, S_{n_{k-1}} - S\,|.
\end{aligned}$$

Since the two right hand terms of the above inequality have been shown to converge almost surely to 0 as $k \to \infty$, it follows that $S_n - S + \mu(S_{n_k} - S_n) \to 0$ a.s. as $n \to \infty$ where for each $n$, $n_k$ satisfies $n_{k-1} < n \leq n_k$. Since $S_n - S \to 0$ in probability, it is easily seen that $\mu(S_{n_k} - S_n) \to 0$. Thus $S_n - S \to 0$ a.s., establishing the result. ∎

**Exercise 2.13.3.**    Let $T_n \to 0$ in probability and $T_n - a_n \to 0$ in probability where the $a_n$ are constants. Prove that $a_n \to 0$. ∎

Theorem 2.13.2 suggests the question whether $\{T_n, n \geq 1\}$ a martingale with $T_n \to T$ in probability implies $T_n \to T$ a.s. The answer is no.

**Example 2.13.1.**    (Personal communication from Gordon Simons). Let $\{X_i, i \geq 1\}$ be independent with

$$P[X_i = 1] = P[X_i = -1] = (2i)^{-1}$$

and

$$P[X_i = 0] = 1 - i^{-1} \qquad \text{for each } i \geq 1.$$

Let $T_0 = 0$ and

$$T_n = \begin{cases} X_n & \text{if } T_{n-1} = 0 \\ n T_{n-1} |\, X_n\,| & \text{if } T_{n-1} \neq 0 \end{cases}$$

for each $n \geq 1$. Then elementary computations show that

   (i)   $\{T_n, n \geq 1\}$ is a martingale,
   (ii)  for each $n \geq 1$, $T_n = 0$ if and only if $X_n = 0$,
   (iii) $P[T_n = 0] = P[X_n = 0] = (n - 1)/n \to 1$, and
   (iv)  $P[T_n \neq 0 \text{ i.o.}] = P[X_n \neq 0 \text{ i.o.}] = 1$

since $\sum_{n=1}^{\infty} P[|X_n| = 1] = \infty$. (iii) implies that $T_n \to 0$ in probability. (iv) implies $P[T_n \to 0] = 0$ since the $T_n$ are integer valued.  ∎

The inequality given in Exercise 2.13.2 could have been used to prove Theorem 2.13.2. In Loève's survey paper [1951] on almost sure convergence, the idea of centering at medians is extended to centering at conditional medians.

We have seen that

$$\sum_{i=1}^{\infty} \text{var}(X_i) < \infty$$

for independent $X_i$ implies that

$$\sum_{i=1}^{\infty} (X_i - EX_i) \qquad \text{converges a.s.,}$$

that is, that the partial sums converge *when* the $X_i$ are centered at their means. At the same time $\sum_{i=1}^{\infty} X_i$ may be almost surely divergent. This suggests an analysis of which centering constants $\{b_i, i \geq 1\}$ (if any) cause $\sum_{i=1}^{\infty} (X_i - b_i)$ to converge almost surely.

**Definition 2.13.1.**  $\sum_{i=1}^{\infty} X_i$ is said to converge almost surely when centered (or said to be essentially convergent almost surely) if there exists constants $\{b_i, i \geq 1\}$ such that

$$\sum_{i=1}^{\infty} (X_i - b_i) \qquad \text{converges almost surely.}  ∎$$

In order to study essential convergence it is most useful to introduce the concept of symmetrization of random variables.

**Definition 2.13.2.**  If $Y$ is a random variable and $Y'$ is independent of $Y$ with the same distribution, then

$$Y^s = Y - Y'$$

is said to be a symmetrized version of $Y$. More generally if $\{Y_i, i \leq 1\}$ is a sequence of random variables and $(Y_i', i \geq 1)$ is a sequence of random variables independent of $(Y_i, i \geq 1)$ with $(Y_i', i \geq 1)$ having the same distribution as $(Y_i, i \geq 1)$, then

$$(Y_i^s, i \geq 1) = (Y_i - Y_i', i \geq 1)$$

is said to be a symmetrized version of $(Y_i, i \geq 1)$. The superscript $s$ will always denote symmetrization. ∎

**Theorem 2.13.3.** Let $\{Y_i, i \geq 1\}$ be any sequence of random variables. Then there exists a symmetrized version $(Y_i^s, i \geq 1)$ (defined on a possibly different probability space).

**Proof.** Application of the Kolmogorov extension theorem. ∎

Note that symmetrized random variables by their definition are indeed symmetric random variables.

The first application of symmetrization will be the establishment of the three series theorem for independent random variables. Recall by Theorem 2.12.2 (iii) that $\{X_i, i \geq 1\}$ independent with $\sum_{i=1}^{\infty} P[|X_i| \geq C] < \infty$, $\sum_{i=1}^{\infty} E[X_i I(|X_i| \leq C)]$ convergent, and $\sum_{i=1}^{\infty} \text{var}[X_i I(|X_i| \leq C)] < \infty$ for some constant $C > 0$ implies that $\sum_{i=1}^{\infty} X_i$ converges a.s. This result was obtained by specializing a martingale result to the case of independence. The content of the three series theorem is that the convergence of the above three series is a necessary as well as a sufficient condition for the almost sure convergence of $\sum_{i=1}^{\infty} X_i$ when the $X_i$ are independent.

The proof of the necessity will depend on the notion of symmetrization.

**Lemma 2.13.1.** Let $\{X_i, i \geq 1\}$ be independent with $\sum_{i=1}^{\infty} X_i$ convergent almost surely and $E \sup X_i^2 < \infty$. Then

$$\sum_{i=1}^{\infty} EX_i \quad \text{converges and} \quad \sum_{i=1}^{\infty} \text{var } X_i < \infty.$$

**Proof.** $\sum_{i=1}^{\infty} X_i$ converges a.s. implies that $\sum_{i=1}^{\infty} X_i^s$ converges a.s.

$$\tfrac{1}{2} \sup(X_i^s)^2 \leq \sup X_i^2 + \sup(X_i')^2.$$

Thus $E[\sup(X_i^s)^2] < \infty$. Since $\{\sum_{i=1}^{n} X_i^s, n \geq 1\}$ is an almost surely convergent martingale with $E \sup(X_i^s)^2 < \infty$, it follows by Theorem 2.11.2 that $\sum_{i=1}^{\infty} \text{var } X_i^s < \infty$. $\text{var } X_i^s = 2 \text{ var } X_i$. Thus $\sum_{i=1}^{\infty} \text{var } X_i < \infty$ as desired. Hence, by Theorem 2.12.2 (ii), $\sum_{i=1}^{\infty} (X_i - EX_i)$ converges a.s. Since $\sum_{i=1}^{\infty} X_i$ converges a.s. by hypothesis it follows that $\sum_{i=1}^{\infty} EX_i$ converges, as desired.  ∎

**Theorem 2.13.4.**   (Kolmogorov's three series theorem). Let $\{X_i, i \geq 1\}$ be independent. Then $\sum_{i=1}^{\infty} X_i$ converges a.s. if and only if

(i)   $\displaystyle\sum_{i=1}^{\infty} P[|X_i| > C] < \infty,$

(ii)   $\displaystyle\sum_{i=1}^{\infty} E[X_i I(|X_i| \leq C)]$ converges, and

(iii)   $\displaystyle\sum_{i=1}^{\infty} \mathrm{var}[X_i I(|X_i| \leq C)] < \infty$ for every constant $C > 0$.

Moreover, if (i), (ii), and (iii) hold for some constant $C > 0$, they hold for all $C > 0$.

**Proof.**   It has already been shown that (i), (ii) and (iii) converging for some $C > 0$ implies that $\sum_{i=1}^{\infty} X_i$ converges a.s. Fix $C > 0$ and assume $\sum_{i=1}^{\infty} X_i$ converges a.s. It suffices to show that (i), (ii), and (iii) are convergent. $\sum_{i=1}^{\infty} X_i$ converging a.s. implies $X_i \to 0$ a.s. Hence, applying the partial converse of the Borel–Cantelli lemma to the independent events $[|X_i| \geq C]$, $\sum_{i=1}^{\infty} P[|X_i| \geq C] < \infty$. Since $P[|X_i| \geq C \text{ i.o.}] = 0$, $\sum_{i=1}^{\infty} X_i I(|X_i| \leq C)$ converges a.s. Applying Lemma 2.13.1 to $\{X_i I(|X_i| \leq C), i \geq 1\}$, it follows that $\sum_{i=1}^{\infty} E[X_i I(|X_i| \leq C)]$ converges and $\sum_{i=1}^{\infty} \mathrm{var}[X_i I(|X_i| \leq C)] < \infty$.  ∎

A comparison of Theorem 2.8.8 and Theorem 2.13.4 suggests that for an arbitrary adapted stochastic sequence $\{X_i, \mathscr{F}_i, i \geq 1\}$ that the convergence of the three conditional series (Eq. (2.8.4), Eq. (2.8.5) and Eq. (2.8.6)) might be necessary as well as sufficient (we know they are sufficient) for the convergence of $\sum_{i=1}^{\infty} X_i$. The following example due to Dvoretzky and to Gilat [1971] shows this conjecture to be false.

**Example 2.13.2.**   Let $\{U_n, n \geq 1\}$ be independent with

$$P[U_n = n^{-1/2}] = P[U_n = -n^{-1/2}] = \tfrac{1}{2}$$

for each $n \geq 1$. Let $X_i = U_i - U_{i-1}$ for $i \geq 2$ and $X_1 = U_1$. Let $\mathscr{F}_n = \mathscr{B}(X_1, X_2, \ldots, X_n)$ for each $n \geq 1$. Then $\sum_{i=1}^{\infty} X_i$ converges almost surely.

$$\sum_{i=1}^{n} E[X_i I(|X_i| \leq 2) \mid \mathscr{F}_{i-1}] = -\sum_{i=2}^{n} U_{i-1} \qquad \text{a.s.,}$$

which is divergent almost surely by the three series theorem for independent random variables.

$$\sum_{i=1}^{n} E[X_i^2 I(|X_i| \leq 2) | \mathscr{F}_{i-1}] \geq \sum_{i=1}^{n} i^{-1} \qquad \text{a.s.}$$

which diverges.

One could ask whether some other condition on the first and second conditional moments is necessary and sufficient for convergence when the $X_i$ are uniformly bounded. Let $Y_i = -(U_i + U_{i-1})$ for $i \geq 2$ and $Y_1 = -U_1$. Then $\{X_i, i \geq 1\}$ and $\{Y_i, i \geq 1\}$ have the *same* first and second conditional moments yet

$$\sum_{i=1}^{n} Y_i = -U_1 - \sum_{i=2}^{n} (U_i + U_{i-1})$$

$$= -U_n - 2\sum_{i=1}^{n-1} U_i$$

*diverges* almost surely. Thus, $\{X_i, i \geq 1\}$ and $\{Y_i, i \geq 1\}$ have the same first and second moment structure, both are uniformly bounded and yet $\sum_{i=1}^{\infty} X_i$ converges almost surely and $\sum_{i=1}^{\infty} Y_i$ diverges almost surely. It follows that there cannot exist a necessary and sufficient condition for the convergence of $S_n$ in terms of first and second conditional moments even for uniformly bounded $X_i$. ∎

The three series theorem is the most important result of Section 2.13. It is often useful in establishing the almost sure convergence or almost sure divergence of partial sums of independent random variables. As a simple application of the three series theorem, we analyze the relationship between the convergence of the partial sums and the partial sums of squares of independent random variables.

**Theorem 2.13.5.** Let $\{X_i, i \geq 1\}$ be a sequence of independent random variables.

(i)  Suppose $\sum_{i=1}^{\infty} X_i^2 < \infty$ a.s. Then $S_n$ converges almost surely if and only if

$$\sum_{i=1}^{\infty} EX_i I(|X_i| \leq 1) \qquad \text{converges.}$$

(ii)  Suppose $S_n$ converges almost surely. Then

$$\sum_{i=1}^{\infty} X_i^2 < \infty \qquad \text{a.s.}$$

if and only if

$$\sum_{i=1}^{\infty} E^2 |X_i| I(|X_i| \leq 1) < \infty$$

if and only if

$$\sum_{i=1}^{\infty} E^2 X_i I(|X_i| \leq 1) < \infty.$$

**Proof.** (i) Suppose $\sum_{i=1}^{\infty} X_i^2 < \infty$ a.s. Thus by the three series theorem, $\sum_{i=1}^{\infty} EX_i^2 I(|X_i| \leq 1) < \infty$ and hence $\sum_{i=1}^{\infty} \text{var } X_i I(|X_i| \leq 1) < \infty$. Also, by the three series theorem $\sum_{i=1}^{\infty} P[|X_i| \geq 1] = \sum_{i=1}^{\infty} P[X_i^2 \geq 1] < \infty$. Thus (i) follows by the three series theorem.

(ii) Suppose $S_n$ converges almost surely and that $\sum_{i=1}^{\infty} X_i^2 < \infty$ a.s. Then by the three series theorem $\sum_{i=1}^{\infty} EX_i^2 I(|X_i| \leq 1) < \infty$. $E^2 |X_i| I(|X_i| \leq 1) \leq EX_i^2 I(|X_i| \leq 1)$. Thus $\sum_{i=1}^{\infty} E^2 |X_i| I(|X_i| \leq 1) < \infty$ and trivially $\sum_{i=1}^{\infty} E^2 X_i I(|X_i| \leq 1) < \infty$.

Suppose $S_n$ converges almost surely and

$$\sum_{i=1}^{\infty} E^2 X_i I(|X_i| \leq 1) < \infty. \qquad (2.13.3)$$

By the three series theorem applied to the convergence of $S_n$,

$$\sum_{i=1}^{\infty} P[X_i^2 > 1] < \infty \qquad (2.13.4)$$

and

$$\sum_{i=1}^{\infty} EX_i^2 I(|X_i| \leq 1) - E^2 X_i I(|X_i| \leq 1) < \infty.$$

Applying Eq. (2.13.3), it follows that $\sum_{i=1}^{\infty} EX_i^2 I(|X_i| \leq 1) < \infty$. Thus $\sum_{i=1}^{\infty} X_i^2 I(|X_i| \leq 1) < \infty$ a.s. Using Eq. (2.13.4), $\sum_{i=1}^{\infty} X_i^2 < \infty$ a.s. follows, thus establishing the theorem. ∎

**Exercise 2.13.4.** Let $\{X_i, i \geq 1\}$ be a sequence of independent random variables. Prove that

$$\sum_{i=1}^{\infty} E \frac{X_i^2}{1 + X_i^2} < \infty \qquad \text{if and only if} \qquad \sum_{i=1}^{\infty} X_i^2 < \infty \quad \text{a.s.} \quad \blacksquare$$

Now we return to the problem of essential convergence, characterizing essential convergence by the convergence of two numerical series. This so called "two series" theorem is to essential convergence as the three series theorem is to ordinary convergence.

**Theorem 2.13.6.**   Let $\{X_i, i \geq 1\}$ be a sequence of independent random variables. $S_n$ converges almost surely when centered (recall Definition 2.13.1) if and only if $\sum_{i=1}^{\infty} X_i^s$ converges almost surely if and only if

(i) $\displaystyle\sum_{i=1}^{\infty} P[|X_i - \mu(X_i)| \geq C] < \infty$      and

(ii) $\displaystyle\sum_{i=1}^{\infty} \text{var}\{[X_i - \mu(X_i)]I(|X_i - \mu(X_i)| \leq C)\} < \infty$

for some constant $C > 0$.

Moreover, if (i) and (ii) hold for some constant $C > 0$ they hold for all $C > 0$.

Further, if $S_n$ converges almost surely when centered then

$$\sum_{i=1}^{\infty} [X_i - \{\mu(X_i) + E[(X_i - \mu(X_i))I(|X_i - \mu(X_i)| \leq C)]\}]$$

converges almost surely for all constants $C > 0$.

**Proof.**   Assume that (i) and (ii) hold for some $C > 0$. We will show that $S_n$ converges a.s. when the $X_i$ are centered at

$$b_i = \mu(X_i) + E[(X_i - \mu(X_i))I(|X_i - \mu(X_i)| \leq C)].$$

Since (ii) holds, Theorem 2.12.2 (ii) implies that

$$\sum_{i=1}^{\infty} \{(X_i - \mu(X_i))I(|X_i - \mu(X_i)| \leq C)$$
$$-E[(X_i - \mu(X_i))I(|X_i - \mu(X_i)| \leq C]\}$$

converges almost surely. Since (i) holds,

$$\sum_{i=1}^{\infty} (X_i - \mu(X_i))I(|X_i - \mu(X_i)| > C)$$

converges almost surely by the Borel–Cantelli lemma. Combining the last two statements implies that $\sum_{i=1}^{\infty} (X_i - b_i)$ converges almost surely. Thus $S_n$ converges almost surely when centered with the $b_i$ the centering constants.

Now assume $S_n$ converges almost surely when centered with centering constants $\{c_i, i \geq 1\}$. We will show that $\sum_{i=1}^{\infty} X_i^s$ converges almost surely $X_i^s = X_i - X_i'$ for $i \geq 1$). $\sum_{i=1}^{\infty} X_i'$ converges almost surely when centered with centering constants $\{c_i, i \geq 1\}$. Thus

$$\sum_{i=1}^{n} (X_i - c_i) - \sum_{i=1}^{n} (X_i' - c_i) = \sum_{i=1}^{n} (X_i - X_i') = \sum_{i=1}^{n} X_i^s$$

converges almost surely as desired.

Now to prove the converse assume $\sum_{i=1}^{\infty} X_i^s$ converges almost surely. Fix $C > 0$. We show that (i) and (ii) hold. By the three series theorem, $\sum_{i=1}^{\infty} P[|X_i^s| \geq C] < \infty$. Fix $i \geq 1$.

$$[X_i^s \geq C] = [X_i - X_i' \geq C] \supset [X_i - \mu(X_i) \geq C, X_i' - \mu(X_i) \leq 0].$$

Thus

$$\tfrac{1}{2} P[X_i - \mu(X_i) \geq C] \leq P[X_i^s \geq C]$$

and applying this argument to $-X_i$ also,

$$\tfrac{1}{2} P[|X_i - \mu(X_i)| \geq C] \leq P[|X_i^s| \geq C]. \tag{2.13.5}$$

Hence $\sum_{i=1}^{\infty} P[|X_i - \mu(X_i)| \geq C] < \infty$ as desired. In order to show that (ii) holds, it suffices to show that

$$\sum_{i=1}^{\infty} E[(X_i - \mu(X_i))^2 I(|X_i - \mu(X_i)| \leq C)] < \infty.$$

For any random variable $Y$ with $EY^2 < \infty$,

$$EY^2 = \int_{-\infty}^{\infty} y^2 \, dP[Y \leq y]$$

$$= -\int_{0}^{\infty} y^2 \, dP[|Y| \geq y]$$

$$= -y^2 P[|Y| \geq y]\Big|_{0}^{\infty} + \int_{0}^{\infty} 2y P[|Y| \geq y] \, dy$$

$$= \int_{0}^{\infty} 2y P[|Y| \geq y] \, dy.$$

Fix $i \geq 1$. Thus

$$E[(X_i - \mu(X_i))^2 I(|X_i - \mu(X_i)| \leq C)] = 2 \int_{0}^{C} x P[|X_i - \mu(X_i)| \geq x] \, dx$$

$$\leq 4 \int_{0}^{C} x P[|X_i^s| \geq x] \, dx,$$

using Eq. (2.13.5). Integrating by parts again (in the reverse direction from above)

$$4 \int_0^C xP[|\ X_i^s\ | \geq x]\ dx = 2C^2P[|\ X_i^s\ | \geq C] - 2 \int_0^C x^2\ dP[|\ X_i^s\ | \geq x]$$

$$= 2C^2P[|\ X_i^s\ | \geq C] + 2\ \mathrm{var}[X_i^s I(|\ X_i^s\ | \leq C)].$$

Hence

$$E[(X_i - \mu(X_i))^2 I(|\ X_i - \mu(X_i)\ | \leq C)]$$
$$\leq 2C^2 P[|\ X_i^s\ | \geq C] + 2\ \mathrm{var}[X_i^s I(|\ X_i^s\ | \leq C)].$$

Since $\sum_{i=1}^\infty X_i^s$ converges almost surely, by the three series theorem,

$$\sum_{i=1}^\infty \mathrm{var}[X_i^s I(|\ X_i^s\ | \leq C)] < \infty.$$

Thus, recalling that $\sum_{i=1}^\infty P[|\ X_i^s\ | \geq C] < \infty$,

$$\sum_{i=1}^\infty E[(X_i - \mu(X_i))^2 I(|\ X_i - \mu(X_i)\ | \leq C)] < \infty.$$

Thus

$$\sum_{i=1}^\infty \mathrm{var}\{[X_i - \mu(X_i)]I(|\ X_i - \mu(X_i)\ | \leq C)\} < \infty$$

as desired.  ∎

Remarkably little implies that $\sum_{i=1}^\infty X_i$ converges almost surely when centered as the following theorem shows.

**Theorem 2.13.7.**  [Doob, 1953, p. 121].  Let $(X_i, i \geq 1)$ be independent. Then

$$\limsup P\left[\left|\sum_{i=1}^n X_i\right| < K\right] > 0$$

for some $K < \infty$ implies that $\sum_{i=1}^\infty X_i$ converges almost surely when centered.

**Proof.**  Doob gives an analytic proof depending on transforms. In the spirit of our approach in this book, we give a probabilistic proof using martingale theory. By hypothesis, there exists $\varepsilon > 0$ and integers $n_k \uparrow \infty$ such that

$$P\left[\left|\sum_{i=1}^{n_k} X_i\right| < K\right] \geq \varepsilon \quad \text{and} \quad P\left[\left|\sum_{i=1}^{n_k} X_i'\right| < K\right] \geq \varepsilon,$$

where $\{X_i', i \geq 1\}$ is as hypothesized in the definition of symmetrization. Fix $k \geq 1$. By the independence of $\sum_{i=1}^{n_k} X_i$ and $\sum_{i=1}^{n_k} X_i'$,

$$P\left[\left|\sum_{i=1}^{n_k} X_i\right| < K, \left|\sum_{i=1}^{n_k} X_i'\right| < K\right] \geq \varepsilon^2$$

for all $k \geq 1$.

$$\left[\left|\sum_{i=1}^{n_k} X_i\right| < K, \left|\sum_{i=1}^{n_k} X_i'\right| < K\right] \subset \left[\left|\sum_{i=1}^{n_k} X_i^s\right| < 2K\right].$$

Hence $P[|\sum_{i=1}^{n_k} X_i^s| < 2K] \geq \varepsilon^2$ for all $k \geq 1$. Thus

$$P\left[\left|\sum_{i=1}^{n_k} X_i^s\right| \geq 2K\right] \leq 1 - \varepsilon^2$$

for all $k \geq 1$. By symmetry,

$$P\left[\sum_{i=1}^{n_k} X_i^s \geq 2K\right] \leq (1 - \varepsilon^2)/2$$

for all $k \geq 1$. By the first Levy maximal inequality,

$$P\left[\max_{n \leq n_k} \sum_{i=1}^{n} X_i^s \geq 2K\right] \leq 2P\left[\sum_{i=1}^{n_k} X_i^s \geq 2K\right] \leq 1 - \varepsilon^2$$

for all $k \geq 1$. Using the continuity of $P$,

$$1 - \varepsilon^2 \geq P\left[\bigcup_{k=1}^{\infty} \left(\max_{n \leq n_k} \sum_{i=1}^{n} X_i^s \geq 2K\right)\right]$$

$$\geq P\left[\sup_{n \geq 1} \sum_{i=1}^{n} X_i^s \geq 2K + 1\right]. \qquad (2.13.6)$$

$[\sup_{n \geq 1} \sum_{i=1}^{n} X_i^s = \infty]$ is a tail event and thus has probability 0 or 1 by the Kolmogorov 0–1 law. Thus, by Eq. (2.13.6),

$$P\left[\sup_{n \geq 1} \sum_{i=1}^{n} X_i^s = \infty\right] = 0.$$

By the symmetry of the $X_i$,

$$P\left[\inf_{n \geq 1} \sum_{i=1}^{n} X_i^s = -\infty\right] = 0.$$

Thus $\sup_{n \geq 1} |\sum_{i=1}^{n} X_i^s| < \infty$ a.s. Hence $\sup_{i \geq 1} |X_i^s| < \infty$ a.s. We could

use the results about MZ regularity to quickly complete the proof. Instead
we present an elementary argument. Let

$$T_n = \sum_{i=1}^{n} X_i^s I(|\, X_i^s\,| \le K) \qquad \text{for} \quad n \ge 1.$$

By the symmetry of the $X_i^s$, $(T_n, n \ge 1)$ is a martingale. Its differences
$\{X_i^s I(|\, X_i^s\,| \le K), i \ge 1\}$ satisfy

$$E[\sup_{i \ge 1} X_i^s I(|\, X_i^s\,| \le K)] < \infty.$$

Hence, by Theorem 2.8.1, $\sup T_n < \infty$ implies $T_n$ converges.

Let $A_K = [\sup_{i \ge 1} |\, X_i^s\,| \le K]$. On $A_K$, $T_n = \sum_{i=1}^{n} X_i^s$ for all $n \ge 1$.
Thus $\sup T_n < \infty$ a.s. on $A_K$. Hence $T_n = \sum_{i=1}^{n} X_i^s$ converges a.s. on $A_K$
for all integers $K \ge 1$. Since $\sup |\, X_i^s\,| < \infty$ a.s., $\bigcup_{K=1}^{\infty} A_K = \Omega$. Thus
$\sum_{i=1}^{\infty} X_i^s$ converges a.s. By Theorem 2.13.6 $\sum_{i=1}^{\infty} X_i$ converges a.s. when
centered. ∎

**Remark.**   A proof of Theorem 2.13.7 can also be based on the observation that $\{\exp(iuS_n)/E \exp(iuS_n), n = 1\}$ is a martingale.

# *Stability and Moments*

## 3.1. Introduction

Throughout Section 3.1, $\{T_n, n \geq 1\}$ denotes an arbitrary stochastic sequence.

**Definition 3.1.1.** The real constants $a_n$, $0 < a_n \to \infty$, and $\{b_n, n \geq 1\}$ are said to stabilize $T_n$ if

$$(T_n - b_n)/a_n \to 0 \qquad \text{a.s.}$$

The $a_n$ are called norming constants and the $b_n$ are called centering constants. ∎

In Chapter 3, the stability of $S_n$ is studied. Of course, it is easy to find norming constants and centering constants that stabilize $\{S_n, n \geq 1\}$. The problem of interest is to study which constants stabilize and which constants do not stabilize $\{S_n, n \geq 1\}$. Of particular interest is the study of when the norming constants $a_n = n$ for each $n \geq 1$ stabilize $\{S_n, n \geq 1\}$, since $S_n/n$ is then the arithmetic average of $X_1, X_2, \ldots, X_n$. When there exists $\{b_n, n \geq 1\}$ such that

$$(S_n - b_n)/n \to 0 \qquad \text{a.s.,}$$

$\{S_n, n \geq 1\}$ is said to obey the strong law of large numbers. Absolute moments play a crucial role in the study of stability.

If $S_n - c_n$ converges almost surely, then $b_n = c_n$ for each $n \geq 1$ and *any* $a_n$, $0 < a_n \to \infty$, stabilizes $\{S_n, n \geq 1\}$. Thus the problem is only of interest when $\{S_n, n \geq 1\}$ is almost surely essentially divergent, that

114

is when

$$P[S_n - c_n \text{ converges}] = 0$$

for every choice of $\{c_n, n \geq 1\}$.

**Exercise 3.1.1.** Prove the result stated above: There exist norming constants $\{a_n, n \geq 1\}$ and centering constants $\{b_n, n \geq 1\}$ which stabilize an arbitrary stochastic sequence $\{T_n, n \geq 1\}$. ∎

Related to the concept of stability is the concept of upper and lower class results.

**Definition 3.1.2.** Norming constants $0 < a_n$ and centering constants $\{b_n, n \geq 1\}$ are said to belong to the upper class of $\{T_n, n \geq 1\}$ if

$$P[|\, T_n - b_n\,| > a_n \text{ i.o.}] = 0$$

and to belong to the lower class if

$$P[|\, T_n - b_n\,| > a_n \text{ i.o.}] = 1. \quad ∎$$

Upper and lower class results for $\{T_n, n \geq 1\}$ make statements about the magnitude of fluctuations $T_n$ for large $n$:

**Theorem 3.1.1.** Norming constants $a_n$, $0 < a_n \to \infty$, and centering constants $\{b_n, n \geq 1\}$ stabilize $\{T_n, n \geq 1\}$ if and only if the norming constants $\{\varepsilon a_n, n \geq 1\}$ and the centering constants $\{b_n, n \geq 1\}$ belong to the upper class of $\{T_n, n \geq 1\}$ for each $\varepsilon > 0$.

**Exercise 3.1.2.** Theorem 3.1.1 is easy to prove. Prove it. ∎

Lower class results appear mainly in Chapters 5 and 6. Most of the results in Chapter 3 are stability results, that is, upper class results in the sense stated in Theorem 3.1.1. According to Theorem 3.1.1, stability results give bounds for the magnitude of fluctuations for large $n$.

**Exercise 3.1.3.** Let $\{X_i, i \geq 1\}$ be independent, and let $a_n$, $0 < a_n \to \infty$, and $\{b_n, n \geq 1\}$ be real. Is it true that either $P[|\, S_n - b_n\,| > a_n \text{ i.o.}] = 1$ or $P[|\, S_n - b_n\,| > a_n \text{ i.o.}] = 0$? That is, is it true that the norming constants $\{a_n, n \geq 1\}$ and the centering constants $\{b_n, n \geq 1\}$ must either be in the upper class or in the lower class of $\{S_n, n \geq 1\}$. Prove that $P[|\, S_n - b_n\,|/a_n \to X] = 1$ implies that $X$ is degenerate. (Recall that $X$ is degenerate if there exists a constant $c$ such that $P[X = c] = 1$.) ∎

## 3.2. Stability Results for Independent Identically Distributed Random Variables

Throughout Section 3.2, $\{X_i, i \geq 1\}$ will denote a sequence of independent identically distributed random variables. Symmetrization and centering at medians play an important role in the study of stability just as they did in the study of the essential convergence of $\{S_n, n \geq 1\}$. In order to exhibit this role, we must study symmetrization further. Recall that $\mu(X)$ denotes the median of $X$.

**Lemma 3.2.1.** (The strong symmetrization inequalities). Let $\{T_n, n \geq 1\}$ be a stochastic sequence and $\{T_n^s, n \geq 1\}$ be the symmetrized version. Then, given $\varepsilon > 0$ and a sequence $\{c_n, n \geq 1\}$,

$$P[\sup_{n \geq 1} |T_n - \mu(T_n)| \geq \varepsilon] \leq 4P[\sup_{n \geq 1} |T_n^s| \geq \varepsilon]$$

$$\leq 8P[\sup_{n \geq 1} |T_n - c_n| \geq \varepsilon/2]. \qquad (3.2.1)$$

*Proof.* The proof is similar to that for Levy's maximal inequalities. Choose $0 < \varepsilon_0 < \varepsilon$ and let $T_n' = T_n - T_n^s$ for $n \geq 1$. Let $t$ be the smallest integer $n \geq 1$ such that $T_n - \mu(T_n) \geq \varepsilon_0$ if such an $n$ exists; otherwise let $t = \infty$, noting that it suffices to prove the left hand inequality of Eq. (3.2.1) with the absolute value signs deleted and with 4 replaced by 2.

$$[T_n^s \geq \varepsilon_0 \text{ for some } n] \supset \bigcup_{n=1}^{\infty} [t = n, T_n' - \mu(T_n) \leq 0].$$

For each $n \geq 1$,

$$P[t = n, T_n' - \mu(T_n) \leq 0] \geq P[t = n]/2$$

since $[t = n]$ and $[T_n' - \mu(T_n) \leq 0]$ are independent and $P[T_n' - \mu(T_n) \leq 0] \geq \frac{1}{2}$. Thus

$$P[T_n^s \geq \varepsilon_0 \text{ for some } n] \geq \sum_{n=1}^{\infty} P[t = n, T_n' - \mu(T_n) \leq 0]$$

$$\geq \sum_{n=1}^{\infty} P[t = n]/2$$

$$= P[T_n - \mu(T_n) \geq \varepsilon_0 \text{ for some } n]/2.$$

Thus

$$P[\sup T_n^s \geq \varepsilon_0] \geq P[T_n^s \geq \varepsilon_0 \text{ for some } n]$$
$$\geq P[T_n - \mu(T_n) \geq \varepsilon_0 \text{ for some } n]/2$$
$$\geq P[\sup(T_n - \mu(T_n)) \geq \varepsilon]/2.$$

Letting $\varepsilon_0 \uparrow \varepsilon$, we obtain

$$P[\sup T_n^s \geq \varepsilon] \geq P[\sup(T_n - \mu(T_n)) \geq \varepsilon]/2$$

by the left continuity of

$$f(\varepsilon) \equiv P[\sup T_n^s \geq \varepsilon],$$

thus proving the left hand inequality of Eq. (3.2.1), with the absolute value signs deleted and 4 replaced by 2.

Fix $\varepsilon_0 < \varepsilon$. For each $n \geq 1$,

$$[|T_n^s| \geq \varepsilon_0] = [|(T_n - c_n) - (T_n' - c_n)| \geq \varepsilon_0]$$
$$\subset [|T_n - c_n| \geq \varepsilon_0/2] \cup [|T_n' - c_n| \geq \varepsilon_0/2].$$

Thus

$$P[|T_n^s| \geq \varepsilon_0 \text{ for some } n] \leq 2P[|T_n - c_n| \geq \varepsilon_0/2 \text{ for some } n].$$

Hence

$$P[\sup |T_n^s| \geq \varepsilon] \leq P[|T_n^s| \geq \varepsilon_0 \text{ for some } n]$$
$$\leq 2P[|T_n - c_n| \geq \varepsilon_0/2 \text{ for some } n]$$
$$\leq 2P[\sup |T_n - c_n| \geq \varepsilon_0/2].$$

Letting $\varepsilon_0 \uparrow \varepsilon$ yields

$$P[\sup |T_n^s| \geq \varepsilon] \leq 2P[\sup |T_n - c_n| \geq \varepsilon/2],$$

completing the proof. ∎

It should be noted that no assumption was made about the joint distributions of $\{T_n, n \geq 1\}$ in the above lemma.

**Theorem 3.2.1.** Let $\{T_n, n \geq 1\}$ be a stochastic sequence and $\{c_n, n \geq 1\}$ a sequence of real numbers.

If $T_n - c_n \to 0$ a.s., then $T_n^s \to 0$ a.s. and $c_n - \mu(T_n) \to 0$.

If $T_n{}^s \to 0$ a.s., then $T_n - c_n \to 0$ a.s. for any real sequence $\{c_n, n \geq 1\}$ such that $c_n - \mu(T_n) \to 0$.

**Proof.**   By Lemma 3.2.1,

$$P[\sup_{n \geq m} | T_n - \mu(T_n)| \geq \varepsilon] \leq 4P[\sup_{n \geq m} | T_n{}^s| \geq \varepsilon]$$

$$\leq 8P[\sup_{n \geq m} | T_n - c_n| \geq \varepsilon/2].$$

Theorem 3.2.1 follows immediately from this.   ∎

**Corollary 3.2.1.**   If $(S_n - b_n)/a_n \to 0$ a.s., then

$$\sum_{i=1}^{n} X_i{}^s/a_n \to 0 \qquad \text{a.s.,}$$

$$(S_n - \mu(S_n))/a_n \to 0 \qquad \text{a.s.,}$$

and

$$(b_n - \mu(S_n))/a_n \to 0.$$

If $\sum_{i=1}^{n} X_i{}^s/a_n \to 0$ a.s., then

$$(S_n - \mu(S_n))/a_n \to 0 \qquad \text{a.s.,}$$

and

$$(S_n - b_n)/a_n \to 0 \qquad \text{a.s.}$$

for all $\{b_n, n \geq 1\}$ such that $(b_n - \mu(S_n))/a_n \to 0$.

**Proof.**   Note for each $n \geq 1$ that $S_n{}^s = \sum_{i=1}^{n} X_i{}^s$ and that

$$\{r \mid \mu(S_n/a_n) = r\} = \{r \mid \mu(S_n)/a_n = r\}.$$

(Recall that the median is not necessarily unique.) Thus Corollary 3.2.1 follows from Theorem 3.2.1.   ∎

Corollary 3.2.1 tells us that the question of which norming constants stabilize $\{S_n, n \geq 1\}$ may be studied *independently* of the question of centering constants. Moreover, it tells us that $\{\mu(S_n), n \geq 1\}$ always works as a set of centering constants. The only interesting question about centering constants remaining unanswered is whether other specific candidates for centering constants stabilize $\{S_n, n \geq 1\}$.

For example, $\{ES_n, n \geq 1\}$ seems like a plausible choice for centering constants, and, indeed will work in many situations. One should note that the solution to the problem of centering constants is simpler when the convergence is to zero rather than to a random variable. A comparison of Theorem 3.2.1 and Corollary 3.2.1 with the two-series theorem indicates this. In particular, if $S_n - b_n \to S$ a.s., it does not necessarily follow that $S_n - \mu(S_n) \to S$ a.s.

We now come to one of the most famous results of probability theory, Kolmogorov's strong law of large numbers [1930]. This is one of the two basic classical strong limit theorems (the law of the iterated logarithm being the other) of probability.

**Theorem 3.2.2.**    $(S_n - bn)/n \to 0$ a.s. if and only if $E|X_1| < \infty$ and $EX_1 = b$.

There are three distinctly different proofs available. We first present Kolmogorov's classical proof and then present a martingale proof due to Doob [1953, p. 341]. A third proof, based on the concept of stationarity, will be presented in Section 3.5.

In order to present Kolmogorov's proof, two computational lemmas are needed.

**Lemma 3.2.2.**    $E|X_1| < \infty$ if and only if $\sum_{i=1}^{\infty} P[|X_1| > i] < \infty$.

**Proof.**    $E|X_1| = \sum_{i=1}^{\infty} E[|X_1| \, I(i-1 < |X_1| \le i)]$. Thus

$$\sum_{i=1}^{\infty} (i-1)P[i-1 < |X_1| \le i] \le E|X_1| \le \sum_{i=1}^{\infty} iP[i-1 < |X_1| \le i].$$

$$\sum_{i=1}^{\infty} iP[i-1 < |X_1| \le i] = \sum_{i=1}^{\infty} i(P[|X_1| > i-1] - P[|X_1| > i])$$

$$= \sum_{i=0}^{\infty} P[|X_1| > i].$$

Thus $\sum_{i=1}^{\infty} P[|X_1| > i] < \infty$ implies $E|X_1| < \infty$ as desired.

Since $\sum_{i=1}^{\infty} (i-1)P[i-1 < |X_1| \le i] < \infty$ implies $\sum_{i=1}^{\infty} iP[i-1 < |X_1| \le i] < \infty$, $E|X_1| < \infty$ implies $\sum_{i=1}^{\infty} P[|X_1| > i] < \infty$ as desired. ∎

**Remark.**    Lemma 3.2.2 is typical of a large number of results relating moments and sums of tail probabilities. These results are often called large deviation results.

**Exercise 3.2.1.**    Prove that $EX_1^2 < \infty$ if and only if

$$\sum_{i=1}^{\infty} iP[|X_1| > i] < \infty. \quad \blacksquare$$

**Lemma 3.2.3.** Let $\{a_{ni}\}$ be a matrix of real numbers and $\{x_i\}$ a sequence of real numbers.

(i) (The Toeplitz lemma). Let $x_i \to x$ as $i \to \infty$. Then

$$\sum_{i=1}^{\infty} |a_{ni}| \le M < \infty \qquad \text{for all} \quad n \ge 1,$$

$$\sum_{i=1}^{\infty} a_{ni} \to 1 \qquad \text{as} \quad n \to \infty,$$

and

$$a_{ni} \to 0 \qquad \text{as} \quad n \to \infty \quad \text{for each} \quad i \ge 1$$

imply that

$$\sum_{i=1}^{\infty} a_{ni} x_i \to x \qquad \text{as} \quad n \to \infty.$$

(A matrix satisfying the above three assumptions is called a Toeplitz matrix.)

(ii) Let $x_i \to 0$ as $i \to \infty$. Then

$$\sum_{i=1}^{\infty} |a_{ni}| \le M < \infty \qquad \text{for all} \quad n \ge 1$$

and

$$a_{ni} \to 0 \qquad \text{as} \quad n \to \infty \quad \text{for each} \quad i \ge 1$$

imply that

$$\sum_{i=1}^{\infty} a_{ni} x_i \to 0.$$

(iii) (The Kronecker lemma). Let $\sum_{i=1}^{\infty} x_i/a_i$ converge for given $0 < a_i \uparrow \infty$. Then

$$\sum_{i=1}^{n} x_i/a_n \to 0 \qquad \text{as} \quad n \to \infty.$$

**Proof.** (i) $\sum_{i=1}^{\infty} a_{ni} x_i - x = \sum_{i=1}^{\infty} a_{ni} x_i - \sum_{i=1}^{\infty} a_{ni} x + \sum_{i=1}^{\infty} a_{ni} x - x$. Since $\sum_{i=1}^{\infty} a_{ni} \to 1$ as $n \to \infty$, it suffices to show that

$$\sum_{i=1}^{\infty} a_{ni}(x_i - x) \to 0 \qquad \text{as} \quad n \to \infty.$$

Fix $\varepsilon > 0$ and choose $I$ such that $|x_i - x| < \varepsilon$ for $i \ge I$.

$$\left| \sum_{i=1}^{\infty} a_{ni}(x_i - x) \right| \le \sum_{i=1}^{I-1} |a_{ni}(x_i - x)| + \sum_{i=I}^{\infty} |a_{ni}(x_i - x)|$$

$$\le \max_{i \le I-1} |x_i - x| \sum_{i=1}^{I-1} |a_{ni}| + \sum_{i=I}^{\infty} |a_{ni}| \varepsilon$$

$$\le \max_{i \le I-1} |x_i - x| \sum_{i=1}^{I-1} |a_{ni}| + M\varepsilon \to M\varepsilon \qquad \text{as} \quad n \to \infty.$$

Thus (i) is established. (ii) follows similarly. (iii) Let $a_0 = 0$, $y_1 = 0$, and $y_{n+1} = \sum_{i=1}^{n} x_i/a_i$ for $n \geq 1$. Fix $n \geq 1$. Summing by parts,

$$\sum_{i=1}^{n} x_i/a_n = \sum_{i=1}^{n} a_i(y_{i+1} - y_i)/a_n$$

$$= y_{n+1} - \sum_{i=1}^{n} (a_i - a_{i-1})y_i/a_n. \qquad (3.2.2)$$

Let

$$a_{ni} = \begin{cases} (a_i - a_{i-1})/a_n & \text{if } i \leq n \\ 0 & \text{if } i > n. \end{cases}$$

$\{a_{ni}\}$ is a Toeplitz matrix. Hence

$$\sum_{i=1}^{n} (a_i - a_{i-1})y_i/a_n \to \sum_{i=1}^{\infty} x_i/a_i$$

by (i) and hypothesis. Now

$$y_{n+1} \to \sum_{i=1}^{\infty} x_i/a_i$$

by hypothesis. Hence by Eq. (3.2.2) $\sum_{i=1}^{n} x_i/a_n \to 0$.  ∎

*Remark.*   The Kronecker lemma is an extremely important tool in establishing stability results. For, given $0 < a_n \uparrow \infty$, in order to prove

$$\sum_{i=1}^{n} (X_i - b_i)/a_n \to 0 \qquad \text{a.s.,}$$

it suffices to show that

$$\sum_{i=1}^{\infty} (X_i - b_i)/a_i \qquad \text{converges a.s.}$$

Thus, the Kronecker lemma allows us to reduce the proof of stability results to results given in Chapter 2. Rademacher was apparently the first person to make use of the Kronecker lemma to prove stability results.

*Proof of Theorem 3.2.2.*   Assuming $E \mid X_1 \mid < \infty$ we first show that $\sum_{i=1}^{n} (X_i - EX_1)/n \to 0$ a.s. Without loss of generality, assume $EX_1 = 0$. $E \mid X_1 \mid < \infty$ implies that

$$\sum_{i=1}^{\infty} P[\mid X_i \mid > i] = \sum_{i=1}^{\infty} P[\mid X_1 \mid > i] < \infty.$$

Hence $P[|X_i| > i \text{ i.o.}] = 0$ by the Borel–Cantelli lemma. Thus

$$\sum_{i=1}^{n} X_i I(|X_i| \le i)/n \to 0 \qquad \text{a.s.}$$

if and only if

$$\sum_{i=1}^{n} X_i/n \to 0 \qquad \text{a.s.} \qquad (3.2.3)$$

By the dominated convergence theorem and the fact that $EX_1 = 0$,

$$EX_i I(|X_i| \le i) = EX_1 I(|X_1| \le i) \to 0.$$

Since ordinary convergence implies Cesaro mean convergence (this an application of the Toeplitz lemma),

$$\sum_{i=1}^{n} EX_i I(|X_i| \le i)/n \to 0.$$

Using Eq. (3.2.3), $\sum_{i=1}^{n} X_i/n$ converges a.s. if and only if

$$\sum_{i=1}^{n} \{X_i I(|X_i| \le i) - E[X_i I(|X_i| \le i)]\}/n \to 0 \qquad \text{a.s.}$$

By the Kronecker lemma, it thus suffices to show that

$$\sum_{i=1}^{\infty} \{X_i I(|X_i| \le i) - E[X_i I(|X_i| \le i)]\}/i$$

converges a.s. By the often used Theorem 2.12.2 (ii) ($p = 2$ case) it suffices to show that

$$\sum_{i=1}^{\infty} \text{var}[X_i I(|X_i| \le i)/i] < \infty.$$

Thus it suffices to show that

$$\sum_{i=1}^{\infty} EX_i^2 I(|X_i| \le i)/i^2 < \infty,$$

recalling that var $Y \le EY^2$. This we show by computation.

$$\sum_{i=1}^{\infty} EX_i^2 I(|X_i| \le i)/i^2 = \sum_{i=1}^{\infty} \sum_{j=1}^{i} E[X_i^2 I(j - 1 < X_i \le j)]i^{-2}$$

$$= \sum_{j=1}^{\infty} E[X_1^2 I(j - 1 < |X_1| \le j)] \sum_{i=j}^{\infty} i^{-2}.$$

That $\sum_{i=j}^{\infty} i^{-2} \leq Kj^{-1}$ for all $j \geq 1$ and some $K < \infty$ is easily seen by comparison with the integral of $x^{-2}$. Thus

$$\sum_{i=1}^{\infty} EX_i^2 I(|X_i| \leq i)/i^2 \leq K \sum_{j=1}^{\infty} EX_1^2 I(j-1 < |X_1| \leq j)j^{-1}$$

$$\leq K \sum_{j=1}^{\infty} E[|X_1| I(j-1 < |X_1| \leq j)]$$

$$= KE|X_1| < \infty,$$

proving that $S_n/n \to 0$ a.s. as desired.

Assume $(S_n - bn)/n \to 0$ a.s. for some constant $b$. We must show that $E|X_1| < \infty$ and $EX_1 = b$. Since $S_n/n \to b$ a.s. and $S_{n-1}/n \to b$ a.s., subtraction shows that $X_n/n \to 0$ a.s. By the partial converse to the Borel–Cantelli theorem, $\sum_{n=1}^{\infty} P[|X_n| > n] < \infty$. By Lemma 3.2.2, $E|X_1| < \infty$. By the first half of the proof

$$S_n/n \to EX_1 \qquad \text{a.s.}$$

Hence $b = EX_1$.  ∎

Before plunging into the martingale proof of the strong law, we shall pause and consider some applications. A very famous application is the number theoretic result (referred to in Section 1.4) that almost every number is normal. Let $\Omega = [0, 1]$. Consider the decimal expansion $\omega = 0.x_1 x_2 \cdots$ for each $\omega \in \Omega$. (Some $\omega$'s have two expansions of course; but this will not affect our discussion since the set of such $\omega$ has Lebesgue measure zero.) For $k = 0, 1, \ldots, 9$ let $N_n^{(k)}(\omega)$ denote the number of times $k$ appears among the first $n$ $x_i$ of the decimal expansion of $\omega$. $\omega \in \Omega$ is said to be normal (to the base 10) if

$$N_n^{(k)}(\omega)/n \to 0.1 \qquad \text{for} \quad k = 0, 1, \ldots, 9 \quad \text{as} \quad n \to \infty.$$

In order to study normal numbers from a probabilistic viewpoint, define random variables $X_1, X_2, \ldots$ by letting $X_n(\omega)$ be the $n$th number in the decimal expansion of $\omega$ for each $\omega \in [0, 1]$, taking $\mathscr{F}$ to be the Borel sets of $[0, 1]$, and $P$ to be Lebesgue measure on $[0, 1]$. It is an easy matter to check that the $X_i$ are independent identically distributed with $P[X_1 = k] = 0.1$ for $k = 0, 1, \ldots, 9$. Fix an integer $0 \leq k \leq 9$. Let $f(x) = 1$ if $x = k$ and $f(x) = 0$ if $x \neq k$. Then $\{f(X_i), i \geq 1\}$ is a sequence of independent identically distributed random variables with $Ef(X_1) = 0.1$.

Thus by the strong law,

$$\sum_{i=1}^{n} f(X_i)/n \to 0.1 \qquad \text{a.s.}$$

But $\sum_{i=1}^{n} f(X_i(\omega))/n = N_n^{(k)}(\omega)/n$ and hence we have the fundamental result that almost every (with respect to Lebesgue measure) real number is normal to the base 10. Clearly it follows that almost every real number is normal to all bases since there are only countably many bases.

As discussed in Section 1.4, the strong law has interesting implications in the statistical estimation problem. Here, as a second application in statistics, we look at the problem of estimating an unknown distribution function $F$. Let $X_1, X_2, \ldots, X_n$ be independent identically distributed each with distribution $F$. The statistician is allowed to observe $X_1, X_2, \ldots, X_n$ in forming an estimate of $F$. For each real $x$, let

$$F_n(x) = \sum_{i=1}^{n} J_x(X_i)/n$$

where $J_x(y) = 0$ if $y > x$ and $J_x(y) = 1$ if $y \le x$. $F_n$ is called the empirical distribution function. We look at its behavior as an estimator of $F$. (It is instructive to choose $n \ge 1$ and suppose a specific set of values for $X_1, X_2, \ldots, X_n$ and graph $F_n$.) It is immediate from the strong law that $F_n(x) \to F(x)$ a.s. for each real $x$. Using a little "hard" analysis, it is not too difficult to strengthen this result to obtain the Glivenko–Cantelli theorem, namely,

$$\sup_{-\infty < x < \infty} |F_n(x) - F(x)| \to 0 \qquad \text{a.s. as } n \to \infty.$$

(See Chung [1968, pp. 123–125], for example.)

The following example gives an application to classical real analysis.

***Example 3.2.1.*** Let $f$ and $g$ be measurable real valued functions defined on $[0, 1]$. Suppose $0 < f < Cg < \infty$ a.s. Lebesgue measure for some $0 < C < \infty$. Suppose $g$ is integrable. We will show that

$$\int_0^1 \cdots \int_0^1 \frac{\sum_{i=1}^n f(x_i)}{\sum_{i=1}^n g(x_i)} \, dx_1 \cdots dx_n \to \frac{\int_0^1 f(x)\, dx}{\int_0^1 g(x)\, dx}.$$

The interesting thing about this result is how immediately it follows from the strong law. Let $\Omega = [0, 1] \times [0, 1] \times \cdots$, $\mathscr{F}$ be Borel sets of $\Omega$, and let $P$ be a Lebesgue measure on $\Omega$. For each $\omega = (x_1, x_2, \ldots)$ and $i \ge 1$, let $X_i(\omega) = x_i$. Then $\{X_i, i \ge 1\}$ is a sequence of independent identically distributed random variables. By the strong law

$$\frac{\sum_{i=1}^{n} f(X_i)}{\sum_{i=1}^{n} g(X_i)} = \frac{\sum_{i=1}^{n} f(X_i)/n}{\sum_{i=1}^{n} g(X_i)/n} \to \frac{\int_0^1 f(x)\,dx}{\int_0^1 g(x)\,dx} \qquad \text{a.s.}$$

as $n \to \infty$. Moreover

$$\left| \frac{\sum_{i=1}^{n} f(X_i)}{\sum_{i=1}^{n} g(X_i)} \right| \leq C \qquad \text{a.s.}$$

Hence by the Lebesgue dominated convergence theorem,

$$E\left[ \frac{\sum_{i=1}^{n} f(X_i)}{\sum_{i=1}^{n} g(X_i)} \right] \to \frac{\int_0^1 f(x)\,dx}{\int_0^1 g(x)\,dx}.$$

But this is the desired result since

$$E\left[ \frac{\sum_{i=1}^{n} f(X_i)}{\sum_{i=1}^{n} g(X_i)} \right] = \int_0^1 \cdots \int_0^1 \frac{\sum_{i=1}^{n} f(x_i)}{\sum_{i=1}^{n} g(x_i)}\,dx_1 \cdots dx_n. \quad \blacksquare$$

Another interesting application of the strong law of large numbers is to Monte Carlo simulation. Consider a continuous function $f : [0, 1] \to [0, 1]$. Suppose it is desired to evaluate $\int_0^1 f(x)\,dx$ by numerical methods. Let $X_{11}, X_{12}, X_{21}, X_{22}, \ldots$ be independent identically distributed uniform random variables on $[0, 1]$. Let $Y_i = I(f(X_{i1}) \geq X_{i2})$. Then by the strong law of large numbers

$$\sum_{i=1}^{n} Y_i/n \to EY_1 = \int_0^1 f(x)\,dx \qquad \text{a.s.}$$

Thus, using a random number generator to generate the $X_{ij}$, a numerical method for evaluating $\int_0^1 f(x)\,dx$ is clearly suggested.

We now give a remarkably simple martingale proof of the strong law of large numbers, due to Doob [1953, pp. 341–342]. Let $Z$ be a random variable and $\mathcal{H}_n = \mathcal{B}(S_i, i \geq n)$ for $n \geq 1$. According to Theorem 2.8.6 (ii), $E[Z \mid \mathcal{H}_n]$ converges almost surely and in $L_1$ as $n \to \infty$.

$$S_n = E[S_n \mid \mathcal{H}_n] = \sum_{i=1}^{n} E[X_i \mid \mathcal{H}_n] = nE[X_1 \mid \mathcal{H}_n] \qquad \text{a.s.}$$

since $E[X_1 \mid \mathcal{H}_n] = E[X_i \mid \mathcal{H}_n]$ a.s. for $i \leq n$ by symmetry. Thus

$$E[X_1 \mid \mathcal{H}_n] = S_n/n \qquad \text{a.s.}$$

Applying Theorem 2.8.6 (ii) with $Z = X_1$, $S_n/n$ converges almost surely and in $L_1$ to a random variable $X$. Since $X$ is a tail random variable, that is, measurable with respect to $\bigcap_{n=1}^{\infty} \mathscr{B}(X_n, X_{n+1}, \ldots)$, it follows by the Kolmogorov 0–1 law that $X$ is constant almost surely. Since $ES_n/n = EX_1$ and $S_n/n$ converges in $L_1$, it thus follows that $X = EX_1$ a.s., proving the strong law.

Viewing the strong law of large numbers as a result about the magnitude of the fluctuations of $\{S_n, n \geq 1\}$ when $E\,|\,X_1\,| < \infty$, it seems reasonable to ask whether there are analogous fluctuation results when $E\,|\,X_1\,|^r < \infty$, $r \neq 1$.

**Theorem 3.2.3.**    (i)    If $E\,|\,X_1\,|^r \overset{\cdot}{<} \infty$ for some $0 < r < 1$, then

$$S_n/n^{1/r} \to 0 \qquad \text{a.s.}$$

   (ii)    If $E\,|\,X_1\,|^r < \infty$ for some $1 \leq r < 2$, then

$$(S_n - nEX_1)/n^{1/r} \to 0 \qquad \text{a.s.}$$

   (iii)    Conversely, if $(S_n - b_n)/n^{1/r} \to 0$ a.s. for some $0 < r < 2$ and centering constants $\{b_n, n \geq 1\}$, then $E\,|\,X_1\,|^r < \infty$.

**Remark.**    Besides the generalization to $r \neq 1$, note that $(S_n - bn)/n \to 0$ a.s. implying $E\,|\,X_1\,| < \infty$ has been slightly generalized to

$$(S_n - b_n)/n \to 0 \qquad \text{a.s. implying} \quad E\,|\,X_1\,| < \infty.$$

The generalization of Kolmogorov's strong law of large numbers given in Theorem 3.2.3 is due to Marcinkiewicz and is often called the Marcinkiewicz strong law of large numbers.

**Proof.**    (i) and (ii). We only need prove (ii) for $r > 1$, since (ii) for $r = 1$ is the content of the Kolmogorov strong law of large numbers. The method of proof is the same as that used in the classical proof of Kolmogorov's strong law of large numbers. We give the main steps, omitting some details.

Assume $E\,|\,X_1\,|^r < \infty$ for some $0 < r < 2$ with $r \neq 1$. If $r > 1$, assume without loss of generality that $EX_1 = 0$ by centering at $EX_1$ if necessary. It suffices to show that

$$\sum_{i=1}^{n} X_i I(|\,X_i\,| \leq i^{1/r})/n^{1/r} \to 0 \qquad \text{a.s.}$$

For $0 < r < 1$,

$$\sum_{i=1}^{\infty} E[|X_i| I(|X_i| \leq i^{1/r})]/i^{1/r}$$

$$= \sum_{i=1}^{\infty} \sum_{j=1}^{i} E[X_1 I(j-1 < |X_1|^r \leq j)]/i^{1/r}$$

$$= \sum_{j=1}^{\infty} E[|X_1| I(j-1 < |X_1|^r \leq j)] \sum_{i=j}^{\infty} i^{-1/r}$$

$$\leq K \sum_{j=1}^{\infty} j^{1/r-1} E[|X_1|^r I(j-1 < |X_1|^r \leq j)] j^{1-1/r}$$

$$= K \sum_{j=1}^{\infty} E[|X_1|^r I(j-1 < |X_1|^r \leq j)]$$

$$= KE|X_1|^r < \infty$$

for some constant $K < \infty$ since $|X_1| = |X_1|^r |X_1|^{1-r}$, $|X_1|^{1-r} \leq j^{1/r-1}$ when $j-1 < |X_1|^r \leq j$ for each $j \geq 1$, and $\sum_{i=j}^{\infty} i^{-1/r} \leq K j^{1-1/r}$ for each $j \geq 1$ and some $K < \infty$. For $1 < r < 2$,

$$\left| \sum_{i=1}^{\infty} E[X_i I(|X_i| \leq i^{1/r})]/i^{1/r} \right| = \left| \sum_{i=1}^{\infty} E[X_i I(|X_i| > i^{1/r})]/i^{1/r} \right|$$

$$\leq \sum_{i=1}^{\infty} E[|X_i| I(|X_i| > i^{1/r})]/i^{1/r}.$$

The last term is shown to be finite by an argument similar to that given for

$$\sum_{i=1}^{\infty} E[|X_i| I(|X_i| \leq i^{1/r})]/i^{1/r}$$

in the case $0 < r < 1$.

Thus $E|X_1|^r < \infty$ for some $0 < r < 2$ with $r \neq 1$ implies that

$$\sum_{i=1}^{\infty} EX_i I(|X_i| \leq i^{1/r})]/i^{1/r}$$

converges as $n \to \infty$. Thus it suffices to show that $E|X_1|^r < \infty$ implies that

$$\sum_{i=1}^{\infty} \{X_i I(|X_i| \leq i^{1/r}) - E[X_i I(|X_i| \leq i^{1/r})]\}/i^{1/r} \qquad \text{converges a.s.}$$

Using Theorem 2.12.2 (ii), it suffices to show that

$$\sum_{i=1}^{\infty} E[|X_i|^2 I(|X_i| \le i^{1/r})]/i^{2/r} < \infty.$$

$$\sum_{i=1}^{\infty} E[|X_i|^2 I(|X_i| \le i^{1/r})]/i^{2/r}$$

$$= \sum_{i=1}^{\infty} i^{-2/r} \sum_{j=1}^{i} E[X_1^2 I(j-1 < |X_1|^r \le j)]$$

$$= \sum_{j=1}^{\infty} E[X_1^2 I(j-1 < |X_1|^r \le j)] \sum_{i=j}^{\infty} i^{-2/r}$$

$$\le K \sum_{j=1}^{\infty} j^{-2/r+1} E[X_1^2 I(j-1 < |X_1|^r \le j)]$$

$$\le K \sum_{j=1}^{\infty} j^{-2/r+1} j^{2/r-1} E[|X_1|^r I(j-1 < |X_1|^r \le j)]$$

$$= KE|X_1|^r < \infty$$

for some constant $K < \infty$, proving (i) and (ii). (Note that the above argument fails when $r = 2$ since $\sum_{i=j}^{\infty} i^{-2/r} = \infty$.)

To prove (iii), assume

$$(S_n - b_n)/n^{1/r} \to 0 \qquad \text{a.s.}$$

for some $0 < r < 2$ and centering constants $\{b_n, n \ge 1\}$. This implies that

$$\sum_{i=1}^{n} X_i^s/n^{1/r} \to 0 \qquad \text{a.s.},$$

which implies that

$$X_n^s/n^{1/r} \to 0 \qquad \text{a.s.}$$

as $n \to \infty$ which implies that $E|X_1^s|^r < \infty$. It can be shown that $E|X_1 - \mu(X_1)|^r \le 2E|X_1^s|^r$. (See Exercise 3.2.2 below.) Thus $E(|X_1|^r) < \infty$ since $E|X_1 - \mu(X_1)|^r < \infty$ implies $E|X_1|^r < \infty$ by the numerical inequality

$$|a + b|^r \le c_r(|a|^r + |b|^r)$$

where $c_r = 1$ if $r < 1$ and $c_r = 2^{r-1}$ if $r \ge 1$.  ∎

***Exercise 3.2.2.*** Prove that $E|X - \mu(X)|^r \le 2E|X^s|^r \le 4c_r \times E|X - a|^r$ for all $r > 0$ and all real numbers $a$ where $c_r = 1$ if $r < 1$

and $c_r = 2^{r-1}$ if $r \geq 1$. Hint: To prove the left-hand inequality, note that

$$P[|\, X - \mu(X)\,|^r > x] \leq 2P[|\, X^s\,|^r > x]$$

and use integration by parts. ∎

It is important to emphasize that Theorem 3.2.3 is a result about the asymptotic fluctuations of $S_n$. (i) says that when $E\,|\,X_1\,|^r < \infty$ for some $0 < r < 1$, then the magnitude of the fluctuations of $S_n$ about 0 are asymptotically of no larger order than $n^{1/r}$. (ii) says that when $E\,|\,X_1\,|^r < \infty$ for some $1 \leq r < 2$, then the magnitude of the asymptotic fluctuations of $S_n$ about the line $nEX_1$ are asymptotically of no larger order than $n^{1/r}$. We will shortly see that it is impossible to make the analogous statement for $r \geq 2$.

It is perhaps of interest to look at what Theorem 3.2.3 has to say about statistical estimation. Let $\mu = EX_1$ be unknown and consider the estimators $\bar{X}_{(n)} = \sum_{i=1}^n X_i/n$ for each $n \geq 1$. Here $n$ is the number of observations in the sample. The statistician, of course, wants $\bar{X}_{(n)}$ to be "close" to $\mu$. The strong law of large numbers tells us that $\bar{X}_{(n)} \to \mu$ almost surely as $n \to \infty$, that being the so called "strong consistency" of the estimators $\bar{X}_{(n)}$. If the statistician happens to know that $E\,|\,X_1\,|^r < \infty$ for some $1 < r < 2$, then by Theorem 3.2.3 (ii),

$$n^{1-1/r}\,|\,\bar{X}_{(n)} - \mu\,| \to 0 \qquad \text{a.s.}$$

as $n \to \infty$. This is clearly a stronger statement than merely saying that $\bar{X}_{(n)}$ is strongly consistent since $n^{1-1/r} \to \infty$ for $1 < r < 2$.

Theorem 3.2.3 characterizes the convergence of $(S_n - b_n)/n^{1/r}$ to zero as the following corollary indicates.

**Corollary 3.2.2.** Let $0 < r < 2$.

$$(S_n - b_n)/n^{1/r} \to 0 \qquad \text{a.s.}$$

if and only if

$$E\,|\,X_1\,|^r < \infty \qquad \text{and} \qquad (b_n - na)/n^{1/r} \to 0$$

where $a = 0$ if $0 < r < 1$ and $a = EX_1$ if $1 \leq r < 2$.

**Proof.** By Theorem 3.2.3 (iii),

$$(S_n - b_n)/n^{1/r} \to 0 \qquad \text{a.s.}$$

implies $E \mid X_1 \mid^r < \infty$. By Theorem 3.2.3 (i) and (ii), $E \mid X_1 \mid^r < \infty$ implies

$$(S_n - an)/n^{1/r} \to 0 \qquad \text{a.s.}$$

Thus the desired result follows. $\blacksquare$

For each $0 < r < 2$, the asymptotic fluctuation behavior of $\{X_n/n^{1/r}, n \geq 1\}$ is essentially the same as that of $\{S_n/n^{1/r}, n \geq 1\}$ as the following theorem shows.

**Theorem 3.2.4.** Let $0 < r < 2$ and $EX_1 = 0$ if $r \geq 1$. Then either

$$\lim S_n/n^{1/r} = \lim X_n/n^{1/r} = 0 \qquad \text{a.s.}$$

or

$$\lim \sup \mid S_n \mid/n^{1/r} = \lim \sup \mid X_n \mid/n^{1/r} = \infty \qquad \text{a.s.}$$

**Proof.** Theorem 3.2.4 is really a corollary of the Kolmogorov and Marcinkiewicz strong laws of large numbers.

$$S_n/n^{1/r} \to 0 \qquad \text{a.s.}$$

if and only if $E \mid X_1 \mid^r < \infty$ by these results. $E \mid X_1 \mid^r < \infty$ if and only if

$$\sum_{i=1}^{\infty} P[\mid X_i \mid > i^{1/r}] < \infty$$

if and only if

$$\sum_{i=1}^{\infty} P[\mid X_i \mid > \varepsilon i^{1/r}] < \infty$$

for each $\varepsilon > 0$. Thus

$$\lim S_n/n^{1/r} = 0 \qquad \text{a.s.}$$

if and only if

$$\lim X_n/n^{1/r} = 0 \qquad \text{a.s.}$$

Assume $P[X_n/n^{1/r} \nrightarrow 0] > 0$. Thus for some $\varepsilon > 0$, $P[\mid X_n \mid/n^{1/r} > \varepsilon \text{ i.o.}] = 1$, using the Kolmogorov 0–1 law. Thus by the Borel–Cantelli lemma

$$\sum_{n=1}^{\infty} P[\mid X_n \mid^r/\varepsilon^r > n] = \sum_{n=1}^{\infty} P[\mid X_1 \mid^r/\varepsilon^r > n] = \infty.$$

Thus $E \mid X_1 \mid^r = \infty$. Applying Lemma 3.2.2 to $\mid X_1/M \mid^r$,

$$\sum_{n=1}^{\infty} P[\mid X_n \mid > Mn^{1/r}] = \infty$$

for each $M > 0$. Hence

$$\limsup |X_n|/n^{1/r} = \infty \qquad \text{a.s.}$$

by the converse to the Borel–Cantelli lemma.

$$\limsup |X_n|/n^{1/r} = \limsup |S_n - S_{n-1}|/n^{1/r}$$
$$\leq \limsup |S_n|/n^{1/r} + \limsup |S_{n-1}|/n^{1/r}$$
$$= 2 \limsup |S_n|/n^{1/r}.$$

Hence $\limsup |S_n|/n^{1/r} = \infty$ a.s.  ∎

   Can we make a statement similar to Theorem 3.2.4 when $a_n \uparrow \infty$ sufficiently rapidly but $\{a_n, n \geq 1\}$ does not necessarily equal $\{n^{1/r}, n \geq 1\}$ for some $0 < r < 2$ as we have assumed previously? The following simple lemma is central to exhibiting such a generalization of Theorem 3.2.4.

   **Lemma 3.2.4.** Let $0 < a_i/i^{1/r}$ be nondecreasing for some $r > 0$. Then either

$$\sum_{i=1}^{\infty} P[|X_i| > Ka_i] = \infty \qquad \text{for all} \quad K > 0$$

or

$$\sum_{i=1}^{\infty} P[|X_i| > Ka_i] < \infty \qquad \text{for all} \quad K > 0.$$

   **Proof.**   Assume

$$\sum_{i=1}^{\infty} P[|X_i| > K_0 a_i] = \infty$$

for some $K_0 > 0$. Clearly

$$\sum_{i=1}^{\infty} P[|X_i| > Ka_i] = \infty$$

for all $K \leq K_0$. Fix $K > K_0$. Choose an integer $n > K^r/K_0^r$. By hypothesis, $a_{in}^r/in \geq a_i^r/i$. Thus

$$a_{in}^r/a_i^r \geq n > K^r/K_0^r.$$

Thus

$$\sum_{i=1}^{\infty} P[|X_i| > Ka_i] \geq \sum_{i=1}^{\infty} P[|X_i| > K_0 a_{in}]$$

$$\geq \sum_{i=1}^{\infty} P[|X_i| > K_0 a_{in+j}]$$

for $j = 0, 1, \ldots, n - 1$. Thus

$$n \sum_{i=1}^{\infty} P[|X_i| > Ka_i] \geq \sum_{j=0}^{n-1} \sum_{i=1}^{\infty} P[|X_i| > K_0 a_{in+j}]$$

$$= \sum_{i=1}^{\infty} P[|X_i| > K_0 a_i] = \infty.$$

Thus divergence for some $K_0 > 0$ implies divergence for all $K > 0$, establishing the lemma. ∎

Lemma 3.2.4 should be interpreted as a statement about the asymptotic behavior of $X_n/a_n$:

**Corollary 3.2.3.** Let $0 < a_i/i^{1/r}$ be nondecreasing for some $r > 0$. Then either

$$\limsup |X_n|/a_n = \infty \qquad \text{a.s.}$$

or

$$\lim X_n/a_n = 0 \qquad \text{a.s.}$$

*Proof.* Immediate from Lemma 3.2.3, the Borel–Cantelli lemma, and the converse to the Borel–Cantelli lemma. ∎

**Theorem 3.2.5.** [Feller, 1946a]. Let $0 < a_n/n^{1/r} \uparrow \infty$ for some $r < 2$. Suppose $EX_1 = 0$ if $E|X_1| < \infty$. Suppose that either $a_n/n$ is nondecreasing or that $a_n/n$ is nonincreasing. Then either

$$\limsup |S_n|/a_n = \limsup |X_n|/a_n = \infty$$

almost surely (and $\sum_{n=1}^{\infty} P[|X_n| > a_n] = \infty$) or

$$\lim S_n/a_n = \lim X_n/a_n = 0$$

almost surely (and $\sum_{n=1}^{\infty} P[|X_n| > a_n] < \infty$).

*Proof.* By Corollary 3.2.3 either

$$\limsup |X_n|/a_n = \infty$$

almost surely (and $\sum_{n=1}^{\infty} P[|X_n| > a_n] = \infty$) or

$$\lim X_n/a_n = 0$$

almost surely (and $\sum_{n=1}^{\infty} P[|X_n| > a_n] < \infty$). If $\limsup |X_n|/a_n = \infty$, then it follows very easily (see the proof of Theorem 3.2.4) from $0 < a_n \uparrow \infty$ that $\limsup |S_n|/a_n = \infty$. Thus it remains to show that

$$\lim X_n/a_n = 0 \quad \text{a.s. implies} \quad \lim S_n/a_n = 0 \quad \text{a.s.}$$

There are two cases. Either $EX_1 = 0$ or $E|X_1| = \infty$. The rest of the proof proceeds in much the same way as the proof of the Kolmogorov and Marcinkiewicz laws of large numbers. We omit the computations. ∎

**Exercise 3.2.3.**  Finish the proof of Theorem 3.2.5. Some of the computations here are somewhat tricky—one may wish to refer to Feller's paper for help. ∎

Chow and Robbins [1961] prove a result related to Theorem 3.2.5. They show that $E|X_1| = \infty$ implies for any positive sequence $\{a_i, i \geq 1\}$ that either $\limsup |S_n|/a_n = \infty$ a.s. or $\liminf |S_n|/a_n = 0$ a.s. This result has an interesting gambling interpretation. For each $i \geq 1$, let $b_i > 0$ be the gambler's entry fee to play round $i$ and $X_i$ be the gambler's winnings at trial $i$. The game will be called fair if there exists $\{b_i, i \geq 1\}$ such that $S_n/\sum_{i=1}^n b_i \to 1$ a.s. Chow and Robbin's result shows that $E|X_1| = \infty$ implies that the game cannot be made fair in this sense for any choice of $b_i$.

Derman and Robbins [1955] also prove a result closely connected with Theorem 3.2.5. Let $E|X_1| = \infty$. By Theorem 3.2.5, $\limsup |S_n|/n = \infty$ a.s. Are there hypotheses which imply $S_n/n \to \infty$ a.s.; that is, when $P[S_n < Mn \text{ i.o.}] = 0$ for all constants $M$?

First, an easy to prove result is left as an exercise:

**Exercise 3.2.4.**  Let $EX_1^+ = \infty$ and $EX_1^- < \infty$. Prove by truncation and an appeal to Kolmogorov's strong law of large numbers that $S_n/n \to \infty$ a.s. ∎

**Theorem 3.2.6.**  [Derman and Robbins, 1955].  Let $0 < \alpha < \beta < 1$. Suppose

$$\liminf x^\alpha P[X_1 > x] > 0 \tag{3.2.4}$$

(which implies $E(X_1^+)^\alpha = \infty$) and $E(X_1^-)^\beta < \infty$. Then

$$S_n/n \to \infty \quad \text{a.s.}$$

Indeed

$$S_n/n^{1/\beta} \to \infty \quad \text{a.s.}$$

**Proof.** Let $Y_i = X_i I(X_i \geq 0)$ and $Z_i = X_i I(X_i < 0)$ for each $i \geq 1$. $\{Y_i, i > 1\}$ and $\{Z_i, i \geq 1\}$ are each sequences of independent distributed sequences of random variables. Also $E \mid Z_1 \mid^\beta = E(X_1^-)^\beta < \infty$. Hence by the Marcinkiewicz strong law of large numbers,

$$\sum_{i=1}^{n} Z_i/n^{1/\beta} \to 0 \qquad \text{a.s.} \qquad (3.2.5)$$

Fix $K > 0$.

$$P\left[\sum_{i=1}^{n} Y_i \leq Kn^{1/\beta}\right] \leq P[\max_{i \leq n} Y_i \leq Kn^{1/\beta}] = P^n[Y_1 \leq Kn^{1/\beta}]$$

$$= P^n[X_1 \leq Kn^{1/\beta}]$$

$$\leq \left(1 - \frac{c}{K^\alpha n^{\alpha/\beta}}\right)^n$$

for $n$ sufficiently large and some $c > 0$ by Eq. (3.2.4).

$$\sum_{n=1}^{\infty} \left(1 - \frac{c}{K^\alpha n^{\alpha/\beta}}\right)^n < \infty,$$

since

$$\left(1 - \frac{c}{K^\alpha n^{\alpha/\beta}}\right)^n \leq \left[\exp\left(-\frac{c}{K^\alpha n^{\alpha/\beta}}\right)\right]^n$$

$$= \exp\left(-\frac{c}{K^\alpha} n^{1-\alpha/\beta}\right).$$

Thus by the Borel–Cantelli lemma,

$$P\left[\sum_{i=1}^{n} Y_i \leq Kn^{1/\beta} \text{ i.o.}\right] = 0.$$

That is,

$$P\left[\sum_{i=1}^{n} Y_i \geq Kn^{1/\beta} \text{ for all } n \text{ sufficiently large}\right] = 1$$

for each $K > 0$. Hence

$$\lim \sum_{i=1}^{n} Y_i/n^{1/\beta} = \infty \qquad \text{a.s.}$$

Thus, using Eq. (3.2.5),

$$\lim \sum_{i=1}^{n} X_i/n^{1/\beta} = \lim \sum_{i=1}^{n} Y_i/n^{1/\beta} + \lim \sum_{i=1}^{n} Z_i/n^{1/\beta} = \infty \qquad \text{a.s.}$$

Clearly this implies that $\lim S_n/n = \infty$ a.s.  ∎

All the preceding results of Section 3.2 concern norming constants for which $a_n/n^{1/r} \uparrow \infty$ for some $r < 2$. We now consider norming constants which do not increase this rapidly. As a first observation, the asymptotic fluctuation behavior of $\{X_n/n^{1/r}, n \geq 1\}$ is no longer the same as that of $\{S_n/n^{1/r}, n \geq 1\}$ when $r \geq 2$.

**Theorem 3.2.7.**  Suppose $X_1$ is nondegenerate and $r \geq 2$. Then

$$\limsup |S_n - b_n|/n^{1/r} = \infty \qquad \text{a.s.}$$

(regardless of the behavior of $\limsup |X_n|/n^{1/r}$!) for every choice of centering constants $\{b_n, n \geq 1\}$.

Theorem 3.2.7 is an easy consequence of the central limit theorem for independent identically distributed random variables. For easy reference we state this fundamental result. The proof of this central limit theorem can be found in any graduate level introductory probability text.

**Theorem 3.2.8.**  (Central limit theorem). If $X_1$ is nondegenerate (which implies var $X_1 > 0$) with var $X_1 < \infty$, then

$$P\left[\frac{S_n - nEX_1}{(n \text{ var } X_1)^{1/2}} \leq x\right] \to \frac{1}{(2\pi)^{1/2}} \int_{-\infty}^{x} \exp(-y^2/2) \, dy$$

as $n \to \infty$ for each real $x$.

**Proof of Theorem 3.2.7.**  Suppose $X_1$ is not degenerate. We assume $P[\limsup |S_n - b_n|/n^{1/r} < \infty] > 0$ and produce a contradiction. By the Kolmogorov 0–1 law, there then exists $K < \infty$ such that

$$\limsup |S_n - b_n|/n^{1/r} \leq K \qquad \text{a.s.}$$

This implies that

$$\limsup \left|\sum_{i=1}^{n} X_i^s\right| \Big/ n^{1/r} \leq 2K \qquad \text{a.s.}$$

and that

$$\limsup |X_i^s|/i^{1/r} \leq 4K \qquad \text{a.s.}$$

Thus

$$P\left[\sup_{m \geq n} \left|\sum_{i=1}^{m} X_i^s\right| \Big/ m^{1/r} > 2K + 1\right] \to 0 \qquad \text{as } n \to \infty \qquad (3.2.6)$$

and

$$P[|X_i^s| > (4K + 1)i^{1/r} \text{ i.o.}] = 0. \qquad (3.2.7)$$

Using Eq. (3.2.7), it follows by the Borel–Cantelli lemma and Lemma 3.2.2 that $E\,|\,X_1{}^s\,|^r < \infty$. Thus $E(X_1{}^s)^2 < \infty$. $X_1$ nondegenerate implies $X_1{}^s$ nondegenerate. Thus, we may apply the central limit theorem to obtain that

$$P\left[\left|\sum_{i=1}^{n} X_i{}^s\right|\Big/n^{1/2} > 2K + 1\right] \to a > 0$$

as $n \to \infty$. But this contradicts Eq. (3.2.6), thus establishing the theorem. ∎

Theorem 3.2.7 can be generalized to norming constants $\{a_n, n \geq 1\}$ not necessarily equal to $\{n^{1/r}, n \geq 1\}$ for some $r \geq 2$. It can be shown that $X_1$ nondegenerate and $\limsup a_n/n^{1/2} < \infty$ implies that $\limsup |\,S_n - b_n\,|/a_n = \infty$ a.s.

**Exercise 3.2.5.**   Prove the preceding remark. ∎

The study of norming constants which increase so slowly that they do *not* satisfy $a_n/n^{1/r} \uparrow \infty$ for some $r < 2$ is radically more difficult and produces unexpected and beautiful results. For example:

**Theorem 3.2.9.**   (The Hartman-Wintner law of the iterated logarithm [Hartman and Winter, 1941]).   If $EX_1{}^2 < \infty$ and $EX_1 = 0$, then

$$\limsup S_n/(2nEX_1{}^2 \log_2 n)^{1/2} = 1 \qquad \text{a.s.}$$

(Here and in the rest of the book, "$\log_2$" denotes "log log.")

The proof of this result is very involved and we do not yet have the tools at our disposal to tackle it. The proof is presented in Chapter 5 where laws of the iterated logarithm and related results are studied in detail. It is an easy consequence of the Kolmogorov 0–1 law that

$$\limsup S_n/(2nEX_1{}^2 \log_2 n)^{1/2} = c \qquad \text{a.s.}$$

for some constant $-\infty \leq c \leq \infty$. Thus, what is striking about Theorem 3.2.9 is that $c = 1$ regardless of the distribution of $X_1$ when $EX_1 = 0$ and $EX_1{}^2 < \infty$.

**Exercise 3.2.6.**   Let $0 < a_n \to \infty$. Prove that there exists $c \in [-\infty, \infty]$ such that

$$P[\limsup S_n/a_n = c] = 1$$

and that there exists $c \in [-\infty, \infty]$ such that

$$P[\limsup | S_n |/a_n = c] = 1.$$

Note that

$$\limsup | S_n |/(2nEX_1^2 \log_2 n)^{1/2} = 1 \qquad \text{a.s.}$$

is a weakening of the conclusion of Theorem 3.2.9 (substitute $\{-X_i, i \geq 1\}$ in the statement of Theorem 3.2.8). If $EX_1^2 < \infty$ and $EX_1 = 0$, it follows easily that

$$\lim X_n/(2n\ EX_1^2 \log_2 n)^{1/2} = 0 \qquad \text{a.s.}$$

Thus Theorem 3.2.9 provides us with an example of norming constants $\{a_n, n \geq 1\}$ which increase to infinity slowly enough such that with probability one

$$\limsup | S_n |/a_n \neq \limsup | X_n |/a_n. \quad \blacksquare$$

**Exercise 3.2.7.**   Let $X_1$ be nondegenerate normal with $EX_1 = 0$. Prove that $\sum_{i=1}^{\infty} S_i/i$ diverges a.s.   $\blacksquare$

## 3.3. Stability Results for Martingales

Throughout Section 3.3, $\{X_i, \mathscr{F}_i, i \geq 1\}$ will be assumed to be a martingale difference sequence. The martingale results of Chapter 2 and the Kronecker lemma immediately yield interesting stability results for martingales.

**Theorem 3.3.1.**   [Chow, 1965].   Let $a_i$ be $\mathscr{F}_{i-1}$ measurable for each $i \geq 1$ and $0 < a_i \uparrow \infty$ a.s. Then

$$\sum_{i=1}^{\infty} E[| X_i |^p | \mathscr{F}_{i-1}]/a_i^p < \infty \qquad \text{for some} \quad 0 < p \leq 2$$

implies

$$S_n/a_n \to 0 \qquad \text{a.s.}$$

**Proof.**   Immediate from Corollary 2.8.5, Exercise 2.8.8, and the Kronecker lemma.   $\blacksquare$

There is a clever application of Theorem 3.3.1 to mixing stochastic sequences. Loosely speaking, a stochastic sequence is mixing if its random variables with indices far apart are almost independent. This idea is formalized in a variety of ways. Blum *et al.* [1963] introduce:

**Definition 3.3.1.** Given a stochastic sequence $\{Y_i, i \geq 1\}$, let

$$\mathcal{G}_{ij} = \mathcal{B}(Y_k, i \leq k \leq j)$$

for all $1 \leq i \leq j < \infty$. $\{Y_i, i \geq 1\}$ is said to be *-mixing if there exists an integer $M$ and a function $\phi$ for which

$$\phi(m) \to 0 \qquad \text{as} \quad m \to \infty$$

and $A \in \mathcal{G}_{1n}$, $B \in \mathcal{G}_{m+n, m+n}$ implies

$$| P(A \cap B) - PA\,PB | \leq \phi(m)PA\,PB \qquad (3.3.1)$$

for all $m \geq M$ and all $n \geq 1$. ∎

We first translate Eq. (3.3.1) into a statement about expectations.

**Lemma 3.3.1.** Let $\{Y_i, i \geq 1\}$ be *-mixing with respect to a function $\phi$ and an integer $M$. Suppose $E\,|\,Y_i\,| < \infty$ for each $i \geq 1$. Then

$$| E[Y_{n+m} \,|\, \mathcal{G}] - EY_{n+m} | \leq \phi(m)E\,|\,Y_{n+m}\,| \qquad \text{a.s.}$$

for each $\sigma$ field $\mathcal{G} \subset \mathcal{G}_{1n}$, each $n \geq 1$, and each $m \geq M$.

**Proof.** Fix $n \geq 1$, $m \geq M$. Choose $Z_k$ simple and $\mathcal{G}_{n+m, n+m}$ measurable such that

$$0 \leq Z_k \uparrow Y_{n+m}^+ \qquad \text{as} \quad k \to \infty.$$

Fix $k \geq 1$. Then there exists $j \geq 1$ such that

$$Z_k = \sum_{i=1}^{j} b_i I(B_i),$$

where $b_i \geq 0$ and $B_i \in \mathcal{G}_{n+m, m+n}$ for $1 \leq i \leq j$. Choose $A \in \mathcal{G}$ such that $P(A) > 0$.

$$
\begin{aligned}
| E[Z_k \,|\, A] - EZ_k | &= \left| \sum_{i=1}^{j} b_i P(B_i \cap A)/PA - \sum_{i=1}^{j} b_i PB_i \right| \\
&= \left| \sum_{i=1}^{j} b_i [P(B_i \cap A)/PA - PB_i] \right| \\
&\leq \sum_{i=1}^{j} b_i \,|\, P(B_i \cap A)/PA - PB_i \,| \\
&\leq \phi(m) \sum_{i=1}^{j} b_i PB_i = \phi(m)EZ_k.
\end{aligned}
$$

Thus $| E[Z_k \mid A] - EZ_k | \le \phi(m)EZ_k$. Letting $k \to \infty$ implies $| E[Y_{n+m}^+ \mid A]$ $-EY_{n+m}^+ | \le \phi(m)EY_{n+m}^+$. Thus

$$\left| \int_A (E[Y_{n+m}^+ \mid \mathscr{G}] - EY_{n+m}^+) \, dP/PA \right| \le \phi(m)EY_{n+m}^+ \qquad \text{for all} \quad A \in \mathscr{G}.$$

But this implies $| E[Y_{n+m}^+ \mid \mathscr{G}] - E[Y_{n+m}^+] | \le \phi(m)EY_{n+m}^+$ a.s. Similarly $| E[Y_{n+m}^- \mid \mathscr{G}] - E[Y_{n+m}^-] | \le \phi(m)E[Y_{n+m}^-]$ a.s. Combining these last two statements proves the lemma. ∎

**Theorem 3.3.2.** [Blum *et al.*, 1963]. Let $\{Y_i, i \ge 1\}$ be *-mixing with $EY_i = 0$ and $E | Y_i | \le K < \infty$ for each $i \ge 1$ and

$$\sum_{i=1}^{\infty} EY_i^2/i^2 < \infty. \tag{3.3.2}$$

Then

$$\sum_{i=1}^{n} Y_i/n \to 0 \qquad \text{a.s.}$$

**Proof.** The clever and simple martingale proof that we give is due to Bartfai (see Revesz [1968]). Fix $\varepsilon > 0$. Referring to Lemma 3.3.1, choose $M$ sufficiently large so that

$$| E[Y_{nM+k} \mid Y_{(n-1)M+k}, Y_{(n-2)M+k}, \ldots, Y_{M+k}] | \le \varepsilon E[| Y_{nM+k} |] \le \varepsilon K \tag{3.3.3}$$

for each $n \ge 2$ and $0 \le k \le M - 1$. Note that we have used the facts that $EY_i = 0$ and $E | Y_i | \le K$ for each $i \ge 1$.

It suffices to show that $\sum_{n=2}^{N} Y_{nM+k}/N \to 0$ a.s. as $N \to \infty$ for each $0 \le k < M$. Fix $0 \le k < M$. Let $\mathscr{H}_n = \mathscr{B}(Y_{nM+k}, Y_{(n-1)M+k}, \ldots, Y_{M+k})$ for each $n \ge 1$ and $\mathscr{H}_0 = \{\varnothing, \Omega\}$. $\{Y_{nM+k} - E[Y_{nM+k} \mid \mathscr{H}_{n-1}], \mathscr{H}_n, n \ge 2\}$ is a martingale difference sequence. Eq. (3.3.2) implies that $\sum_{n=2}^{\infty} E\{(Y_{nM+k} - E[Y_{nM+k} \mid \mathscr{H}_{n-1}])^2 \mid \mathscr{H}_{n-1}\}/n^2 < \infty$ a.s. Thus, by Theorem 3.3.1 with $p = 2$ and $a_i = i$ for $i \ge 1$,

$$\sum_{n=2}^{N} (Y_{nM+k} - E[Y_{nM+k} \mid \mathscr{H}_{n-1}])/N \to 0 \qquad \text{a.s.}$$

By Eq. (3.3.3),

$$\left| \sum_{n=2}^{N} E[Y_{nM+k} \mid \mathscr{H}_{n-1}]/N \right| \le \varepsilon K \qquad \text{a.s.}$$

for each $0 \leq k < M$ and each $N \geq 2$. Thus, since $\varepsilon > 0$ was arbitrarily chosen,

$$\sum_{n=2}^{N} Y_{nM+k}/N \to 0 \qquad \text{a.s. as} \quad N \to \infty,$$

as desired. ∎

The above proof is a nice illustration of a remark made in Chapter 2; namely, that centering random variables at their conditional expectations and then resorting to martingale theory can be useful.

Mixing sequences are quite important in probabilistic number theory. For example, Theorem 3.3.2 has an immediate application to the theory of normal numbers. The previously given definition of a normal number (limiting relative frequency of $1/10$ for each digit in the decimal expansion of the number) can be strengthened by considering the limiting relative frequency of blocks of digits. Fix $M \geq 1$, and consider successive overlapping blocks of $M$ consecutive digits of the decimal expansion of a number $\omega \in [0, 1]$. There are $n - M + 1$ such blocks in the first $n$ places. Let $N_n(\omega)$ denote the number of such blocks that are identical to a given block. For example, for $\omega = 0.015013\ldots$, if the given block is 01, then $N_6(\omega) = 2$. We say that a number $\omega \in [0, 1]$ is normal if $\lim_{n \to \infty} N_n(\omega)/n = 10^{-M}$ for all $M \geq 1$ and all possible given blocks of length $M$. For a given $M \geq 1$ and a given block of length $M$ let $X_i(\omega) = 1$ if the $i$th digit in the decimal expansion of $\omega$ completes a block identical to the given block; it is equal to zero otherwise. Let $P$ be Lebesgue measure and $\mathscr{F}$ be the Borel sets of $[0, 1]$. Clearly $\{X_i, i \geq 1\}$ is *-mixing and satisfies the other hypotheses of Theorem 3.3.2 as well. Indeed $\{X_i, i \geq 1\}$ is an example of an "$m$-dependent" sequence of random variables; that is, $\{X_1, \ldots, X_n\}$ is independent of $\{X_{n+m+1}, X_{n+m+2}, \ldots\}$ for each $n \geq 1$. Here $m = M - 1$. $EX_i = 10^{-M}$ for each $i \geq 1$. Thus, by Theorem 3.3.2,

$$\sum_{i=1}^{n} X_i/n \to 10^{-M} \qquad \text{a.s.}$$

Since the total number of possible given blocks is countable it follows that almost every $\omega \in [0, 1]$ is a normal number.

The study of continued fraction expansions provides an example where the theory of mixing random variables is of vital importance.

***Example 3.3.1.*** Continued fraction expansions of real numbers when viewed properly provide an example of a *-mixing sequence. Let the mapping $T : (0, 1) \rightarrow (0, 1)$ be defined by

$$Tx = x^{-1} - [x^{-1}]$$

where $[y]$ denotes the largest integer $\leq y$. Let $a_1(x) = [x^{-1}]$ and $a_{n+1}(x) = a_1(T^n x)$ for $n \geq 1$ and each $x \in (0, 1)$. Here $T^n$ is the $n$th iterate of $T$. Then $(a_1(x), a_2(x), \ldots)$ is referred to as the continued fraction expansion of $x$. This is sometimes written

$$x = \cfrac{1}{a_1(x) + \cfrac{1}{a_2(x) + \cfrac{1}{a_3(x) + \cdots}}}$$

Let $\Omega = (0, 1)$, $\mathscr{F}$ be the Borel sets, and $P$ be defined by

$$PA = \log^{-1} 2 \int_A (1 + x)^{-1} \, dx$$

for each $A \in \mathscr{F}$. Then, analysis shows that $(a_1, a_2, \ldots, a_n)$ and $(a_{1+m}, a_{2+m}, \ldots, a_{n+m})$ have the same joint distribution for each $n \geq 1$ and $m \geq 1$. (Such random variables are called stationary and are systematically studied in Section 3.5.) Moreover, analysis that we omit shows that $\{a_1, a_2, \ldots\}$ is a *-mixing sequence with $\phi(m) = Kq^{\sqrt{m}}$ for some $K < \infty$ and $0 < q < 1$. These observations are most useful in stating almost everywhere limit theorems (with respect to Lebesgue measure) about $\{a_i, i \geq 1\}$. For example for fixed integer $p \geq 1$, let

$$I_i(x) = \begin{cases} 1 & \text{if } a_i(x) = p \\ 0 & \text{otherwise} \end{cases}$$

for each $i \geq 1$ and $x \in (0, 1)$. Then by Theorem 3.3.2 with $Y_i = I_i$ for each $i \geq 1$,

$$\sum_{i=1}^{n} I_i / n \rightarrow EI_1 \qquad \text{a.s.}$$

Elementary computation shows

$$EI_1 = \frac{\log\left(\dfrac{p+1}{p}\right) - \log\left(\dfrac{p+2}{p+1}\right)}{\log 2}.$$

Thus

$$\sum_{i=1}^{n} I_i/n \to \frac{\log\left(\dfrac{p+1}{p}\right) - \log\left(\dfrac{p+2}{p+1}\right)}{\log 2} \qquad \text{a.s.}$$

Since Lebesgue measure is absolutely continuous with respect to $P$, it follows that the above convergence is almost everywhere with respect to Lebesgue measure also. This convergence result is clearly the continuous fraction analog to the result that almost every number is normal. ∎

*Exercise 3.3.1.* Prove that $(a_1 a_2 \ldots a_n)^{1/n}$ converges almost everywhere (with respect to a Lebesgue measure) to a finite number. For a thorough study of the use of mixing sequences in probabilistic number theory, one should consult the article by Philipp [1971]. ∎

Mixing sequences arise in other ways also. Let $\{Z_i, i \geq 1\}$ be a sequence of independent random variables. Fix an integer $m \geq 1$. Define a sequence $\{Y_i, i \geq 1\}$ of so called moving averages by

$$Y_i = \sum_{j=i}^{i+m} Z_j/(m+1).$$

Clearly $Y_{j+n+1}$ is independent of $\{Y_i, i \leq n\}$ for each $n \geq 1$ and $j \geq m$. Thus, in a trivial way, $\{Y_i, i \geq 1\}$ is *-mixing. $\{Y_i, i \geq 1\}$ is another example of an "$m$-dependent" sequence of random variables. If the $Z_i$ are identically distributed as well with $EZ_1 = 0$ and $EZ_1^2 < \infty$, then it follows from Theorem 3.3.2 that

$$\sum_{i=1}^{n} Y_i/n \to 0 \qquad \text{a.s. as} \quad n \to \infty.$$

For further examples of *-mixing sequences, see Blum *et al.* [1963].

We return now to the general theory of stability for martingales. Theorem 2.8.9 and the Kronecker lemma immediately yield the analog of Theorem 3.3.1 for absolute moments higher than 2.

**Theorem 3.3.3.** [Chow, 1965]. Let $a_i$ be $\mathscr{F}_{i-1}$ measurable for each $i \geq 1$ and $a_i \uparrow \infty$ a.s. Let $\{b_i, i \geq 1\}$ be a sequence of positive constants such that $\sum_{i=1}^{\infty} b_i < \infty$. Then

$$\sum_{i=1}^{\infty} E[|X_i|^p \mid \mathscr{F}_{i-1}](b_i)^{1-p/2} a_i^{-p} < \infty$$

for some $p \geq 2$ implies

$$\sum_{i=1}^{n} X_i/a_n \to 0.$$

**Corollary 3.3.1.** $\sum_{i=1}^{\infty} E[|X_i|^p]/i^{1+p/2-\varepsilon} < \infty$ for some $\varepsilon > 0$ and $p \geq 2$ implies that

$$\sum_{i=1}^{n} X_i/n \to 0 \qquad \text{a.s.}$$

**Proof.** Let $b_i = i^{-1-\varepsilon/(p/2-1)}$ and $a_i = i$ for $i \geq 1$. Apply Theorem 3.3.3. ∎

It is tempting to try to set $\varepsilon = 0$ in Corollary 3.3.1. Indeed, this strengthened version of Corollary 3.3.1 is true. However, the result lies surprisingly deep and depends on two new inequalities: the Marcinkiewicz and Zygmund inequality extended to the martingale case by Burkholder [1966] and the Hájek and Rényi inequality extended to the martingale case by Chow [1960]. We present this analysis not only because the strengthened Corollary 3.3.1 is interesting but also because the two above mentioned inequalities are rather important additions to one's bag of computational tricks.

We first establish the martingale version of the Marcinkiewicz and Zygmund inequality, following a new proof given recently by Burkholder. Some preliminary results from martingale theory are required first.

**Theorem 3.3.4.** [Doob, 1953, p. 314].  Let $\{T_k, k \geq 1\}$ be a martingale or nonnegative submartingale. Then

$$P[\max_{k \leq n} |T_k| > \varepsilon] \leq E[|T_n| \, I(\max_{k \leq n} |T_k| > \varepsilon)]/\varepsilon$$

for each $\varepsilon > 0$ and each $n \geq 1$. (Recall Exercise 2.9.1.)

**Proof.** Fix $\varepsilon > 0$ and $n \geq 1$. Let $t$ be the smallest integer $k \leq n$ such that $|T_k| > \varepsilon$; otherwise let $t = \infty$.

$$P[\max_{k \leq n} |T_k| > \varepsilon] = \sum_{k=1}^{n} P[t = k, |T_k| > \varepsilon]$$

$$\leq \sum_{k=1}^{n} E[|T_k| \, I(t = k)]/\varepsilon.$$

By Corollary 2.6.1, $\{|T_k|, k \geq 1\}$ is a submartingale. Hence

$$E[|T_k| \, I(t = k)] \leq E[|T_n| \, I(t = k)]$$

and

$$P[\max_{k \leq n} | T_k | > \varepsilon] \leq E[| T_n | I(t \leq n)]/\varepsilon$$
$$= E[| T_n | I(\max_{k \leq n} | T_k | > \varepsilon)]/\varepsilon. \quad \blacksquare$$

The following simple lemma is often useful in converting a probability inequality into a moment inequality.

**Lemma 3.3.1.**  Let $X$ and $Y$ be nonnegative random variables satisfying

$$P[X > \beta\varepsilon] \leq CE[YI(X > \varepsilon)]/\varepsilon$$

for every $\varepsilon > 0$ and fixed $\beta > 0$. Fix $p > 1$. Then

$$EX^p \leq [Cp\beta^p/(p - 1)]^p EY^p.$$

**Proof.**

$$EX^p = \int_0^\infty px^{p-1}P[X > x] \, dx$$
$$\leq \beta Cp \int_0^\infty x^{p-2}\left[\int_{\{\omega | X(\omega) > x/\beta\}} Y(\omega) \, dP(\omega)\right] dx$$
$$= \beta Cp \int_\Omega Y(\omega) \int_0^{\beta X(\omega)} x^{p-2} \, dx \, dP(\omega)$$
$$= [C\beta^p p/(p - 1)] \int_\Omega Y(\omega)X^{p-1}(\omega) \, dP(\omega)$$
$$\leq [C\beta^p p/(p - 1)]E^{1/p}(Y^p)E^{1-1/p}(X^p)$$

by the Holder inequality. Hence

$$EX^p \leq [C\beta^p p/(p - 1)]^p EY^p$$

as desired.  $\blacksquare$

**Corollary 3.3.2.**  [Doob, 1953, p. 317].  Let $\{T_k, k \geq 1\}$ be a martingale or nonnegative submartingale.

For fixed $p > 1$, suppose

$$E | T_k |^p < \infty \qquad \text{for each} \quad k \geq 1.$$

Then

$$E \max_{k \leq n} | T_k |^p \leq C_p E | T_n |^p$$

for each $n \geq 1$. Here $C_p = [p/(p - 1)]^p$.

**Proof.** Immediate from Theorem 3.3.4 and Lemma 3.3.1 with $\beta = 1$ and $C = 1$. ∎

**Exercise 3.3.2.** Use Corollary 3.3.2 and the method of subsequences to show that $\sum_{i=1}^{\infty} EX_i^2 < \infty$ implies $S_n$ converges a.s. (a fact we already know, of course). ∎

Corollary 3.3.2 plays a central role in proving an important result in the study of the convergence of orthogonal series. In terms of content, the discussion that follows should be viewed as an addendum to Section 2.3. Let $\{f_i, i \geq 1\}$ be an arbitrary orthonormal system ($f_i$ orthogonal and $\int_\Omega f_i^2(x)\,d\mu(x) = 1$ for each $i \geq 1$). Let constants $\{c_i, i \geq 1\}$ satisfy $\sum_{i=1}^{\infty} c_i^2 < \infty$. $\sum_{i=1}^{\infty} c_i f_i$ is, as we know, not necessarily convergent almost surely unless $\{f_i, i \geq 1\}$ is one of the "nice" orthonormal systems such as the trigonometric system, the Haar system, or $f_i$ independent with mean 0. However, Garsia [1967] has shown most interestingly that there exists a rearrangement (dependent on the choice of $c_i$) $\{f_{\sigma_1}, f_{\sigma_2}, \ldots\}$ (that is, $(\sigma_1, \sigma_2, \ldots)$ is a permutation of $(1, 2, \ldots)$) such that $\sum_{i=1}^{\infty} c_{\sigma_i} f_{\sigma_i}$ converges almost surely. It is Corollary 3.3.2 that provides the key to proving this striking result. Clearly, there exists $n_k \uparrow \infty$ such that $\sum_{i=1}^{n_k} c_i f_i$ converges a.s. Consider those permutations $\sigma$ which permute the integers only within the "blocks" $\{1, 2, \ldots, n_1\}$, $\{n_1 + 1, \ldots, n_2\}$, $\ldots$. Let $n_0 = 1$. Then Garsia's result easily is proved if we can show for some such $\sigma \equiv (\sigma_1, \sigma_2, \ldots)$ that

$$E \max_{n_k \leq n < n_{k+1}} \left( \sum_{i=n_k}^{n} c_{\sigma_i} f_{\sigma_i} \right)^2 \leq K \sum_{i=n_k}^{n_{k+1}-1} c_i^2 \qquad (3.3.4)$$

for each $k \geq 0$ and $K < \infty$.

**Exercise 3.3.3.** Assume Eq. (3.3.4) and prove Garsia's convergence result. ∎

Fix $n \geq 2$. Let $\sigma = \{\sigma_1, \ldots, \sigma_n\}$ denote an arbitrary permutation of $\{1, \ldots, n\}$. The heart of the proof of Eq. (3.3.4) is that there exists a constant $C$ such that for every choice of reals $\{x_1, \ldots, x_n\}$.

$$\sum_{\sigma} \max_{1 \leq j \leq n} \left( \sum_{i=1}^{j} x_{\sigma_i} \right)^2 \Big/ n! \leq C \left( \sum_{i=1}^{n} x_i \right)^2 + C \sum_{i=1}^{n} x_i^2. \qquad (3.3.5)$$

**Exercise 3.3.4.** Assume Eq. (3.3.5) and prove Eq. (3.3.4). Hint: Identify the $x_i$ with the $c_i f_i$ and integrate with respect to $\mu$. ∎

It is in the proof of Eq. (3.3.5) that Corollary 3.3.2 is useful. We construct a probability space. Let $\Omega'$ consist of all permutations of $\{1, \ldots, n\}$ and let $P'$ assign probability $1/n!$ to each element of $\Omega'$. Suppose $\{x_1, \ldots, x_n\}$ are real numbers satisfying $\sum_{i=1}^{n} x_i = 0$. (We will drop this restriction presently by centering.) For $k \geq 1$, let (for fixed $\sigma = \{\sigma_1, \ldots, \sigma_n\}$)

$$U_k(\sigma) = \sum_{j=1}^{k} \frac{x_{\sigma_j}}{(n-k)} \quad \text{if} \quad 1 \leq k < n.$$

**Exercise 3.3.5.** Prove that $\{U_k, 1 \leq k < n\}$ is a martingale with respect to $(\Omega', P')$. ∎

Let $m = [(n+1)/2]$. Thus, by Corollary 3.3.2 applied to $\{U_1, \ldots, U_m\}$,

$$\sum_{\sigma} \max_{1 \leq k \leq m} U_k^2/n! \leq C_2 \sum_{\sigma} U_m^2/n!$$

Thus, centering at $\bar{x} = \sum_{i=1}^{n} x_i/n$, for arbitrary $\{x_1, \ldots, x_n\}$

$$\left[\sum_{\sigma} \max_{1 \leq j \leq m} \left(\sum_{i=1}^{j} x_{\sigma_i} - j\bar{x}\right)^2\right]\Big/(2n!)$$

$$\leq \left[(n-m) \sum_{\sigma} \max_{1 \leq j \leq m} \frac{(\sum_{i=1}^{j} x_{\sigma_i} - j\bar{x})^2}{n-j}\right]\Big/n!$$

$$\leq C_2 \sum_{\sigma} \left(\sum_{i=1}^{m} x_{\sigma_i} - m\bar{x}\right)^2\Big/n!.$$

Let $y_{\sigma_i} = x_{\sigma_i} - \bar{x}$ for $1 \leq i \leq m$.

$$\frac{\sum_{\sigma} (\sum_{i=1}^{m} y_{\sigma_i})^2}{n!} = \sum_{i=1}^{m} \frac{\sum_{\sigma} y_{\sigma_i}^2}{n!} + \frac{\sum_{1 \leq i \neq j \leq m} \sum_{\sigma} y_{\sigma_i} y_{\sigma_j}}{n!}.$$

But

$$\frac{1}{n!} \sum_{\sigma} y_{\sigma_i}^2 = \frac{\sum_{i=1}^{n} y_i^2}{n}$$

and

$$\frac{1}{n!} \sum_{\sigma} y_{\sigma_i} y_{\sigma_j} = \frac{1}{n(n-1)} \sum_{1 \leq i \neq j \leq n} y_i y_j.$$

Also

$$0 = \left(\sum_{i=1}^{n} y_i\right)^2$$

$$= \sum_{i=1}^{n} y_i^2 + \sum_{1 \leq i \neq j \leq n} y_i y_j.$$

Combining,

$$\frac{\sum_\sigma (\sum_{i=1}^m y_{\sigma_i})^2}{n!} = \frac{m}{n} \sum_{i=1}^n y_i^2 - \frac{m(m-1)}{n(n-1)} \sum_{i=1}^n y_i^2$$

$$= \frac{m(n-m)}{n(n-1)} \sum_{i=1}^n y_i^2.$$

Thus

$$C_2 \sum_\sigma \left( \sum_{i=1}^m x_{\sigma_i} - m\bar{x} \right)^2 \Big/ n! = C_2 \left( \frac{m(n-m)}{n(n-1)} \right) \sum_{i=1}^n (x_i - \bar{x})^2$$

$$\leq C_2 \sum_{i=1}^n x_i^2.$$

For fixed $j$ and $\sigma$, using $(a+b)^2 \leq 2a^2 + 2b^2$,

$$\left( \sum_{i=1}^j x_{\sigma_i} \right)^2 - 2(j\bar{x})^2 \leq 2\left( \sum_{i=1}^j x_{\sigma_i} - j\bar{x} \right)^2.$$

Thus

$$\sum_\sigma \max_{1 \leq j \leq m} \left( \sum_{i=1}^j x_{\sigma_i} - j\bar{x} \right)^2 \Big/ (2n!)$$

$$\geq \sum_\sigma \max_{1 \leq j \leq m} \left[ \left( \sum_{i=1}^j x_{\sigma_i} \right)^2 - 2(j\bar{x})^2 \right] \Big/ (4n!)$$

$$\geq \sum_\sigma \max_{1 \leq j \leq m} \left( \sum_{i=1}^j x_{\sigma_i} \right)^2 \Big/ (4n!) - m^2 \left( \sum_{i=1}^n x_i \right)^2 \Big/ (2n^2).$$

Combining inequalities,

$$C_2 \sum_{i=1}^n x_i^2 + m^2 \left( \sum_{i=1}^n x_i \right)^2 \Big/ (2n^2) \geq \sum_\sigma \max_{1 \leq j \leq m} \left( \sum_{i=1}^j x_{\sigma_i} \right)^2 \Big/ [4n!]. \qquad (3.3.6)$$

Consider $m < j \leq n$.

$$\sum_{i=1}^j x_{\sigma_i} = \sum_{i=1}^m x_{\sigma_i} + \sum_{i=m+1}^j x_{\sigma_i}$$

$$= \sum_{i=1}^m x_{\sigma_i} + \sum_{i=1}^{j-m} x_{\sigma_i'}$$

for

$$\sigma' = (\sigma_{m+1}, \ldots, \sigma_n, \sigma_1, \sigma_2, \ldots, \sigma_m).$$

Thus

$$\sum_\sigma \max_{1 \leq j \leq n} \left( \sum_{i=1}^j x_{\sigma_i} \right)^2 \leq 4 \sum_\sigma \max_{1 \leq j \leq m} \left( \sum_{i=1}^j x_{\sigma_i} \right)^2.$$

Combining this with Eq. (3.3.6), it follows that there exists $C < \infty$ such that

$$\sum_{\sigma} \max_{1 \leq j \leq n} \left( \sum_{i=1}^{j} x_{\sigma_i} \right)^2 \bigg/ n! \leq C \sum_{i=1}^{n} x_i^2 + C \left( \sum_{i=1}^{n} x_i \right)^2,$$

proving Eq. (3.3.5) as desired.

We now return to the derivation of the basic martingale inequalities needed to sharpen Corollary 3.3.1. A new notation turns out to be convenient. Given a stochastic sequence $\{T_k, k \geq 1\}$, let $T_0 = 0$ a.s. and for each $n \geq 1$ let

$$Q_n(T) = \left[ \sum_{k=1}^{n} (T_k - T_{k-1})^2 \right]^{1/2},$$

$$M_n(T) = \max_{k \leq n} |T_k|,$$

and

$$Q(T) = \lim Q_n(T).$$

**Theorem 3.3.5.** [Burkholder, 1966]. Let $\{R_k, k \geq 1\}$ be a martingale or nonnegative submartingale. Then there exists a constant $D$ such that $P[Q_n(R) > \varepsilon] \leq DE |R_n|/\varepsilon$ for every $n \geq 1$ and every $\varepsilon > 0$.

**Proof.**  Fix $\varepsilon > 0$ and $n \geq 1$. We apply Gundy's decomposition theorem for martingales to $(R_k, 1 \leq k \leq n)$ (Theorem 2.9.1). Let $R_k = T_k + U_k + V_k$ denote the decomposition, taking $K = \varepsilon$.

$$Q_n(R) \leq Q_n(T) + Q_n(U) + Q_n(V)$$

by the Minkowski inequality. Using the Chebyshev inequality and the decomposition theorem,

$$P[Q_n(R) > \varepsilon] \leq P[\max_{1 \leq k \leq n} |T_k| > 0] + P\left[ \sum_{k=1}^{n} |U_k - U_{k-1}| > \varepsilon/3 \right]$$

$$+ P[Q_n(V) > \varepsilon/3]$$

$$\leq P[\max_{1 \leq k \leq n} |T_k| > 0] + 3E \sum_{k=1}^{n} |U_k - U_{k-1}|/\varepsilon$$

$$+ 9E[Q_n^2(V)]/\varepsilon^2$$

$$\leq 31 \max E |R_k|/\varepsilon.$$

Take $D = 31$.  ∎

An alternate proof of Theorem 3.3.5 based on Lemma 2.9.2 can be given (see Burkholder [1973]).

**Theorem 3.3.6.** (Martingale version of Marcinkiewicz and Zygmund inequality [Burkholder, 1966]). Let $\{T_n, \mathscr{G}_n, n \geq 1\}$ be a martingale and $p > 1$. Then there exists constants $c_p$ and $C_P$ such that

$$c_p E[Q^p(T)] \leq \sup_{n \geq 1} E \mid T_n \mid^p \leq C_p E[Q^p(T)]. \qquad (3.3.7)$$

**Proof.** The somewhat involved proof given here due to Burkholder represents a considerable simplication over earlier and even more involved proofs. Fix $n \geq 1$. Let

$$U_k = E[T_n{}^+ \mid \mathscr{G}_k] \qquad \text{and} \qquad V_k = E[T_n{}^- \mid \mathscr{G}_k]$$

for $1 \leq k \leq n$. $\{U_k, 1 \leq k \leq n\}$ and $\{V_k, 1 \leq k \leq n\}$ are nonnegative martingales.

$$T_k = E[T_n \mid \mathscr{G}_k] = E[T_n{}^+ \mid \mathscr{G}_k] - E[T_n{}^- \mid \mathscr{G}_k] = U_k - V_k$$

for $1 \leq k \leq n$. By Minkowski's inequality,

$$Q_n(T) \leq Q_n(U) + Q_n(V). \qquad (3.3.8)$$

We shall analyze the right hand terms of Eq. (3.3.8) separately. First, we will show that there exists a constant $C > 0$ for which

$$P[\max(Q_n(U), M_n(U)) > 2\varepsilon]$$
$$\leq CE\{U_n I[\max(Q_n(U), M_n(U)) > \varepsilon]\}/\varepsilon \qquad \text{for every } \varepsilon > 0. \qquad (3.3.9)$$

Fix $\varepsilon > 0$. Let $W_k = U_k I[\max(Q_k(U), M_k(U)) > \varepsilon]$ for $1 \leq k \leq n$. Fix $k$.

$$[\max(Q_k(U), M_k(U)) > \varepsilon] \supset [\max(Q_{k-1}(U), M_{k-1}(U)) > \varepsilon].$$

Thus

$$E[W_k \mid \mathscr{G}_{k-1}] = E\{U_k I[\max(Q_k(U), M_k(U)) > \varepsilon] \mid \mathscr{G}_{k-1}\}$$
$$\geq E\{U_k I[\max(Q_{k-1}(U), M_{k-1}(U)) > \varepsilon] \mid \mathscr{G}_{k-1}\}$$
$$= I[\max(Q_{k-1}(U), M_{k-1}(U)) > \varepsilon] E[U_k \mid \mathscr{G}_{k-1}]$$
$$= I[\max(Q_{k-1}(U), M_{k-1}(U)) > \varepsilon] U_{k-1}$$
$$= W_{k-1} \qquad \text{a.s.}$$

That is, $\{W_k, 1 \leq k \leq n\}$ is a nonnegative submartingale

Let $t$ be the smallest integer $k \geq 1$ such that $Q_k(U) > \varepsilon$ if such a $k$ exists; otherwise let $t = \infty$. Consider the event

$$A_n = [Q_n(U) > 2\varepsilon, \, M_n(U) \leq \varepsilon].$$

On this event, $t \leq n$ and

$$4\varepsilon^2 < Q_n{}^2(U) = Q_{t-1}^2(U) + (U_t - U_{t-1})^2 + \sum_{k=t+1}^{n} (U_k - U_{k-1})^2$$

(taking $\sum_{n+1}^{n} (\cdot) = 0$ as a convention). On $A_n$,

$$|U_t - U_{t-1}| = (U_t - U_{t-1})I(U_t - U_{t-1} \geq 0) + (U_{t-1} - U_t)I(U_t - U_{t-1} < 0)$$
$$\leq U_t I(U_t - U_{t-1} \geq 0) + U_{t-1} I(U_t - U_{t-1} < 0) \leq \varepsilon$$

since $U_{t-1} \leq M_n(U) \leq \varepsilon$ and $U_t \leq M_n(U) \leq \varepsilon$. Hence on $A_n$,

$$4\varepsilon^2 < \varepsilon^2 + \varepsilon^2 + \sum_{k=t+1}^{n} (U_k - U_{k-1})^2.$$

On $A_n$, $k > t$ implies $U_k - U_{k-1} = W_k - W_{k-1}$. Hence on $A_n$,

$$\varepsilon^2 < 2\varepsilon^2 < \sum_{k=1}^{n} (W_k - W_{k-1})^2$$

(taking $W_0 = 0$), that is, $Q_n(W) > \varepsilon$. $W_k \leq U_k$ for $k \leq n$. Hence $M_n(W) \leq M_n(U)$. Combining the last two statements,

$$[Q_n(U) > 2\varepsilon, \, M_n(U) \leq \varepsilon] \subset [Q_n(W) > \varepsilon, \, M_n(W) \leq \varepsilon].$$

Thus

$$P[\max(Q_n(U), M_n(U)) > 2\varepsilon]$$
$$\leq P[Q_n(U) > 2\varepsilon, \, M_n(U) \leq \varepsilon] + P[M_n(U) > \varepsilon]$$
$$\leq P[Q_n(W) > \varepsilon, \, M_n(W) \leq \varepsilon] + P[M_n(U) > \varepsilon]$$
$$\leq P[Q_n(W) > \varepsilon] + P[M_n(U) > \varepsilon].$$

Applying Theorems 3.3.4 and 3.3.5, we obtain that

$$P[\max(Q_n(U), M_n(U)) > 2\varepsilon] \leq (D/\varepsilon)E(W_n) + (1/\varepsilon)E[U_n I(M_n(U) > \varepsilon)]$$
$$\leq CE[U_n I(\max(Q_n(U), M_n(U)) > \varepsilon)]/\varepsilon$$

taking $C = D + 1$. Thus Eq. (3.3.9) is established.

By Lemma 3.3.1 with $\beta = 2$,

$$E[\max(Q_n(U), M_n(U))^p] \leq \left(\frac{Cp2^p}{p-1}\right)^p EU_n^p$$

$$\leq \left(\frac{Cp2^p}{p-1}\right)^p E\,|\,T_n\,|^p.$$

Thus

$$EQ_n(U)^p \leq \left(\frac{Cp2^p}{p-1}\right)^p E\,|\,T_n\,|^p. \tag{3.3.10}$$

The same argument shows that Eq. (3.3.10) holds with $\{V_k, 1 \leq k \leq n\}$ substituted for $\{U_k, 1 \leq k \leq n\}$. Using Eq. (3.3.8),

$$E(Q_n(T))^p \leq E(Q_n(U) + Q_n(V))^p$$

$$\leq 2^{p-1}[E(Q_n(V)^p) + E(Q_n(V)^p)]$$

$$\leq 2^p\left(\frac{Cp2^p}{p-1}\right)^p E\,|\,T_n\,|^p.$$

Since this holds for each $n \geq 1$, since $E\,|\,T_n\,|^p$ is increasing in $n$, and since $Q_n(T) \uparrow Q(T)$, the left-hand inequality of Eq. (3.3.7) is established.

To prove the right-hand side of Eq. (3.3.7), we may assume that $EQ_n(T)^p < \infty$ and $E\,|\,T_n\,|^p > 0$ for each $n \geq 1$. The right-hand side follows from the left-hand side by a duality argument. Fix $n \geq 1$. Since $EQ_n(T)^p < \infty$, $\max_{k \leq n} E\,|\,T_k - T_{k-1}\,|^p < \infty$ (taking $T_0 = 0$ a.s.) and hence $E\,|\,T_n\,|^p < \infty$. Consider the $L_p$ space $L_p(\Omega, \mathscr{G}_n, P)$ and its dual $L_q(\Omega, \mathscr{G}_n, P)$. Motivated by this consideration, let

$$T_n' = \operatorname{sgn} T_n\,|\,T_n\,|^{p-1}/(E\,|\,T_n\,|^p)^{(p-1)/p}.$$

Note that $T_n' \in L_q(\Omega, \mathscr{G}_n, P)$ satisfies

$$E T_n T_n' = (E\,|\,T_n\,|^p)^{1/p} \qquad \text{and} \qquad E\,|\,T_n'\,|^q = 1.$$

Let $T_k' = E[T_n'\,|\,\mathscr{G}_k]$ define a martingale $\{T_k', 1 \leq k \leq n\}$. By the martingale property,

$$E T_n T_n' = E\left[\sum_{k=1}^{n}(T_k - T_{k-1})\sum_{j=1}^{n}(T_j' - T_{j-1}')\right]$$

$$= \sum_{k=1}^{n}\sum_{j=1}^{n} E(T_k - T_{k-1})(T_j' - T_{j-1}')$$

$$= E\left[\sum_{k=1}^{n}(T_k - T_{k-1})(T_k' - T_{k-1}')\right]$$

(taking $T_0' = 0$ a.s.). By the Holder inequality for sums,

$$E\left[\sum_{k=1}^{n} (T_k - T_{k-1})(T_k' - T_{k-1}')\right] \leq E[Q_n(T)Q_n(T')].$$

Hence

$$(E \mid T_n \mid^p)^{1/p} = E(T_n T_n') \leq E[Q_n(T)Q_n(T')]$$
$$\leq [EQ_n(T)^p]^{1/p}[EQ_n(T')^q]^{1/q}$$
$$\leq c_q^{-1/q}(EQ_n(T)^p)^{1/p}$$

by the left hand side of Eq. (3.3.7) applied to $\{T_k', 1 \leq k \leq n\}$. Letting $n \to \infty$ thus establishes the right hand side of Eq. (3.3.7). ∎

If $\{X_i, i \geq 1\}$ is a sequence of independent random variables with mean zero, then Kolmogorov's maximal inequality (see Exercise 2.9.1 and Theorem 3.3.4) says that

$$P[\max_{i \leq n} \mid S_i \mid \geq \varepsilon] \leq ES_n^2/\varepsilon^2.$$

Can this be generalized to an inequality for $P[\max_{i \leq n} \mid S_i/a_i \mid \geq \varepsilon]$? For instance, if $a_i = i$ for $i \geq 1$, such an inequality could be useful in establishing $S_i/i \to 0$ a.s. Hájek and Rényi [1955] give such an inequality. We present a martingale version (thus including the case of independence) due to Chow [1960].

**Theorem 3.3.7.** Let $\{T_n, \mathscr{G}_n, n \geq 1\}$ be a submartingale, $\varepsilon > 0$, and $0 < a_n$ increasing and $\mathscr{G}_{n-1}$ measurable for $n \geq 1$. ($\mathscr{G}_0 = \{\varnothing, \Omega\}$). Then

$$\varepsilon P[\max_{1 \leq k \leq n} T_k/a_k \geq \varepsilon] \leq ET_1^+/a_1 + \sum_{k=2}^{n} E[(T_k^+ - T_{k-1}^+)]/a_k \quad (3.3.11)$$

for each $n \geq 1$.

**Proof.** (Due to Tomkins, personal communication)

$$P[\max_{k \leq n} T_k/a_k \geq \varepsilon] = P[\max_{k \leq n} T_k^+/a_k \geq \varepsilon].$$

$\{T_k^+, k \geq 1\}$ is a submartingale. Thus, without loss of generality suppose $T_k \geq 0$ a.s. for $k \geq 1$. Let

$$Z_k = T_1/a_1 + \sum_{i=2}^{k} (T_i - T_{i-1})/a_i.$$

$\{Z_k, k \geq 1\}$ is a nonnegative submartingale.

Let $t$ be the first $n \geq k \geq 1$ such that $T_k/a_k \geq \varepsilon$ if such a $k$ exists, otherwise let $t = n + 1$. Summation by parts shows that

$$Z_k = T_k/a_k + \sum_{i=1}^{k-1} (a_i^{-1} - a_{i+1}^{-1})T_i.$$

On $t = k$, $\varepsilon \leq T_k/a_k \leq Z_k$. Thus, on $t \leq n$, $Z_t \geq \varepsilon$. Hence

$$\varepsilon P[\max_{1 \leq k \leq n} T_k/a_k \geq \varepsilon] \leq \varepsilon P[t \leq n]$$

$$\leq EZ_t I[t \leq n] = \sum_{j=1}^{n} \int_{t=j} Z_j$$

$$\leq \sum_{j=1}^{n} \int_{t=j} Z_n = EZ_n I(t \leq n) \leq EZ_n. \quad \blacksquare$$

**Corollary 3.3.3.** (The Kolmogorov maximal inequality). Let $\{X_i, i \geq 1\}$ be a sequence of independent random variables with mean zero. Then

$$P[\max_{i \leq n} |S_i| \geq \varepsilon] \leq ES_n^2/\varepsilon^2$$

for each $n \geq 1$.

**Proof.** Let $T_n = S_n^2$ and $a_n = 1$ for each $n \geq 1$. $\{T_n, n \geq 1\}$ is a submartingale. By Theorem 3.3.7,

$$\varepsilon^2 P[\max_{1 \leq k \leq n} S_k^2 \geq \varepsilon^2] \leq ES_n^2. \quad \blacksquare$$

Theorem 3.3.7 yields a stability result:

**Corollary 3.3.4.** [Chow, 1960]. Let $(T_n, \mathscr{G}_n, n \geq 1)$ be a nonnegative submartingale with $0 < a_n$ increasing and $\mathscr{G}_{n-1}$ measurable for $n \geq 1$ ($\mathscr{G}_0 = \{\phi, \Omega\}$). Suppose for some $\alpha \geq 1$, $ET_1^\alpha/a_1^\alpha < \infty$ and

$$\sum_{k=2}^{\infty} E[(T_k^\alpha - T_{k-1}^\alpha)/a_k^\alpha] < \infty. \tag{3.3.12}$$

Then

$$\lim T_n/a_n = 0 \qquad \text{a.s.}$$

**Proof.** Fix $n \geq 1$. $\{T_k^\alpha, k \leq n\}$ is a submartingale by Theorem 2.6.1. Using Chow's inequality (Eq. (3.3.11)),

$$\varepsilon^\alpha P[\sup_{k \geq n} T_k/a_k \geq \varepsilon] = \varepsilon^\alpha P[\sup_{k \geq n} T_k^\alpha/a_k^\alpha \geq \varepsilon^\alpha]$$

$$\leq ET_n^\alpha/a_n^\alpha + \sum_{k=n+1}^{\infty} E[(T_k^\alpha - T_{k-1}^\alpha)/a_k^\alpha]. \tag{3.3.13}$$

By Eq. (3.3.12) and the Kronecker lemma,

$$\sum_{k=2}^{n} E[(T_k{}^{\alpha} - T_{k-1}^{\alpha})/a_n{}^{\alpha}] = ET_n{}^{\alpha}/a_n{}^{\alpha} - ET_1{}^{\alpha}/a_n{}^{\alpha} \to 0$$

as $n \to \infty$. Thus $ET_n{}^{\alpha}/a_n{}^{\alpha} \to 0$. Hence by Eq. (3.3.13), $P[\sup_{k \ge n} T_k/a_k \ge \varepsilon] \to 0$ as $n \to \infty$. This implies that $T_n/a_n \to 0$ a.s. ∎

At last we can state and prove the main result, the strengthened version of Corollary 3.3.1 discussed previously.

**Theorem 3.3.8.** [Chow, 1960, 1967].

$$\sum_{i=1}^{\infty} E \mid X_i \mid^p / i^{1+p/2} < \infty$$

for some $p \ge 2$ implies $S_n/n \to 0$ a.s.

**Proof.** The result is known when $p = 2$. Thus suppose $p > 2$. Throughout, $C$ will denote a constant, although not necessarily the same constant each time. Fix $n \ge 1$. By Burkholder's inequality,

$$E \left| \sum_{i=1}^{n} X_i \right|^p \le CE \left| \sum_{i=1}^{n} X_i^2 \right|^{p/2}.$$

By the Holder inequality,

$$\sum_{i=1}^{n} (1) X_i^2 \le \left( \sum_{i=1}^{n} \mid X_i \mid^p \right)^{2/p} n^{1-2/p}$$

and hence

$$E \left| \sum_{i=1}^{n} X_i^2 \right|^{p/2} \le E \sum_{i=1}^{n} \mid X_i \mid^p n^{p/2-1}.$$

Combining,

$$E \left| \sum_{i=1}^{n} X_i \right|^p \le C n^{p/2-1} E \sum_{i=1}^{n} \mid X_i \mid^p. \tag{3.3.14}$$

(To understand the usefulness of Burkholder's inequality, apply Holder's inequality directly to $\mid \sum_{i=1}^{n} X_i \mid^p$ and note that the resulting inequality is weaker.)

Now, by Corollary 3.3.4, it suffices to show that

$$\sum_{n=2}^{\infty} E \left[ \left| \sum_{i=1}^{n} X_i \right|^p - \left| \sum_{i=1}^{n-1} X_i \right|^p \right] / n^p < \infty.$$

Thus, summing by parts, it suffices to show that

$$\lim E \left| \sum_{i=1}^{n} X_i \right|^p \Big/ n^p \to 0$$

and that

$$\sum_{n=1}^{\infty} \left( \frac{1}{n^p} - \frac{1}{(n+1)^p} \right) E \left| \sum_{i=1}^{n} X_i \right|^p < \infty.$$

By Eq. (3.3.14),

$$E \left| \sum_{i=1}^{n} X_i \right|^p \Big/ n^p \le C n^{-p/2-1} \sum_{i=1}^{n} E \, | \, X_i \, |^p$$

which approaches zero by the Kronecker lemma, since

$$\sum_{i=1}^{\infty} E \, | \, X_i \, |^p / i^{p/2+1} < \infty$$

by hypothesis. Using Eq. (3.3.14),

$$\sum_{n=1}^{\infty} \left( \frac{1}{n^p} - \frac{1}{(n+1)^p} \right) E \left| \sum_{i=1}^{n} X_i \right|^p$$

$$\le C \sum_{n=1}^{\infty} \left( \frac{1}{n^p} - \frac{1}{(n+1)^p} \right) n^{p/2-1} \sum_{i=1}^{n} E \, | \, X_i \, |^p$$

$$\le C \sum_{n=1}^{\infty} n^{-p/2-2} \sum_{i=1}^{n} E \, | \, X_i \, |^p$$

$$= C \sum_{i=1}^{\infty} E \, | \, X_i \, |^p \sum_{n=i}^{\infty} n^{-p/2-2}$$

$$\le C \sum_{i=1}^{\infty} E \, | \, X_i \, |^p \, i^{-p/2-1} < \infty,$$

completing the proof.  ∎

    Recall that Feller's Theorem 3.2.5 characterized the stability of $S_n/a_n$ in terms of the convergence of $\sum_{n=1}^{\infty} P[| \, X_n \, | > a_n]$ where $\{X_i, i \ge 1\}$ was assumed to be an independent sequence and the $a_n$ converged to infinity sufficiently rapidly. We now extend this result to the martingale case.

    **Lemma 3.3.2.**  Let $X$ be a random variable and $\{a_i, i \ge 1\}$ be constants increasing to infinity. Let $\psi$ be any even nondecreasing function satisfying $\psi(a_i) = i$ for all $i \ge 1$. ($\psi$ is essentially the inverse of the function

$i \to a_i$.) Then $E\psi(X) < \infty$ if and only if

$$\sum_{n=1}^{\infty} P[|X| > a_n] < \infty.$$

**Exercise 3.3.6.** Prove Lemma 3.3.2. ∎

Therefore, Feller's result characterizes the stability of $S_n/a_n$ by the finiteness of $E\psi(X_1)$, $\psi$ appropriately chosen. The corresponding martingale result reads in a similar manner. It is based on work of Chung [1947] for the case of independence and is implicit in a paper of Loève's [1951].

**Theorem 3.3.9.** Let $\psi$ be a positive even function and let $a_i$ be $\mathscr{F}_{i-1}$ measurable for $i \geq 2$ with $a_i \uparrow \infty$ a.s. Let $\{Y_i, \mathscr{F}_i, i \geq 1\}$ be a stochastic sequence.

(i)   If $\psi(x)/x$ is nondecreasing and $\psi(x)/x^2$ is nonincreasing, then

$$\sum_{i=2}^{\infty} E[\psi(X_i)/\psi(a_i) \mid \mathscr{F}_{i-1}] < \infty$$

implies that $\sum_{i=1}^{\infty} X_i/a_i$ converges and hence that $\sum_{i=1}^{n} X_i/a_n \to 0$.

(ii)   If $\psi(x)/x$ is nonincreasing then

$$\sum_{i=2}^{\infty} E[\psi(Y_i)/\psi(a_i) \mid \mathscr{F}_{i-1}] < \infty$$

implies that $\sum_{i=1}^{\infty} Y_i/a_i$ converges and hence that $\sum_{i=1}^{n} Y_i/a_n \to 0$.

(iii)   (Sharpness of (i) and (ii)). Let $\{a_i, i \geq 1\}$ be constants increasing to infinity. Let $\psi$ be a function such that either $\psi(x)/x$ is nondecreasing and $\psi(x)/x^2$ nonincreasing or $\psi(x)/x$ is nonincreasing. Given positive constants $m_i$ such that $\sum_{i=1}^{\infty} m_i/\psi(a_i) = \infty$, then there exists independent symmetric $X_i$ with $EX_i = 0$ and $E\psi(X_i) = m_i$ for all $i \geq 1$ such that $\sum_{i=1}^{\infty} X_i/a_n$ diverges almost surely.

**Exercise 3.3.7.** Prove Theorem 3.3.9. (The proof is similar to a number of previously proven results.) ∎

**Corollary 3.3.5.** (i)   If

$$E[|X_i|^p (\log^+ |X_i|)^{1+\varepsilon}] \leq K < \infty$$

for all $i \geq 1$, some $1 \leq p < 2$, and some $\varepsilon > 0$, then $\sum_{i=1}^{n} X_i/n^{1/p} \to 0$ a.s.

(ii)   Let $\{Y_i, i \geq 1\}$ be a stochastic sequence. If

$$E[|\,Y_i\,|^p \,(\log^+ |\,Y_i\,|)^{1+\varepsilon}] \leq K < \infty$$

for all $i \geq 1$, some $0 < p < 1$, and some $\varepsilon > 0$, then $\sum_{i=1}^n Y_i/n^{1/p} \to 0$ a.s.

**Exercise 3.3.8.**   Prove Corollary 3.3.5.   ∎

**Exercise 3.3.9.**   Let $\{Z_i, i \geq 1\}$ be identically distributed with $E\,|\,Z_1\,| < \infty$. Let $\mathscr{B}_n = \mathscr{B}(Z_1, \ldots, Z_n)$ for $n \geq 1$. Prove there exists $\{Y_n, n \geq 1\}$ such that each $Y_n$ is $\mathscr{B}_{n-1}$ measurable and $\sum_{i=1}^n (Z_i - Y_i)/n \to 0$ a.s.   ∎

One should contrast Corollary 3.3.5 with the Kolmogorov and the Marcinkiewicz laws of large numbers for independent identically distributed random variables (Theorems 3.2.2 and 3.2.3). The inclusion of the logarithmic factor in Corollary 3.3.5 compensates for the weakening of the hypothesis that the $X_i$ are independent identically distributed.

By Theorems 3.2.2 and 3.2.3 we know that $\{X_i, i \geq 1\}$ independent identically distributed with $EX_1 = 0$ and $EX_1^2 < \infty$ implies that

$$\sum_{i=1}^n X_i/n^{1/p} \to 0 \qquad \text{a.s.}$$

for each $0 < p < 2$. Stating this in a way to suggest a generalization, $EX_1 = 0$ and $EX_1^2 < \infty$ implies that

$$\sum_{i=1}^n X_i \Big/ \Big(\sum_{i=1}^n \operatorname{var} X_i\Big)^{1/p} \to 0 \qquad \text{a.s.}$$

for each $0 < p < 2$. Lévy [1954] generalizes this to the martingale case.

**Theorem 3.3.10.**   Let $f$ be a positive increasing function defined on $[0, \infty)$ such that

$$\int_0^\infty f^{-2}(t)\, dt < \infty.$$

Let $s_n^2 = \sum_{i=1}^n E[X_i^2\,|\,\mathscr{F}_{i-1}]$ for $n \geq 1$ and $s_0^2 = 0$. Then $s_n^2 \to \infty$ implies $\lim \sum_{i=1}^n X_i/f(s_n^2) \to 0$.

**Proof.**

$$\sum_{i=1}^\infty E[X_i^2 f^{-2}(s_i^2)\,|\,\mathscr{F}_{i-1}] = \sum_{i=1}^\infty (s_i^2 - s_{i-1}^2)f^{-2}(s_i^2)$$

$$\leq \sum_{i=1}^\infty \int_{s_{i-1}^2}^{s_i^2} f^{-2}(t)\, dt < \infty \qquad \text{a.s.}$$

by hypothesis. Hence by Theorem 2.8.7,

$$\sum_{i=1}^{\infty} X_i/f(s_i^2) \qquad \text{converges a.s.}$$

By the Kronecker lemma,

$$\sum_{i=1}^{n} X_i/f(s_n^2) \to 0 \qquad \text{a.s.}$$

on the event $[s_n^2 \to \infty]$, using fact the $f(x) \uparrow \infty$ as $x \uparrow \infty$. ∎

### 3.4. Stability Results for Independent Random Variables

Clearly a number of results asserting the stability of partial sums of independent random variables follow immediately from results stated in Section 3.3. Statements of these results are omitted.

Throughout Section 3.4, $\{X_i, i \geq 1\}$ denotes a sequence of independent (not necessarily identically distributed) random variables. We first examine some of Prokhorov's [1949] and Loève's [1963, p. 252] work on the characterization of the stability of $\{S_n/a_n, n \geq 1\}$.

**Theorem 3.4.1.** Let $0 < a_n \uparrow \infty$. Let $\{n_k, k \geq 1\}$ be an integer subsequence such that there exists a constant $c > 1$ for which $c \leq a_{n_{k+1}}/a_{n_k}$ for each $k \geq 1$ and such that there exists a constant $d > c$ for which either $a_{n_{k+1}}/a_{n_k} \leq d$ or $n_{k+1} - n_k = 1$ for each $k \geq 1$. Let

$$T_k = (S_{n_k} - S_{n_{k-1}})/a_{n_k}$$

for all $k \geq 1$ (taking $n_0 = 0$ and $S_0 = 0$). Then

(i)   $[S_n - \mu(S_n)]/a_n \to 0 \qquad$ a.s.

if and only if

(ii)   $T_k - \mu(T_k) \to 0 \qquad$ a.s.

if and only if

(iii)   $\sum_{k=1}^{\infty} P[|T_k - \mu(T_k)| > \varepsilon] < \infty \qquad$ for each   $\varepsilon > 0$.

**Remark.**   Given any sequence $0 < a_n \uparrow \infty$, it is not always possible to choose $n_k$ as in the hypothesis of Theorem 3.4.1. For example, $10^1, 10^1,$

$10^2, 10^2, 10^4, 10^4, 10^8, 10^8, \ldots$, is such a sequence. However, it should be realized that if the sequence $0 < a_n \uparrow \infty$ is reasonably well behaved then the required $n_k$ exist.

**Proof of Theorem 3.4.1.**  The equivalence of (ii) and (iii) is immediate by the Borel–Cantelli lemma and its converse, since the $T_k$ are independent.

In proving the equivalence of (i) and (ii), we assume without loss of generality that the $X_i$ are symmetric. Assume (i), namely that $S_n/a_n \to 0$ a.s. For each $k \geq 1$,

$$T_k = \frac{S_{n_k}}{a_{n_k}} - \frac{S_{n_{k-1}}}{a_{n_{k-1}}} \frac{a_{n_{k-1}}}{a_{n_k}} \quad \text{and} \quad \frac{a_{n_{k-1}}}{a_{n_k}} \leq 1.$$

Hence $T_k \to 0$ a.s.; that is, (ii) holds.

Assume (ii) holds; that is, $T_k \to 0$ a.s. We shall use the method of subsequences to show that $S_n/a_n \to 0$ a.s. Fix $k \geq 1$.

$$\frac{S_{n_k}}{a_{n_k}} = \sum_{i=1}^{n_k} \frac{X_i}{a_{n_k}} = \sum_{j=1}^{k} \frac{a_{n_j} T_j}{a_{n_k}}.$$

Let $c_{kj} = a_{n_j}/a_{n_k}$ if $j \leq k$, $= 0$ if $j > k$. We want to show that $\sum_{j=1}^{k} c_{kj} T_j \to 0$ a.s. By Lemma 3.2.3 (ii), it suffices to show that $\sum_{j=1}^{k} c_{kj} \leq M < \infty$ and $c_{kj} \to 0$ as $k \to \infty$ for each $j \geq 1$.

$$\sum_{j=1}^{k} c_{kj} = \sum_{j=1}^{k} \frac{a_{n_j}}{a_{n_k}}$$

$$= \frac{a_{n_1}}{a_{n_2}} \frac{a_{n_2}}{a_{n_3}} \cdots \frac{a_{n_{k-1}}}{a_{n_k}} + \frac{a_{n_2}}{a_{n_3}} \frac{a_{n_3}}{a_{n_4}} \cdots \frac{a_{n_{k-1}}}{a_{n_k}} + \cdots + 1$$

$$\leq (c^{-1})^{k-1} + (c^{-1})^{k-2} + \cdots + 1$$

$$\leq \sum_{j=0}^{\infty} c^{-j} \equiv M < \infty.$$

Trivially $c_{kj} \to 0$ as $k \to \infty$. Thus $S_{n_k}/a_{n_k} \to 0$ a.s.

Fix $n \geq 1$. Choose $k$ such that $n_{k-1} < n \leq n_k$.

$$\frac{|S_n|}{a_n} \leq \frac{|S_n - S_{n_{k-1}}|}{a_n} + \frac{|S_{n_{k-1}}|}{a_n} \leq \frac{|S_n - S_{n_{k-1}}|}{a_n} + \frac{|S_{n_{k-1}}|}{a_{n_{k-1}}}.$$

There are two possibilities. If $n_k = n_{k-1} + 1$,

$$\frac{|S_n - S_{n_{k-1}}|}{a_n} = \frac{|X_{n_k}|}{a_{n_k}} = \frac{|S_{n_k} - S_{n_{k-1}}|}{a_{n_k}} \leq \frac{|S_{n_k}|}{a_{n_k}} + \frac{|S_{n_{k-1}}|}{a_{n_{k-1}}}.$$

If $n_k \neq n_{k-1} + 1$,

$$\frac{|S_n - S_{n_{k-1}}|}{a_n} + \frac{|S_{n_{k-1}}|}{a_{n_{k-1}}} \leq \frac{|S_n - S_{n_{k-1}}|}{a_{n_k}} \frac{a_{n_k}}{a_{n_{k-1}}} + \frac{|S_{n_{k-1}}|}{a_{n_{k-1}}}$$

$$\leq d \frac{|S_n - S_{n_{k-1}}|}{a_{n_k}} + \frac{|S_{n_{k-1}}|}{a_{n_{k-1}}}.$$

We know that $S_{n_k}/a_{n_k} \to 0$ a.s. as $k \to \infty$. Thus to show that $S_n/n \to 0$ a.s. $n \to \infty$, it only remains to show in the second case that

$$\max_{n_{k-1} < n \leq n_k} \frac{|S_n - S_{n_{k-1}}|}{a_{n_k}} \to 0 \qquad \text{a.s. as} \quad k \to \infty.$$

Using the Lévy inequality,

$$\sum_{k=1}^{\infty} P[\max_{n_{k-1} < n \leq n_k} |S_n - S_{n_{k-1}}|/a_{n_k} > \varepsilon] \leq 2 \sum_{k=1}^{\infty} P[|T_k| > \varepsilon] < \infty$$

for $\varepsilon > 0$. Hence $\max_{n_{k-1} < n \leq n_k} |S_n - S_{n_{k-1}}|/a_{n_k} \to 0$ a.s., establishing the theorem. ∎

**Corollary 3.4.1.** [Chung, 1951]. $S_n/n \to 0$ a.s. if and only if $S_{2^k}/2^k \to 0$ a.s. and $S_n/n \to 0$ in probability.

**Proof.** Trivially, $S_n/n \to 0$ a.s. implies $S_{2^k}/2^k \to 0$ a.s. and $S_n/n \to 0$ in probability.

Assume $S_{2^k}/2^k \to 0$ a.s. Then $(S_{2^k} - S_{2^{k-1}})/2^k \to 0$ a.s. By the preceding theorem, $S_n/n - \mu(S_n/n) \to 0$ a.s. Since $S_n/n \to 0$ in probability, $\mu(S_n/n) \to 0$. Hence $S_n/n \to 0$ a.s. ∎

It is trivial that $S_{2^k}/2^k \to 0$ a.s. if and only if $T_k = (S_{2^k} - S_{2^{k-1}})/2^k \to 0$ a.s. Thus Corollary 3.4.1 shows that convergence in probability can play the role that centering at medians does in Theorem 3.4.1.

From an aesthetic and perhaps practical viewpoint (iii) of Theorem 3.4.1 is a somewhat unsatisfactory condition since it is not a condition in terms of properties of individual summands. Recently Nagaev [1972] has given a solution to the well-known and challenging (see Chung [1951] for a discussion) problem of finding necessary and sufficient conditions for the strong law of large numbers ($a_n = n$). Prokhorov had already shown that a necessary and sufficient condition in terms of absolute moments of individual summands was unlikely (see Section 5.3). After proving the impossibility of such a necessary and sufficient condition, Nagaev gives a reasonably simply expressed necessary and sufficient condition under the (without loss of generality) assumption that the $X_i$ are symmetric. The condition is:

$$\sum_{n=1}^{\infty} P[|X_n| > \varepsilon n] < \infty \qquad \text{for each} \quad \varepsilon > 0$$

and

$$\sum_{r=1}^{\infty} \exp[-\varepsilon h_r(\varepsilon) 2^{r+1}] < \infty \qquad \text{for each} \quad \varepsilon > 0.$$

Here $f_n(h, \varepsilon) = E \exp(h X_n) I(|X_n| \leq n\varepsilon)$. Also, $h_r(\varepsilon)$ is the solution of the equation

$$\Psi_r(h, \varepsilon) \equiv \sum_{n=2^r+1}^{2^{r+1}} \frac{d}{dh} [f_n(h, \varepsilon)]/f_n(h, \varepsilon) = \varepsilon 2^{r+1}$$

provided

$$\sup_h \Psi_r(h, \varepsilon) \geq \varepsilon 2^{r+1}.$$

Otherwise, $h_r(\varepsilon) = \infty$. $h_r(\varepsilon)$ is well defined since $\Psi_r(h, \varepsilon)$ is monotone in $h$.

The proof, which we do not give, consists, using the notation of Theorem 3.4.1, in carefully computing

$$\sum_{k=1}^{\infty} P[|T_k| > \varepsilon],$$

where $a_n = n$.

When first moments have been finite, they have up until now served as centering constants. Thus, it seems reasonable to ask whether a result analogous to Theorem 3.4.1 holds with random variables centered at means. Let $a_n \uparrow \infty$ and $\varepsilon_n \to 0$. Note that $S_n/a_n \to 0$ a.s. if and only if $\sum_{i=1}^{n} X_i I(|X_i| \leq \varepsilon_i a_i)/a_n \to 0$ a.s. and $\sum_{i=1}^{\infty} P[|X_i| > \varepsilon_i a_i] < \infty$. Thus if we can show that

$$\sum_{i=1}^{\infty} P[|X_i| > \varepsilon_i a_i] < \infty,$$

the problem of establishing $S_n/a_n \to 0$ a.s. is reduced to one of bounded random variables $|X_i| \leq \varepsilon_i a_i$ a.s. for each $i \geq 1$. For random variables so bounded we do obtain an analog to Theorem 3.4.1. We first state two preparatory lemmas.

**Lemma 3.4.1.**   (Kolmogorov).   Let $\{Y_i, 1 \leq i \leq n\}$ be independent and $|Y_i| \leq c < \infty$ a.s. for $1 \leq i \leq n$. Let $T_k = \sum_{i=1}^{k}(Y_i - EY_i)$. Then

$$P[\max_{1 \leq k \leq n} |T_k| < \varepsilon] \leq (\varepsilon + 2c)^2 \Big/ \sum_{i=1}^{n} \text{var}(Y_i)$$

for all $\varepsilon > 0$.

**Proof.**   Let $t$ be the smallest integer $k \geq 1$ such that $|T_k| \geq \varepsilon$ if such a $k$ exists, otherwise, let $t = \infty$. Let $T_0 = 0$ a.s. Fix $k \geq 1$.

$$T_k I(t = k) + T_k I(t > k) = T_k I(t \geq k)$$
$$= T_{k-1} I(t \geq k) + (Y_k - EY_k) I(t \geq k). \qquad (3.4.1)$$

$$E(Y_k - EY_k)(T_{k-1} I(t \geq k)) = E(Y_k - EY_k) E[T_{k-1} I(t \geq k)] = 0$$

by independence. Squaring both sides of Eq. (3.4.1) and taking expectations yields, using $I(t = k)I(t > k) = 0$ a.s.,

$$ET_k^2 I(t = k) + ET_k^2 I(t > k) = ET_{k-1}^2 I(t \geq k) + \text{var}(Y_k) P[t \geq k]. \qquad (3.4.2)$$

$$|T_k I(t = k)| \leq |T_{k-1} I(t = k)| + |(Y_k - EY_k) I(t = k)|$$
$$\leq (\varepsilon + 2c) I(t = k).$$

Hence, by Eq. (3.4.2),

$$(\varepsilon + 2c)^2 P[t = k] + ET_k^2 I(t > k) \geq ET_{k-1}^2 I(t > k - 1) + \text{var}(Y_k) P[t \geq k].$$

Summing over $1 \leq k \leq n$ yields

$$(\varepsilon + 2c)^2 P[t \leq n] + ET_n^2 I(t > n) \geq \sum_{k=1}^{n} \text{var}(Y_k) P[t > n].$$

$t > n$ implies $T_n^2 < \varepsilon^2$. Hence

$$(\varepsilon + 2c)^2 P[t \leq n] + \varepsilon^2 P[t > n] \geq \sum_{k=1}^{n} \text{var}(Y_k) P[t > n].$$

This implies

$$(\varepsilon + 2c)^2 \geq \sum_{k=1}^{n} \text{var}(Y_k) P[t > n]$$

as desired.   ∎

**Lemma 3.4.2.**   $|EX - \mu(X)| \leq (2 \text{ var } X)^{1/2}.$

*Exercise 3.4.1.*  Prove Lemma 3.4.2.  ∎

**Theorem 3.4.2.**  Let constants $0 < a_i \uparrow \infty$. Let $\{n_k, k \geq 1\}$ be an integer subsequence such that there exists a constant $c > 1$ for which $c \leq a_{n_{k+1}}/a_{n_k}$ for each $k \geq 1$ and such that there exists a constant $d > c$ for which either $a_{n_{k+1}}/a_{n_k} \leq d$ or $n_{k+1} - n_k = 1$ for each $k \geq 1$. Assume $|X_i| \leq \varepsilon_i a_i$ a.s. for each $i \geq 1$ where $\varepsilon_i \to 0$. Let $T_k = (S_{n_k} - S_{n_{k-1}})/a_{n_k}$ for all $k \geq 1$ taking $n_0 = 0$ and $S_0 = 0$. Then the following statements are equivalent:

(i)   $(S_n - ES_n)/a_n \to 0$ a.s.

(ii)  $T_k - ET_k \to 0$ a.s.

(iii) $\sum\limits_{k=1}^{\infty} P[|\, T_k - ET_k\,| > \varepsilon] < \infty$ for all $\varepsilon > 0$.

**Proof.**  The equivalence of (ii) and (iii) is trivial.

(i) implying (ii) is proved exactly as (i) implying (ii) was proved in Theorem 3.4.1.

In order to complete the proof, we must show that (ii) implies (i). Assume $T_k - ET_k \to 0$ a.s. This implies that $T_k - \mu(T_k) \to 0$ a.s. By Theorem 3.4.1, $T_k - \mu(T_k) \to 0$ a.s. implies that $(S_n - \mu(S_n))/a_n \to 0$ a.s. Hence it suffices to show that

$$[ES_n - \mu(S_n)]/a_n \to 0.$$

By Lemma 3.4.2, it suffices to show that

$$\text{var}(S_n/a_n) \to 0.$$

Fix $n \geq 1$. Choose $k$ such that $n_{k-1} < n \leq n_k$.

$$\text{var}(S_n/a_n) \leq \text{var}(S_{n_k}/a_{n_k})\, d^2.$$

Hence it suffices to show that $\text{var}(S_{n_k}/a_{n_k}) \to 0$. $\text{Var}(S_{n_k}^s/a_{n_k}) \to 0$ implies $\text{var}(S_{n_k}/a_{n_k}) \to 0$. Thus it suffices to show that

$$\text{var}(S_{n_k}^s/a_{n_k}) \to 0.$$

Fix $k \geq 1$.

$$\text{var}\left(\frac{S_{n_k}^s}{a_{n_k}}\right) = \sum_{i=1}^{k} \text{var}\left(T_i^s\, \frac{a_{n_i}}{a_{n_k}}\right) = \sum_{i=1}^{k} \text{var}(T_i^s)\, \frac{a_{n_i}^2}{a_{n_k}^2}.$$

By Lemma 3.2.3 (ii), it suffices to show that

$$\text{var}(T_k^s) = \sum_{j=n_{k-1}+1}^{n_k} \text{var}(X_j^s/a_{n_k}) \to 0.$$

Fix $\varepsilon > 0$.

$$|X_j^s/a_j| \leq |X_j/a_j| + |X_j'/a_j| \leq 2\varepsilon_j \qquad \text{a.s.}$$

for each $j \geq 1$ by hypothesis. Thus by Lemma 3.4.1

$$P\left[\max_{n_{k-1}<j\leq n_k}\left|\sum_{i=n_{k-1}+1}^{j} \frac{X_i^s}{a_{n_k}}\right| < \varepsilon\right] \leq \frac{(\varepsilon + 2\max_{n_{k-1}<j\leq n_k}\varepsilon_j)^2}{\sum_{j=n_{k-1}+1}^{n_k}\text{var}(X_j^s/a_{n_k})}. \qquad (3.4.3)$$

$S_n^s/a_n \to 0$ a.s. since $(S_n - \mu(S_n))/a_n \to 0$ a.s. This implies

$$\max_{n_{k-1}<j\leq n_k}\left|\sum_{i=n_{k-1}+1}^{j} X_i^s/a_{n_k}\right| \to 0 \qquad \text{a.s.}$$

Thus the left hand side of Eq. (3.4.3) converges to one as $k \to \infty$. Thus

$$\limsup_{k\to\infty} \sum_{j=n_{k-1}+1}^{n_k} \text{var}(X_j^s/a_{n_k})/(\varepsilon + 2\max_{n_{k-1}<j\leq n_k}\varepsilon_j)^2 \leq 1.$$

But

$$\max_{n_{k-1}<j\leq n_k}\varepsilon_j \to 0.$$

Hence

$$\limsup_{k\to\infty} \sum_{j=n_{k-1}+1}^{n_k} \text{var}(X_j^s/a_{n_k}) \leq \varepsilon^2$$

for all $\varepsilon > 0$. Hence

$$\sum_{j=n_{k-1}+1}^{n_k} \text{var}(X_j^s/a_{n_k}) \to 0$$

as $k \to \infty$, as desired. ∎

Theorem 3.4.2 can be slightly strengthened (at the expense of a more complicated proof) by replacing the hypothesis that $|X_i| \leq \varepsilon_i a_i$ a.s. for each $i \geq 1$ by the hypothesis $|X_i| \leq a_i$ a.s. for each $i \geq 1$. A proof of this is given by Loève [1963, p. 253]. The special case $a_i = i$ for each $i \geq 1$ was proved by Prokhorov [1949].

There is a deeper question closely related to Theorems 3.4.1 and 3.4.2 discussed in detail in Chapter 5: Can the stability of $\{S_n, n \geq 1\}$ with $a_n = n$ for each $n \geq 1$ (that is, the strong law of large numbers) be characterized by the magnitudes of the variances of $\{X_i, i \geq 1\}$? We do have the simply stated result of Kolmogorov giving a sufficient condition:

$$\sum_{i=1}^{\infty} \text{var}(X_i)/i^2 < \infty \quad \text{implies} \quad (S_n - ES_n)/n \to 0 \quad \text{a.s.}$$

This important strong law of large numbers follows immediately from Theorem 3.3.1 with $p = 2$ and $a_i = i$ for each $i \geq 1$. Recent work by Teicher [1968] sharpens this result.

**Theorem 3.4.3.**   Suppose $EX_i = 0$ for $i \geq 1$. Then the three conditions

(i)   $\sum_{i=2}^{\infty} (EX_i^2/i^4) \sum_{j=1}^{i-1} EX_j^2 < \infty$,

(ii)   $\sum_{i=1}^{n} EX_i^2/n^2 \to 0$, and

(iii)   there exists constants $c_i$ such that $\sum_{i=1}^{\infty} P[|X_i| > c_i] < \infty$ and $\sum_{i=1}^{\infty} c_i^2 EX_i^2/i^4 < \infty$ together imply that $S_n/n \to 0$ a.s.

**Proof.**   The proof is dependent upon standard martingale results. Fix $n \geq 1$.

$$(S_n/n)^2 = \sum_{i=1}^{n} X_i^2/n^2 + 2\sum_{i=2}^{n} X_i S_{i-1}/n^2.$$

$T_n = \sum_{i=2}^{n} X_i S_{i-1}/i^2$ is easily shown to be a martingale.

$$ET_n^2 = \sum_{i=2}^{n} E[X_i^2 S_{i-1}^2]/i^4$$

$$= \sum_{i=2}^{n} EX_i^2 ES_{i-1}^2/i^4$$

$$= \sum_{i=2}^{n} EX_i^2 \sum_{j=1}^{i-1} EX_j^2/i^4 \leq K < \infty$$

for all $n \geq 2$ by (i). Hence $\sup E|T_n| < \infty$ follows and thus $T_n$ converges a.s. by Doob's basic martingale convergence theorem. By the Kronecker lemma,

$$\sum_{i=2}^{n} X_i S_{i-1}/n^2 \to 0 \quad \text{a.s.}$$

It remains to be shown that $\sum_{i=1}^{n} X_i^2/n^2 \to 0$ a.s. Let $Y_i = X_i I(|X_i| \leq c_i)$ and $Z_i = Y_i^2 - EY_i^2$ for $i \geq 1$. Note that $\{Z_i/i^2, i \geq 1\}$ is a sequence of independent mean zero random variables.

$$EZ_i^2 \leq EY_i^4 = EX_i^4 I(|X_i| \leq c_i) \leq c_i^2 EX_i^2$$

for $i \geq 1$. Hence

$$\sum_{i=1}^{\infty} E(Z_i/i^2)^2 \leq \sum_{i=1}^{\infty} c_i^2 EX_i^2/i^4 < \infty$$

by (iii). Thus

$$\sum_{i=1}^{\infty} (Y_i^2 - EY_i^2)/i^2 \qquad \text{converges a.s.}$$

by Theorem 2.12.2 (ii). This implies that

$$\sum_{i=1}^{n} (Y_i^2 - EY_i^2)/n^2 \to 0 \qquad \text{a.s.} \qquad (3.4.4)$$

by the Kronecker lemma. Now $EY_i^2 \leq EX_i^2$ for each $i \geq 1$ and hence by (ii), $\sum_{i=1}^{n} EY_i^2/n^2 \to 0$. Thus by Eq. (3.4.4), $\sum_{i=1}^{n} Y_i^2/n^2 \to 0$ a.s. Since $\sum_{i=1}^{\infty} P[|X_i| > c_i] < \infty$, it follows that $\sum_{i=1}^{n} X_i^2/n^2 \to 0$ a.s., thus establishing the theorem. ∎

**Corollary 3.4.2.** $\sum_{i=1}^{\infty} \text{var}(X_i)/i^2 < \infty$ implies (i), (ii), and (iii) of Theorem 3.4.3 (and thus Theorem 3.4.3 does sharpen Kolmogorov's strong law of large numbers).

*Exercise 3.4.2.* Prove Corollary 3.4.2. ∎

It is rather easy to construct examples where Theorem 3.4.2 applies and Kolmogorov's law of large numbers does not:

*Example 3.4.1.* Let $\{X_i, i \geq 1\}$ be a sequence of independent random variables with

$$P[X_{i-1} = i^{1/2}/\log^{1/2} i] = P[X_{i-1} = -i^{1/2}/\log^{1/2} i] = \tfrac{1}{2}$$

for each $i \geq 2$. Does $S_n/n \to 0$ a.s.? $EX_{i-1}^2 = i/\log i$ for $i \geq 2$. Thus

$$\sum_{i=1}^{\infty} EX_i^2/i^2 = \sum_{i=2}^{\infty} (i \log i)^{-1} = \infty.$$

Hence no conclusion can be reached using Kolmogorov's strong law. Let $c_i = i^{1/2}$ for each $i \geq 1$. We verify (i)–(iii) of Theorem 3.4.3. (iii) is clearly satisfied. We verify (ii). Let $N = [n^{1/2}]$.

$$\sum_{i=1}^{n} EX_i^2/n^2 = \sum_{i=1}^{N} EX_i^2/n^2 + \sum_{i=N+1}^{n} EX_i^2/n^2$$

$$\leq 2 \sum_{i=2}^{N+1} i/n^2 + \log^{-1} N \sum_{i=N+2}^{n+1} i/n^2 \to 0 \qquad \text{a.s.} \quad n \to \infty.$$

The above computation shows that

$$\sum_{j=1}^{i-1} EX_j^2 \le i + (i^2 + i) \log^{-1} i \le 2i^2 \log^{-1} i$$

for $i$ sufficiently large. Thus, verifying (i),

$$\sum_{i=2}^{\infty} \frac{EX_i^2}{i^4} \sum_{j=1}^{i-1} EX_j^2 \le C \sum_{i=2}^{\infty} \frac{1}{(i^3 \log i)} \frac{i^2}{\log i} < \infty.$$

Thus, by Theorem 3.4.3, $S_n/n \to 0$ a.s.  ∎

Teicher also shows that there is an infinite hierarchy of successfully weaker and more complicated conditions that can replace (i) in the statement of Theorem 3.4.3. These conditions are rather sharp as Theorem 3.4.4 suggests:

**Theorem 3.4.4.**   [Revesz, 1968, p. 65].   Let $|X_i| \le i$ a.s. and $EX_i = 0$ for all $i \ge 1$. Then $S_n/n \to 0$ a.s. implies that

$$\lim \left( \sum_{i=1}^{n} EX_i^2/i^2 \right) \Big/ \log n = 0.$$

**Proof.**   Omitted.  ∎

### 3.5.   Stability Results for Stationary Random Variables

Recall that $\{X_i, i \ge 1\}$ independent identically distributed with $EX_1$ finite implies that $\sum_{i=1}^{n} X_i/n \to EX_1$ a.s., this being Kolmogorov's strong law of large numbers. A sequence $\{X_i, i \ge 1\}$ of independent identically distributed random variables is stationary in the sense that $X_1, X_2, \ldots,$ $X_n, \ldots$ has the same distribution as $X_k, X_{k+1}, \ldots, X_{k+n-1}, \ldots$ for each $k \ge 1$. From a physical viewpoint it seems very natural to consider stochastic sequences whose probability behavior stays stationary over time, irrespective of the dependence structure. We isolate this stationarity and study stability results which follow from it.

In Section 3.5, $A = B$ will not mean $P[A \triangle B] = 0$ as it has meant by our convention. Rather

$$A = B$$

will mean that $A$ and $B$ are identical events. Also, in Section 3.5 random variables will be assumed to be finite for every $\omega \in \Omega$.

**Definition 3.5.1.** Recall that $R_\infty$ denotes the Cartesian product of a countable infinity of copies of the real line. Let $\mathscr{C}_\infty$ be the Borel sets of $R_\infty$; that is, the smallest $\sigma$ field of sets of $R_\infty$ which contains all the measurable finite dimensional product cylinders $\prod_{i=1}^\infty C_i$. ($\prod_{i=1}^\infty C_i$ is a measurable finite dimensional product cylinder if each $C_i$ is a one-dimensional Borel set and $C_i = R_1$ except for finitely many $i$.)

A stochastic sequence $\{X_i, i \geq 1\}$ is said to be stationary if $X_1, X_2, \ldots, X_n, \ldots$ has the same distribution as $X_{k+1}, X_{k+2}, \ldots, X_{k+n}, \ldots$ for each $k \geq 1$; that is, if for each $k \geq 1$

$$P[(X_1, X_2, \ldots, X_n, \ldots) \in C] = P[(X_{k+1}, X_{k+2}, \ldots, X_{k+n}, \ldots) \in C]$$

for every $C \in \mathscr{C}_\infty$. ∎

The main result of Section 3.5 is a rather striking generalization of the Kolmogorov strong law of large numbers; namely, that $E \mid X_1 \mid < \infty$ for $\{X_i, i \geq 1\}$ stationary implies $S_n/n$ is almost surely convergent. The natural mathematical framework for this result, known as the pointwise ergodic theorem, is that of "measure-preserving" transformations on a probability space. We need to consider a sizable amount of background material before we can establish the pointwise ergodic theorem. However, the elegance and importance of the pointwise ergodic theorem makes the effort worthwhile.

Consider a measurable transformation $T$ from $\Omega$ to $\Omega$ defined on the probability space $(\Omega, \mathscr{F}, P)$. By "measurable" we simply mean that $T^{-1}(\mathscr{F}) \subset \mathscr{F}$.

**Definition 3.5.2.** A transformation $T$ from $\Omega$ to $\Omega$ is measure preserving (alternatively "$T$ preserves $P$") if it is measurable and if $P[T^{-1}A] = P(A)$ for all $\mathscr{A} \in \mathscr{F}$. ∎

Every measure-preserving transformation generates a stationary sequence and any stationary sequence can be represented by means of a measure-preserving transformation. Throughout Section 3.5, $\mathscr{C}_\infty$ will denote the Borel sets of $R_\infty$.

**Theorem 3.5.1.** Let $X$ be a random variable and $T$ a measure-preserving transformation. Then setting

$$X_1(\omega) = X(\omega), \quad X_2(\omega) = X(T\omega), \ldots, X_n(\omega) = X(T^{n-1}\omega), \ldots$$

for each $\omega \in \Omega$ defines a stationary sequence $\{X_i, i \geq 1\}$.

**Proof.**  It suffices to show that

$$X_1, X_2, \ldots, X_n, \ldots$$

has the same distribution as

$$X_2, X_3, \ldots, X_{n+1}, \ldots$$

in order to verify that $\{X_i, i \geq 1\}$ is stationary.

Choose $C \in \mathscr{C}_\infty$. Let $Y \equiv (X, XT, \ldots)$ and $A = Y^{-1}(C)$.

$$T^{-1}A = T^{-1}Y^{-1}(C) = (YT)^{-1}C$$
$$= \{(\omega \mid XT(\omega), XT^2(\omega), \ldots, XT^n\omega, \ldots) \in C\}.$$

Thus

$$P[(X_1, X_2, \ldots, X_n, \ldots) \in C]$$
$$= P[\omega \mid (X(\omega), X(T\omega), \ldots, X(T^{n-1}\omega), \ldots) \in C]$$
$$= P[A] = P[T^{-1}A]$$
$$= P[\omega \mid (X(T\omega), X(T^2\omega), \ldots, X(T^n\omega), \ldots) \in C]$$
$$= P[(X_2, X_3, \ldots, X_{n+1}, \ldots) \in C]. \quad \blacksquare$$

In order to go the other way; that is, to represent a given stationary sequence by a random variable and a measure-preserving transformation, is slightly more involved. Let $\{X_i, i \geq 1\}$ be any stochastic sequence. Let

$$\hat{P}(C) = P[(X_1, X_2, \ldots, X_n, \ldots) \in C]$$

define a probability space $(R_\infty, \mathscr{C}_\infty, \hat{P})$. For each $\omega = (x_1, x_2, \ldots, x_n, \ldots) \in R_\infty$, let $\hat{X}_i(\omega) = x_i$ define the stochastic sequence $\{\hat{X}_i, i \geq 1\}$. $\{\hat{X}_i, i \geq 1\}$ is said to be the coordinate representation of the stochastic sequence $\{X_i, i \geq 1\}$. Clearly $\{\hat{X}_i, i \geq 1\}$ and $\{X_i, i \geq 1\}$ have the same distribution, justifying the use of the word "representation". Since in the study of sequences of random variables we are only interested in properties of $\{X_i, i \geq 1\}$ expressible in terms of its distribution, it is irrelevant whether we work with $(\Omega, \mathscr{F}, P)$, $\{X_i, i \geq 1\}$ or $(R_\infty, \mathscr{C}_\infty, \hat{P})$, $\{\hat{X}_i, i \geq 1\}$.

**Theorem 3.5.2.**  Let $\{X_i, i \geq 1\}$ be the coordinate representation of a stationary sequence. There exists a measure-preserving transformation $S$ on $(R_\infty, \mathscr{C}_\infty, \hat{P})$ such that $X_1(\omega) = X_1(\omega)$, $X_2(\omega) = X_1(S\omega), \ldots, X_n(\omega) = X_1(S^{n-1}\omega), \ldots$ for all $\omega \in R_\infty$.

**Proof.** Let $S(x_1, x_2, \ldots, x_n, \ldots) = (x_2, x_3, \ldots, x_{n+1}, \ldots)$ define a transformation $S : R_\infty \to R_\infty$. For each $n \geq 1$,

$$X_1(S^{n-1}(x_1, x_2, \ldots)) = X_1(x_n, x_{n+1}, \ldots)$$
$$= x_n = X_n(x_1, x_2, \ldots).$$

Hence $X_1(S^{n-1}(\omega)) = X_n(\omega)$ for each $n \geq 1$ and all $\omega \in R$ as desired.

We must show that $S$ is measure preserving. Let $C$ be a measurable finite dimensional product cylinder. Clearly $S^{-1}C$ is also such a cylinder set. Since these cylinder sets generate $\mathscr{C}_\infty$, it follows that $S$ is measurable. Let $C \in \mathscr{C}_\infty$.

$$\hat{P}(S^{-1}C) = \hat{P}[\omega \mid S\omega \in C] = \hat{P}[(x_1, x_2, \ldots) \mid (x_2, x_3, \ldots) \in C]$$
$$= \hat{P}[(X_2, X_3, \ldots) \in C]$$
$$= \hat{P}[(X_1, X_2, \ldots) \in C]$$

by stationarity. But

$$\hat{P}[(X_1, X_2, \ldots) \in C] = \hat{P}(C). \quad \blacksquare$$

For obvious reasons, $S$ is called the shift transformation.

We mention some examples of stationary sequences. Let $\{X_i, i \geq 1\}$ be a sequence of independent identically distributed random variables. $\{X_i, i \geq 1\}$ is clearly stationary. For fixed $m \geq 1$, let $Y_n = \sum_{i=n}^{n+m-1} X_i/m$ define $\{Y_n, n \geq 1\}$. This "moving average" is clearly stationary. More generally the following theorem shows how to construct new stationary sequences from a given stationary sequence.

**Theorem 3.5.3.** Let $\{X_i, i \geq 1\}$ be stationary. Let $\phi : R_\infty \to R_1$ be measurable (that is, $\phi^{-1}(B) \in \mathscr{C}_\infty$ for all Borel sets $B$ of $R_1$). Let

$$Y_i = \phi(X_i, X_{i+1}, \ldots)$$

for $i \geq 1$. Then $\{Y_i, i \geq 1\}$ is stationary.

**Exercise 3.5.1.** Prove Theorem 3.5.3. $\quad \blacksquare$

Recall the discussion of continued fractions in Example 3.3.1 yielded a stationary sequence. As already stated, another way to exhibit stationary sequences is to start with a measure-preserving transformation and a random variable. For example, let $\Omega$ be the unit disk, $\mathscr{F}$ be the Borel sets of $\Omega$, and $\pi P$ be two-dimensional Lebesgue measure. Let $(r, \theta)$ denote a typical element of $\Omega$ with $0 \le r \le 1$, $0 \le \theta < 2\pi$. For fixed $a$, let $T(r, \theta)$ $= (r, \theta + a)$ (addition modulo $2\pi$) define a transformation $T : \Omega \to \Omega$. Clearly $T$ is measure preserving since $T$ merely rotates measurable sets of $\Omega$, not changing their measures. Then for any random variable $X$, $\{XT^{i-1}, i \ge 1\}$ is stationary.

A random sequence $\{X_i, i \ge 1\}$ is said to be exchangeable or symmetrically dependent if for all $n \ge 1$ and all distinct positive integers $(i_1, i_2, \ldots, i_n)$ the distribution of $(X_{i_1}, X_{i_2}, \ldots, X_{i_n})$ depends only on $n$. Clearly $\{X_i, i \ge 1\}$ is a stationary sequence. Indeed the assumption of exchangeability seems and is much stronger (see Loève [1963, pp. 364–365]).

An important class of examples of stationary sequences will be given in Section 3.6 where the theory of stationary sequences is applied to stochastic sequences satisfying the Markov property.

***Exercise 3.5.2.*** Let $A$ and $B$ be independent normal mean zero variance one random variables. Let

$$X_n = A \sin 2\pi n\alpha + B \cos 2\pi n\alpha$$

for a fixed constant $\alpha$ define the stochastic sequence $\{X_n, n \ge 1\}$. Verify that $\{X_n, n \ge 1\}$ is stationary.  ∎

***Exercise 3.5.3.*** Let $\Omega = \{1, 2, \ldots, n\}$ and $T$ be a permutation of $\Omega$. Find necessary and sufficient conditions on $P$ such that $T$ is measure preserving.  ∎

If $\{X_i, i \ge 1\}$ are independent identically distributed, then $\sum_{i=1}^{n} X_i/n$ must converge to a constant if it converges. When the $X_i$ are merely stationary, convergence to a nonconstant limit is possible. In order to be able to study the limiting random variable in the stationary case we need to introduce the related concepts of an invariant set and ergodicity.

**Definition 3.5.3.** Given a measure preserving transformation $T$, a measurable event $A$ is said to be invariant if $T^{-1}A = A$. If

$$P[T^{-1}A \bigtriangleup A] = 0,$$

$A$ is said to be almost invariant.  ∎

**Lemma 3.5.1.** The collection of invariant events forms a $\sigma$ field $\mathscr{I}$. The collection of almost invariant events $\bar{\mathscr{I}}$ forms a $\sigma$ field which is the completion of $\mathscr{I}$ with respect to $\mathscr{F}$ and $P$ (that is, every almost invariant event differs from an invariant event by a measurable event of probability 0.)

**Exercise 3.5.4.**  Prove Lemma 3.5.1.  ∎

**Definition 3.5.4.**  A measure preserving transformation is ergodic if for all $A \in \mathscr{I}$ either $P(A) = 0$ or $P(A) = 1$.  ∎

**Lemma 3.5.2.**  Let $T$ be a measure-preserving transformation. $T$ is ergodic if and only if for all $A \in \bar{\mathscr{I}}$, either $P(A) = 0$ or $P(A) = 1$.

**Proof.**  Clearly $P(A) = 0$ or $P(A) = 1$ for all $A \in \bar{\mathscr{I}}$ implies that $T$ is ergodic since $\mathscr{I} \subset \bar{\mathscr{I}}$.

Assume $T$ is ergodic. Let $A \in \bar{\mathscr{I}}$. Then $A = B \cup N$ where $N$ is an event of probability zero and $B$ is an invariant event. Since $T$ is ergodic, either $P(B) = 0$ or $P(B) = 1$. Hence $P(A) = 0$ or $P(A) = 1$. Thus in the definition of ergodicity, $\mathscr{I}$ can be replaced by $\bar{\mathscr{I}}$.  ∎

Ergodicity has a geometric and, hence, a physical interpretation. Let $T$ be measure preserving and $1 : 1$ but not ergodic. Then there exists $A$ such that $0 < P(A) < 1$ and $A = T^{-1}A$. If we think of $\Omega$ as a geometric (or physical) space and $T$ as a transformation acting on points (or particles) of that space, then iterates of the transformation can never thoroughly mix points of the space. This is because $T^n A = A$ for all $n \geq 1$. Hence points of $A$ can never be moved outside of $A$ by iterates of $T$. As an illustration, recall the example of the unit disk: Let $\Omega$ be the unit disk, $\mathscr{F}$ the Borel sets of the disk, and $\pi P$ be two-dimensional Lebesgue measure restricted to the disk. Represent points of the disk by $(r, \theta)$ with $0 \leq r \leq 1$ and $0 \leq \theta < 2\pi$. For fixed $a$, let $T(r, \theta) = (r, \theta + a)$ (addition modulo $2\pi$) define $T$. Iterates of $T$ merely rotate the disk and hence do not thoroughly mix points. It is clear that $T$ is not ergodic.

**Exercise 3.5.5.**  Construct a random variable $X$ for the example discussed above such that $\sum_{i=1}^{n} X(T^{i-1}\omega)/n$ converges for all $\omega \in \Omega$, but to a nondegenerate random variable. (This exercise is important! It will be seen later that ergodicity forces the limit to be constant almost surely.)  ∎

As just stated, the ergodicity of $T$ is related to the ability of iterates of $T$ to thoroughly mix the points of $\Omega$. The following definition formalizes this notion of mixing.

**Definition 3.5.5.**  Let $T$ be a measure preserving transformation. $T$ is said to be mixing if for all $A, B \in \mathscr{F}$,

$$\lim P[A \cap T^{-n}B] = P[A]P[B]. \quad \blacksquare$$

Halmos [1956] gives a colorful example to help motivate this definition. Consider a container $(\Omega)$ filled with 90% gin and 10% vermouth. The contents are steadily stirred by a swizzle stick. The condition of the container is observed at times $t = 0, 1, 2, \ldots, n, \ldots$ . The movement of a particle $\omega$ under the influence of the stirring at times $t = 0, 1, 2, \ldots, n, \ldots$ is given by $\omega, T\omega, \ldots, T^n\omega, \ldots$ . Let $\mathscr{F}$ be the Borel subsets of $\Omega$ and $P$ be volume, where units are chosen so that $P(\Omega) = 1$. Let $A$ be the set of the vermouth particles at time $t = 0$. Let $B$ be an arbitrary Borel set of $\Omega$. If we have thorough mixing, then the percent vermouth in $B$ for large $n$ should be close to 10%. The set of vermouth particles in $B$ at time $t = n$ is

$$\{\omega \mid \omega \in A, \, T^n\omega \in B\} = A \cap T^{-n}B.$$

Hence thorough mixing means that

$$P(A \cap T^{-n}B)/P(B) \to P(A) = 0.1,$$

thus motivating Definition 3.5.5.

**Theorem 3.5.4.**  Let $T$ be mixing. Then $T$ is ergodic.

**_Proof._**  Let $B$ be invariant and $A \in \mathscr{F}$. Since $B = T^{-n}B$,

$$P[A \cap T^{-n}B] = P[A \cap B] \qquad \text{for all} \quad n \geq 1.$$

Thus, letting $n \to \infty$ and applying the mixing condition, $P[A \cap B] = P[A]P[B]$. Letting $A = B$, we obtain $P[B] = \{P[B]\}^2$, and hence $P[B] = 0$ or $P[B] = 1$. $\quad \blacksquare$

We will not use the concept of mixing; it was presented as an intuitive aid to the understanding of ergodicity. The converse to Theorem 3.5.4 is false. Thus mixing is a more restrictive assumption than ergodicity.

**Definition 3.5.6.**  Let $T$ be measure preserving. A random variable $X$ is invariant if $X(\omega) = X(T\omega)$ for all $\omega \in \Omega$. $\quad \blacksquare$

**Lemma 3.5.3.**  A random variable $X$ is invariant if and only if $X$ is $\mathscr{I}$ measurable.

**Proof.** Assume $X$ is invariant. Fix $a \in R_1$.

$$T^{-1}[X < a] = T^{-1}[\omega \mid X(\omega) < a] = [\omega \mid X(T\omega) < a]$$
$$= [\omega \mid X(\omega) < a] = [X < a].$$

Since $T^{-1}[X < a] = [X < a]$ for all $a \in R_1$, it follows that $T^{-1}[X \in B]$ $= [X \in B]$ for all Borel sets $B$. Hence $[X \in B] \in \mathscr{I}$ for all Borel sets $B$; that is, $X$ is $\mathscr{I}$ measurable as desired.

Assume $X = I(A)$ for $A \in \mathscr{I}$. $X(T\omega) = I(A)(T\omega) = I(T^{-1}A)(\omega) = I(A)(\omega) = X(\omega)$. Passing to simple $\mathscr{I}$ measurable random variables, non-negative $\mathscr{I}$ measurable random variables, and arbitrary $\mathscr{I}$ measurable random variables, the result follows. ∎

Actually only half of the definition of almost invariance suffices to establish almost invariance:

**Lemma 3.5.4.** Let $T$ be measure preserving. An event $A$ is almost invariant if *either*

$$P[A \cap (T^{-1}A)^c] = 0 \qquad \text{or} \qquad P[A^c \cap T^{-1}A] = 0.$$

**Proof.**

$$P[A \cap (T^{-1}A)^c] = P[A] - P[A \cap T^{-1}A]$$
$$= P[T^{-1}A] - P[A \cap T^{-1}A]$$
$$= P[A^c \cap T^{-1}A].$$

Hence

$$P[A \triangle T^{-1}A] = P[A \cap (T^{-1}A)^c] + P[A^c \cap T^{-1}A]$$
$$= 2P[A \cap (T^{-1}A)^c] = 2P[A^c \cap T^{-1}A]. \quad ∎$$

**Lemma 3.5.5.** Let $T$ be a measure-preserving transformation and $X$ a random variable with $E \mid X \mid < \infty$. Then

$$\int_A X \, dP = \int_{T^{-1}A} X(T\omega) \, dP(\omega)$$

and hence $A$ almost invariant implies that

$$\int_A X \, dP = \int_A XT \, dP.$$

**Proof.**   Suppose $X$ is an indicator random variable; that is, $X = I(B)$ for $B \in \mathscr{F}$.

$$\int_A X \, dP = P(A \cap B) = PT^{-1}(A \cap B) = P(T^{-1}A \cap T^{-1}B)$$
$$= \int_{T^{-1}A} I(T^{-1}B) \, dP = \int_{T^{-1}A} X(T\omega) \, dP(\omega)$$

as desired. The proof is completed by supposing $X$ simple, $X$ nonnegative, and $X$ integrable.   ∎

This completes the necessary preliminaries. The proof of the pointwise ergodic theorem depends on an important maximal inequality known as the maximal ergodic theorem.

**Theorem 3.5.5.**   (Maximal ergodic theorem [Riesz, 1945]).   Let $X$ be a random variable with $E \mid X \mid < \infty$ and $T$ a measure-preserving transformation. Let

$$S_k(\omega) = X(\omega) + X(T\omega) + \cdots + X(T^{k-1}\omega)$$

define $S_k$ for $k \geq 1$ and let $M_n = \max_{0 \leq k \leq n} S_k$ for $n \geq 1$ taking $S_0 = 0$. Then

$$\int_{[M_n > 0]} X \, dP \geq 0 \qquad (3.5.1)$$

(that is, $E[X \mid M_n > 0] \geq 0$ if $P[M_n > 0] > 0$) for each $n \geq 1$.

**Proof.**   (Proof due to Garsia [1965]).   Fix $n \geq 1$. For any $0 \leq k \leq n$, $M_n(T\omega) \geq S_k(T\omega)$. Thus

$$X(\omega) + M_n(T\omega) \geq X(\omega) + S_k(T\omega) = S_{k+1}(\omega)$$

for $0 \leq k \leq n$. It follows that

$$X(\omega) \geq \max_{1 \leq k \leq n} S_k(\omega) - M_n(T\omega).$$

Hence

$$\int_{[M_n > 0]} X \, dP \geq \int_{[\omega \mid M_n(\omega) > 0]} [\max_{1 \leq k \leq n} S_k(\omega) - M_n(T\omega)] \, dP(\omega).$$

On the set $[\omega \mid M_n(\omega) > 0]$, $\max_{1 \leq k \leq n} S_k(\omega) = M_n(\omega)$. Hence

$$\int_{[M_n > 0]} X \, dP \geq \int_{[\omega \mid M_n(\omega) > 0]} [M_n(\omega) - M_n(T\omega)] \, dP(\omega).$$

But $M_n T \geq 0$ a.s. and hence

$$\int_{[M_n > 0]} M_n \, dP - \int_{[\omega | M_n(\omega) > 0]} M_n(T\omega) \, dP(\omega)$$

$$\geq \int_{\Omega} M_n \, dP - \int_{\Omega} M_n(T\omega) \, dP(\omega).$$

By Lemma 3.5.5,

$$\int_{\Omega} M_n \, dP - \int_{\Omega} M_n(T\omega) \, dP(\omega) = 0. \quad\blacksquare$$

**Exercise 3.5.6.** (To make the meaning of the maximal ergodic theorem more clear). Assume the notation of the maximal ergodic theorem. Fix $\varepsilon > 0$. Assume $P[\max_{1 \leq k \leq n} S_k / k > \varepsilon] > 0$ for each $n \geq 1$. Prove that $E[X \mid \max_{1 \leq k \leq n} S_k / k > \varepsilon] \geq \varepsilon$ for each $n \geq 1$.   $\blacksquare$

**Theorem 3.5.6.** (Birkoff's pointwise ergodic theorem [1931]). Let $T$ be measure preserving. Then $E \mid X \mid < \infty$ implies that

$$P\left[\omega \,\Big|\, \lim_{n \to \infty} \sum_{k=0}^{n-1} X(T^k \omega)/n = E(X \mid \mathcal{S})(\omega)\right] = 1.$$

**Proof.** $E[X \mid \mathcal{S}]$ being invariant, it follows that

$$\sum_{k=0}^{n-1} E[X \mid \mathcal{S}](T^k \omega)/n = E[X \mid \mathcal{S}](\omega)$$

for each $n \geq 1$ and $\omega \in \Omega$. Hence it suffices to prove the result for $X - E[X \mid \mathcal{S}]$; that is, we may assume $E[X \mid \mathcal{S}] = 0$ a.s. without loss of generality and prove that $\sum_{k=0}^{n-1} XT^k / n \to 0$ a.s.

Fix $n \geq 1$. Let $S_n(\omega) = \sum_{k=0}^{n-1} X(T^k \omega)$ and for fixed $\varepsilon > 0$, $D = [\limsup S_n / n > \varepsilon]$. We would like to show $P[D] = 0$. Since

$$(\limsup S_n/n)(\omega) = (\limsup S_n/n)(T\omega)$$

for all $\omega$, it follows that $\limsup S_n / n$ is an invariant random variable and, therefore, $D$ is an invariant set. Let $X^* = (X - \varepsilon)I(D)$, $S_n^*(\omega) = \sum_{k=0}^{n-1} X^*(T^k \omega)$ for all $\omega \in \Omega$, and $M_n^* = \max_{0 \leq k \leq n} S_k^*$ for all $n \geq 1$, where $S_0^* = 0$. By the maximal ergodic theorem,

$$\int_{[M_n^* > 0]} X^* \, dP \geq 0.$$

We now show that $\int_D X^* \, dP \geq 0$ follows. As $n \to \infty$, the events $[M_n^* > 0]$ increase to

$$[\sup S_k^* > 0] = [\sup S_k^*/k > 0]$$
$$= [\sup S_k/k > \varepsilon] \cap D.$$

But

$$\sup S_k/k \geq \lim \sup S_k/k,$$

which implies that $D \subset [\sup S_k/k > \varepsilon]$. Thus the events $[M_n^* > 0]$ increase to $D$. $E \mid X^* \mid \leq E \mid X \mid + \varepsilon < \infty$. Thus by the Lebesgue dominated convergence theorem,

$$\int_{[M_n^* > 0]} X^* \, dP \to \int_D X^* \, dP.$$

Since $\int_{[M_n^* > 0]} X^* \, dP \geq 0$,

$$\int_D X^* \, dP \geq 0 \qquad \text{for each} \quad n \geq 1. \tag{3.5.2}$$

But, using the fact that $D$ is invariant,

$$\int_D X^* \, dP = \int_D X \, dP - \varepsilon P(D)$$
$$= \int_D E[X \mid \mathscr{I}] \, dP - \varepsilon P(D) = -\varepsilon P(D).$$

Thus, using Eq. (3.5.2), $P(D) = 0$ for each $\varepsilon > 0$ and hence $\lim \sup S_n/n \leq 0$ a.s. Similarly $\lim \inf S_n/n \geq 0$ a.s., establishing the theorem.  ∎

**Remark.**    It is easy to lose the thrust of the proof of the pointwise ergodic theorem: Either $P(D) = 0$ or $P(D) > 0$. If $P(D) > 0$, then the maximal ergodic theorem implies that $X$ is large on $D$ in the sense that $E[X \mid D] \geq \varepsilon$. Yet since $D$ is invariant, the magnitude of $X$ on $D$ is controlled in the sense that $E[X \mid D] = 0$. Thus $P(D) = 0$, from which the theorem follows.

**Corollary 3.5.1.**    Let $T$ be measure preserving and ergodic. Then $E \mid X \mid < \infty$ implies that

$$\lim_{n \to \infty} \sum_{k=0}^{n-1} X(T^k \omega)/n = EX \qquad \text{a.s.}$$

**Proof.** Each event in $\mathscr{S}$ has probability zero or one. Hence $E[X \mid \mathscr{S}]$ is almost surely equal to a constant. Hence $E[X \mid \mathscr{S}] = EX$ a.s. ■

Corollary 3.5.1 is often described in physical terms by saying that the time average $(\lim \sum_{k=0}^{n-1} X(T^k\omega)/n)$ equals the ensemble average $(EX)$. Roughly, the idea is that the average of one physical system over time is the same as the average of infinitely many identical systems at a fixed time.

**Corollary 3.5.2.** (The mean ergodic theorem). Let $T$ be measure preserving. Then $E \mid X \mid < \infty$ implies

$$E \left| \sum_{k=0}^{n-1} X(T^k)/n - E[X \mid \mathscr{S}] \right| \to 0$$

as $n \to \infty$ (that is, convergence in $L_1$).

**Proof.** Let $S_n = \sum_{k=0}^{n-1} X(T^k\omega)$ for $n \geq 1$. Without loss of generality, assume $E[X \mid \mathscr{S}] = 0$ a.s. By the pointwise ergodic theorem, $S_n/n \to 0$ a.s. We want to show that $E \mid S_n \mid/n \to 0$. Let $A = [\sup_{n \geq 1} \mid S_n \mid/n > K]$ for fixed $K > 0$. By the Lebesgue dominated convergence theorem $\int_{A^c} (\mid S_n \mid/n)$ $dP \to 0$ as $n \to \infty$. Hence

$$\lim \sup \int_{\Omega} (\mid S_n \mid/n) \, dP = \lim \sup \int_{A} (\mid S_n \mid/n) \, dP$$

$$\leq \lim \sup \sum_{i=1}^{n} \int_{A} \mid X_i \mid dP/n$$

where we define $X_i(\omega) = XT^{i-1}\omega$ for all $\omega \in \Omega$ and $i \geq 1$.

$$\int_{A} \mid X_i \mid dP = \int_{A \cap (\mid X_i \mid > N)} \mid X_i \mid dP + \int_{A \cap (\mid X_i \mid \leq N)} \mid X_i \mid dP$$

$$\leq \int_{\mid X_1 \mid > N} \mid X_1 \mid dP + NP(A)$$

for $i \geq 1$ and $N$ positive. In this case Theorem 3.5.3 was used to conclude that $\{X_i I(\mid X_i \mid > N), i \geq 1\}$ is stationary. Fix $\varepsilon > 0$. Choose $N$ so that $\int_{\mid X_1 \mid > N} \mid X_1 \mid dP \leq \varepsilon/2$. Then choose $K$ such that $NP(A) \leq \varepsilon/2$, this being possible since $S_n/n \to 0$ a.s. Combining,

$$\lim \sup \int_{\Omega} (\mid S_n \mid/n) \, dP \leq \varepsilon$$

for all $\varepsilon > 0$; that is, $E \mid S_n \mid/n \to 0$ as desired. ■

One might ask whether the condition $E \mid X \mid < \infty$ is necessary as well as sufficient for the conclusion of the pointwise ergodic theorem holding true, the question being motivated by the case of independent identically distributed random variables. One should refer to Halmos [1956, p. 32] for an example such that $E \mid X \mid = \infty$ and yet $\sum_{k=0}^{n-1} X(T^k \omega)/n$ converges a.s.

**Exercise 3.5.7.** If $EX = \infty$ and $T$ is measure preserving and ergodic, show that

$$\sum_{k=1}^{n} X(T^k \omega)/n \to \infty \qquad \text{a.s.} \quad \blacksquare$$

Recall that our main purpose was to establish the pointwise ergodic theorem for stationary sequences, a task we can now carry out. Let $\{X_i, i \geq 1\}$ be a stationary sequence. The concepts of invariance and ergodicity were defined earlier in terms of a measure-preserving transformation used with the coordinate representation of $\{X_i, i \geq 1\}$. We shall now define these concepts *directly* in terms of the given stationary sequence, with no reference to a measure-preserving transformation.

**Lemma 3.5.6.** Let $\{X_i, i \geq 1\}$ be the coordinate representation of a stationary sequence and let $S$ be the associated measure preserving transformation (the shift transformation). Then an event $A$ is invariant if and only if there exists $C \in \mathscr{C}_\infty$ such that $A = [(X_k, X_{k+1}, \ldots) \in C]$ for all $k \geq 1$.

**Proof.** Assume $A \in \mathscr{C}_\infty$ invariant; that is, $S^{-1}A = A$. Because of the coordinate space representation,

$$A = [(X_1, X_2, \ldots) \in A]$$

and

$$S^{-1}A = [(X_2, X_3, \ldots) \in A].$$

Thus, by the invariance of $A$,

$$[(X_1, X_2, \ldots) \in A] = [(X_2, X_3, \ldots) \in A].$$

$$A = [(X_k, X_{k+1}, \ldots) \in A]$$

follows, proving half the lemma.

To go the other way, assume an event $A$ can be represented $A = [(X_k, X_{k+1}, \ldots) \in C]$ for each $k \geq 1$. Thus $A = [(X_1, X_2, \ldots) \in C] = [(X_2, X_3, \ldots) \in C] = S^{-1}A.$ $\blacksquare$

Lemma 3.5.6 suggests how invariance can be defined directly in terms of a stationary sequence.

**Definition 3.5.7.** Let $\{X_i, i \geq 1\}$ be stationary and $A \in \mathscr{B}(X_i, i \geq 1)$. That is,

$$A = [(X_1, X_2, \ldots) \in C] \qquad \text{for some} \quad C \in \mathscr{C}_\infty.$$

Then $A$ is said to be invariant if

$$A = [(X_k, X_{k+1}, \ldots) \in C] \qquad \text{for all} \quad k \geq 1$$

and $A$ is said to be almost invariant if

$$P\{A \triangle [(X_k, X_{k+1}, \ldots) \in C]\} = 0 \qquad \text{for all} \quad k \geq 1.$$

A random variable $Y$ is invariant if there exists a measurable function $\phi : R_\infty \to R_1$ such that $Y = \phi(X_k, X_{k+1}, \ldots)$ for all $k \geq 1$. ∎

**Remark.** In order to verify that an event $A \in \mathscr{B}(X_i, i \geq 1)$ is invariant it clearly suffices to verify

$$A = [(X_2, X_3, \ldots) \in C].$$

A similar remark applies to almost invariance.

Let $\mathscr{H}$ denote the $\sigma$ field of invariant events for a given stationary sequence $\{X_i, i \geq 1\}$.

**Definition 3.5.8.** A stationary sequence $\{X_i, i \geq 1\}$ is ergodic if every invariant event has probability zero or one. ∎

**Lemma 3.5.7.** Let $\{X_i, i \geq 1\}$ be stationary.

(i) Then a random variable $Y$ is invariant if and only if $Y$ is $\mathscr{H}$ measurable.

(ii) Let $A \in \mathscr{B}(X_i, i \geq 1)$, that is,

$$A = [(X_1, X_2, \ldots) \in C] \qquad \text{for some} \quad C \in \mathscr{C}_\infty.$$

Then if either

$$P\{A \cap [(X_2, X_3, \ldots) \notin C]\} = 0$$

or

$$P\{A^c \cap [(X_2, X_3, \ldots) \in C]\} = 0,$$

$A$ is almost invariant.

**Proof.**   We prove (ii) only. Let $A_i = [(X_i, X_{i+1}, \ldots) \in C]$ for $i \geq 1$.

$$P[A \cap A_2{}^c] = P[A_1 \cap A_2{}^c] = P[A_1] - P[A_1 \cap A_2]$$
$$= P[A_2] - P[A_1 \cap A_2]$$

since $P[A_1] = P[A_2]$ by stationarity.

$$P[A_2] - P[A_1 \cap A_2] = P[A_1{}^c \cap A_2] = P[A^c \cap A_2].$$

Thus

$$P[A \bigtriangleup A_2] = P[A \cap A_2{}^c] + P[A^c \cap A_2]$$
$$= 2P[A \cap A_2{}^c] = 2P[A^c \cap A_2].$$

By hypothesis, either $P[A \cap A_2{}^c] = 0$ or $P[A^c \cap A_2] = 0$. Thus $P[A \bigtriangleup A_2] = 0$. Similarly, $P[A \bigtriangleup A_k] = 0$, as desired. ∎

**Theorem 3.5.7.**   (The pointwise ergodic theorem for stationary sequences).   Let $\{X_i, i \geq 1\}$ be stationary with $E \mid X_1 \mid < \infty$. Then $\sum_{i=1}^n X_i/n \to E[X_1 \mid \mathcal{H}]$ a.s. If in addition $\{X_i, i \geq 1\}$ is ergodic, $\sum_{i=1}^n X_i/n \to EX_1$ a.s.

**Proof.**   Using the coordinate space representation $\{\hat{X}_i, i \geq 1\}$, by the pointwise ergodic theorem and Theorem 3.5.2, it follows that

$$\sum_{i=1}^n \hat{X}_i/n \to E[X_1 \mid \mathcal{I}] \qquad \text{a.s.}$$

where $\mathcal{I}$ is the $\sigma$ field of invariant sets with respect to the transformation $S$. Hence, $\sum_{i=1}^n X_i/n$ converges almost surely to a random variable $Y$. Using the mean ergodic theorem (Corollary 3.5.2)

$$E \left| \sum_{i=1}^n X_i/n - Y \right| \to 0$$

as $n \to \infty$.

We identify $Y$ and $E[X_1 \mid \mathcal{H}]$. Let $Y = \limsup \sum_{i=1}^n X_i/n$. (This defines the limiting random variable for all $\omega \in \Omega$, not just those $\omega$ such that $\sum_{i=1}^n X_i(\omega)/n$ converges.) Since

$$\limsup \sum_{i=1}^n X_i/n = \limsup \sum_{i=k}^{k+n-1} X_i/n,$$

$Y$ is an invariant random variable.

Choose $H \in \mathcal{H}$. By the mean ergodic theorem,

$$\int_H \left( \sum_{i=1}^{n} X_i/n \right) dP \to \int_H Y \, dP. \tag{3.5.3}$$

Since $H$ is invariant, there exists $C \in \mathcal{C}_\infty$ such that

$$H = [(X_k, X_{k+1}, \ldots) \in C]$$

for all $k \geq 1$. $\{X_i I[(X_i, X_{i+1}, \ldots) \in C], i \geq 1\}$ is stationary by Theorem 3.5.3. Thus

$$\int_H X_i \, dP = \int_\Omega X_i I[(X_i, X_{i+1}, \ldots) \in C] \, dP = \int_H X_1 \, dP$$

for $i \geq 1$. Thus by Eq. (3.5.3),

$$\int_H X_1 \, dP = \int_H Y \, dP$$

for all $H \in \mathcal{H}$. Hence $E[X_1 \mid \mathcal{H}] = Y$ a.s.  ∎

**Lemma 3.5.8.**  Let $\{X_i, i \geq 1\}$ be independent identically distributed. Then $\{X_i, i \geq 1\}$ is stationary ergodic.

**Proof.**  Trivially, $\{X_i, i \geq 1\}$ is stationary. Let $A$ be an invariant event. There exists $C \in \mathcal{C}_\infty$ such that $A = [(X_k, X_{k+1}, \ldots) \in C]$ for all $k \geq 1$. Hence $A$ is a tail event. By the Kolmogorov 0–1 law $P(A) = 0$ or $P(A) = 1$. Hence $\{X_i, i \geq 1\}$ is ergodic.  ∎

**Remarks.**  Theorem 3.5.7 and Lemma 3.5.8 yield an alternate proof of the strong law of large numbers for independent identically distributed random variables with finite mean. This is the third proof of the strong law given, the classical truncation approach and the martingale approach having been given previously. In general, it can be difficult to verify that a stationary sequence is ergodic. Of course, one method is to show that the tail $\sigma$ field is trivial, as we did in Lemma 3.5.8. Some general results on the establishment of ergodicity are given by Shiryaev [1963].

**Theorem 3.5.8.**  Let $\{X_i, i \geq 1\}$ be stationary ergodic and $\phi$ be a measurable function $\phi : R_\infty \to R_1$. Let $Y_i = \phi(X_i, X_{i+1}, \ldots)$ define $\{Y_i, i \geq 1\}$. Then $\{Y_i, i \geq 1\}$ is stationary ergodic.

**Proof.**   By Theorem 3.5.3, $\{Y_i, i \geq 1\}$ is stationary. Let

$$\phi_k(x) = \phi(x_k, x_{k+1}, \ldots)$$

for each $x = (x_1, x_2, \ldots) \in R_\infty$ define the function $\phi_k : R_\infty \to R_1$. Let $A$ be an invariant event for the $Y$ sequence. $A = [(Y_k, Y_{k+1}, \ldots) \in C]$ for all $k \geq 1$ and some $C \in \mathscr{C}_\infty$. Thus

$$A = \{[\phi(X_k, X_{k+1}, \ldots), \phi(X_{k+1}, X_{k+2}, \ldots), \ldots] \in C\}$$

for all $k \geq 1$. Let

$$C_1 = [x \mid (\phi_1(x), \phi_2(x), \ldots) \in C] \in \mathscr{C}_\infty.$$

$$A = [X_k, X_{k+1}, \ldots) \in C_1]$$

for all $k \geq 1$. Hence $A$ is also an invariant event for the $X$ sequence. Thus $P(A) = 0$ or $P(A) = 1$ since $\{X_i, i \geq 1\}$ is ergodic.   ∎

This result can be applied to moving averages:

**Corollary 3.5.3.**   Let $\{X_i, i \geq 1\}$ be independent identically distributed with $E \mid X_1 \mid < \infty$. Let

$$Y_k = \sum_{i=k}^{k+n-1} X_i/n$$

define $\{Y_k, k \geq 1\}$ for fixed $n \geq 1$. Then

$$\sum_{i=1}^{m} Y_i/m \to EX_1 \qquad \text{a.s.}$$

**Proof.**   By Theorem 3.5.8, $\{Y_i, i \geq 1\}$ is stationary ergodic. Hence by the pointwise ergodic theorem,

$$\sum_{i=1}^{m} Y_i/m \to EY_1 = EX_1 \qquad \text{a.s.}   ∎$$

Recall we derived this result previously from Theorem 3.3.2 for mixing random variables.

**Example 3.5.1.**   (Equidistribution of points under a measure preserving transformation).   Let $\Omega = [0, 1]$, $\mathscr{F}$ be the Borel sets, and $P$ be

Lebesgue measure. Let addition be mod 1. Define $T : [0, 1] \rightarrow [0, 1]$ by

$$Tx = x + \lambda$$

for each $x \in [0, 1]$ where $\lambda$ is a fixed irrational number. Clearly $T$ is measure preserving since it merely translates sets. Let $J$ be a subinterval of $[0, 1]$ and consider the random variable $I$ defined by

$$I(x) = \begin{cases} 1 & \text{if } x \in J \\ 0 & \text{if } x \notin J. \end{cases}$$

By the pointwise ergodic theorem,

$$\sum_{i=1}^{n} IT^{i-1}/n \rightarrow E[I \mid \mathscr{I}] \qquad \text{a.s.}$$

as $n \rightarrow \infty$.

Note that $\sum_{i=1}^{n} I(T^{i-1}x)/n$ is merely the proportion of the points

$$\{x + \lambda, x + 2\lambda, \ldots, x + n\lambda\}$$

falling in $J$. We show that $T$ is ergodic from which the desired equidistribution result

$$\sum_{i=1}^{n} IT^{i-1}/n \rightarrow P(J) \qquad \text{a.s.}$$

follows.

In order to prove that $T$ is ergodic we prove an equivalent result, namely that an arbitrary invariant bounded random variable $X$ is constant almost surely. Such a bounded $X$ is in $L_2$. Hence it has a Fourier series representation $X(x) = \sum_{n=-\infty}^{\infty} c_n e^{2\pi i n x}$ for almost every $x \in [0, 1]$, the convergence of the sum being $L_2$ convergence. Here the constants $\{c_n, -\infty < n < \infty\}$ satisfy $\sum_{n=-\infty}^{\infty} c_n^2 < \infty$. Now

$$X(Tx) = \sum_{n=-\infty}^{\infty} c_n e^{2\pi i n \lambda} \cdot e^{2\pi i n x} \qquad \text{for almost every} \quad x \in [0, 1].$$

By definition, since $X$ is invariant, $X(x) = X(Tx)$ for all $x \in [0, 1]$. Thus

$$\sum_{n=-\infty}^{\infty} c_n e^{2\pi i n x} = \sum_{n=-\infty}^{\infty} c_n e^{2\pi i n \lambda} e^{2\pi i n x}$$

for almost every $x \in [0, 1]$. But this implies $c_n(e^{2\pi i n \lambda} - 1) = 0$ for each integer $n$. Since $\lambda$ is irrational, this implies $c_n = 0$ for all $n \neq 0$. Thus $X = c_0$ a.s., as desired. ∎

Since it is usually hard to verify that a transformation is ergodic, we were in this sense lucky in Example 3.5.1.

*Example 3.5.2.*  (A class of stationary ergodic sequences). A sequence $\{X_i, i \geq 1\}$ is called a normal sequence if $X_1, X_2, \ldots, X_n$ is jointly normally distributed for each $n \geq 1$. Suppose $EX_i = 0$ for each $i \geq 1$ and that the covariance $r_{ij} = EX_i X_j$ depends only on $|i - j|$ for all integers $i$ and $j$. The joint characteristic function of $X_1, \ldots, X_n$ is given by

$$\phi(u_1, u_2, \ldots, u_n) = \exp\left[ -\tfrac{1}{2} \sum_{i=1}^{n} \sum_{j=1}^{n} r_{ij} u_i u_j \right]$$

for all real $(u_1, u_2, \ldots, u_n)$. Since $r_{ij}$ is a function of $|i - j|$, it follows that $\phi$ is also the characteristic function of $(X_2, \ldots, X_{n+1})$. Hence $\{X_i, i \geq 1\}$ is stationary. Suppose $r_{ij} \to 0$ as $|i - j| \to \infty$. Then $\{X_i, i \geq 1\}$ can, in addition, be shown to be ergodic. This is somewhat intuitive since uncorrelated jointly normal random variables are independent. Thus one would suspect that the tail $\sigma$ field of $\{X_i, i \geq 1\}$ consists only of events of probability zero or one since this would be true if the $X_i$ were independent. Since each invariant event can easily be shown to be a tail event, it then follows that each invariant event should have probability zero or one.

It of course follows from the pointwise ergodic theorem that

$$\sum_{i=1}^{n} X_i/n \to 0 \qquad \text{a.s. as} \quad n \to \infty. \quad \blacksquare$$

For an application of the pointwise ergodic theorem to the proof of the Shannon–McMillan theorem of information theory see Ash [1972, Chapter 4].

It is interesting to ask whether the Marcinkiewicz strong law of large numbers for independent identically distributed random variables generalizes to stationary ergodic sequences of random variables. The answer is no. (See Jain, Jogdeo, and Stout [1973].) It should be remarked that the above treatment followed Breiman [1968] closely.

## 3.6.  Application of Stationarity to Stability Results for Markov Sequences

The pointwise ergodic theorem has an interesting application to the study of stochastic sequences satisfying the Markov property. We must first review the definition and some basic properties of Markov sequences and then some related material on conditioning. The experienced reader may wish to skip some of the following material.

**Definition 3.6.1.** A stochastic sequence $\{X_i, i \geq 1\}$ is said to be a Markov sequence (or to satisfy the Markov property) if

$$P[X_{n+1} \in A \mid X_n, \ldots, X_1] = P[X_{n+1} \in A \mid X_n] \qquad \text{a.s.}$$

for all $n \geq 1$ and all Borel sets $A$. ∎

Given a random variable $X$, a random variable $Y$, and a Borel set $A$ of $R_1$, $P[X \in A \mid Y]$ and $E[X \mid Y]$ are defined as functions from $\Omega$ to $R_1$. The conditional probability of $X \in A$ given $Y$ and the conditional expectation of $X$ given $Y$ can also be defined (in a way consistant with their definitions as functions from $\Omega$ to $R_1$) as measurable functions from $R_1$ to $R_1$. This is more convenient for the study of Markov sequences. Let $\mathscr{B}_1$ denote the Borel sets of $R_1$.

**Definition 3.6.2.** The conditional probability of $X \in A$ given $Y$, denoted by $P[X \in A \mid Y = \cdot]$, is defined as any measurable function $\phi : R \to [0, 1]$ satisfying

$$\int_C \phi(y) \, dP_Y(y) = P[X \in A, Y \in C]$$

for all $C \in \mathscr{B}_1$. Here $P_Y$ is defined by $P_Y(C) = P[Y \in C]$ for each $C \in \mathscr{B}_1$.

Let $E \mid X \mid < \infty$. The conditional expection of $X$ given $Y$, denoted by $E[X \mid Y = \cdot]$, is defined as any measurable function $\phi : R \to R$ satisfying

$$\int_C \phi(y) \, dP_Y(y) = \int_{Y \in C} X \, dP$$

for all $C \in \mathscr{B}_1$. ∎

The standard properties of conditional probability and expectation also hold for $P[X \in A \mid Y = \cdot]$ and $E[X \mid Y = \cdot]$. Lemma 3.6.1 gives the connection between $P[X \in A \mid Y]$ and $P[X \in A \mid Y = \cdot]$. The same result holds for $E[X \mid Y]$ and $E[X \mid Y = \cdot]$ and is left unstated.

**Lemma 3.6.1.** (i) Let $P[X \in A \mid Y]$ be given for $A \in \mathscr{B}_1$. Let $g(\cdot)$ be a measurable function from $R_1$ to $R_1$ such that $g(Y) = P[X \in A \mid Y]$.

(Such a $g$ always exists.) Then letting

$$P[X \in A \mid Y = y] = g(y)$$

for each real $y$ defines a version of $P[X \in A \mid Y = \cdot]$.

(ii)  Let $P[X \in A \mid Y = \cdot]$ be given for $A \in \mathscr{C}_1$. Then

$$P[X \in A \mid Y](\omega) = P[X \in A \mid Y = Y(\omega)]$$

for each $\omega \in \Omega$ defines a version of $P[X \in A \mid Y]$.

**Proof.**  (i)  Fix $C \in \mathscr{C}_1$.

$$\int_C P[X \in A \mid Y = y] \, dP_Y(y) = \int_C g(y) \, dP_Y(y)$$

$$= \int_{[Y \in C]} g(Y) \, dP = \int_{[Y \in C]} P[X \in A \mid Y] \, dP$$

$$= P[X \in A, Y \in C],$$

proving (i).

(ii)  Fix $B \in \mathscr{B}(Y)$. There exists $C \in \mathscr{C}_1$ such that $B = [Y \in C]$.

$$\int_B P[X \in A \mid Y] \, dP = \int_{[\omega \mid Y(\omega) \in C]} P[X \in A \mid Y = Y(\omega)] \, dP(\omega)$$

$$= \int_C P[X \in A \mid Y = y] \, dP_Y(y) = P[X \in A, Y \in C]$$

$$= P[(X \in A) \cap B]. \quad \blacksquare$$

**Definition 3.6.3.**  Let $X$ and $Y$ be random variables. A function $p(\cdot, \cdot) : R_1 \times \mathscr{C}_1 \to [0, 1]$ is said to be a regular version of $P[X \in \cdot \mid Y = \cdot]$ if

(i)  $p(\cdot, A) : R_1 \to [0, 1]$ is a version of $P[X \in A \mid Y = \cdot]$ for each $A \in \mathscr{C}_1$ and if

(ii)  $p(y, \cdot) : \mathscr{C}_1 \to [0, 1]$ is a probability measure for each $y \in R_1$.

Foundational work on the existence of regular versions of conditional probabilities (see Breiman [1968, pp. 77–81]) shows that regular conditional versions always exist. Thus in working with Markov sequences we will assume that conditional probabilities are regular. These regular versions are called transition probabilities.

**Definition 3.6.4.** Let $\{X_i, i \geq 1\}$ be a Markov sequence. If for each $n \geq 1$ $p_n(\cdot, \cdot)$ is a regular version of $P[X_{n+1} \in \cdot \mid X_n = \cdot]$ then each $p_n(\cdot, \cdot)$ is called a transition probability and $\{p_n(\cdot, \cdot), n \geq 1\}$ is the collection of transition probabilities for the Markov sequence. ∎

**Remark.** Let $\{X_i, i \geq 1\}$ be a Markov sequence. The transition probabilities $p_n$ are functions defined on $R_1 \times \mathscr{C}_1$. If the range of $\{X_i, i \geq 1\}$ is a proper set of $R_1$, such as the positive integers, the range of the $p_n$ can be restricted. Let $S$ be a Borel set of $R_1$ such that $\bigcup_{i=1}^{\infty} X_i(\Omega) \subset S$. Then $S$ is called a state space of the Markov sequence. A "minimal" $S$ may not exist since $\bigcup_{i=1}^{\infty} X_i(\Omega)$ need not be a Borel set, but this need not concern us. Usually in applications the choice of $S$ is clear. Let $\mathscr{C}_1(S)$ be the Borel sets of $S$. We may take the $p_n$ with domain restricted to $S \times \mathscr{C}_1(S)$ if desired.

A regular version of a conditional probability $p(\cdot, \cdot)$ has certain nice properties resulting from the fact that $p(y, \cdot)$ is a measure for each $y \in R_1$.

**Lemma 3.6.2.** Let $p(\cdot, \cdot)$ be a regular version of the conditional probability of $X$ given $Y$. Let $g$ be a measurable function from $R_1$ to $R_1$ for which $E \mid g(X) \mid < \infty$. Then

$$E[g(X) \mid Y] = \int_{-\infty}^{\infty} g(x) p(Y, dx) \qquad \text{a.s.}$$

**Exercise 3.6.1.** Prove Lemma 3.6.2. Hint: Let $g$ be an indicator function, then a simple function, etc. ∎

Most of the work on Markov sequences concerns sequences whose transition probabilities are independent of $n$.

**Definition 3.6.5.** Let $\{X_i, i \geq 1\}$ be a Markov sequence. If there exist transition probabilities $\{p_i(\cdot, \cdot), i \geq 1\}$ and $p(\cdot, \cdot)$ such that $p_i(\cdot, \cdot) = p(\cdot, \cdot)$ for each $i \geq 1$, then $\{X_i, i \geq 1\}$ is said to have stationary transition probabilities. The function $p(\cdot, \cdot)$ is said to be the transition probability of the Markov sequence. ∎

We will study only Markov sequences with stationary transition probabilities. Thus, all Markov sequences discussed below are assumed to have stationary transition probabilities.

Roughly speaking, the Markov assumption says that given the present, the past contains no further information concerning the future. For example, in predicting the future behavior of a system of particles not being influenced by any external forces, it seems reasonable to assume that it is sufficient to know the present coordinates and velocities of the particles and that any information about the past evolution of the system is irrelevant. Nature abounds with phenomena where the Markov assumption seems appropriate. We have already studied a class of Markov sequences when we studied partial sums $\{S_n, n \geq 1\}$ of independent random variables $\{X_i, i \geq 1\}$. If, in addition, the $X_i$ are identically distributed, $\{S_n, n \geq 1\}$ can be shown to have stationary transition probabilities. These two rather intuitive facts require formal verification (see Chung [1960, pp. 286–287]).

The transition probability of a Markov sequence $\{X_i, i \geq 1\}$ does not by itself determine the joint distributions of $\{X_i, i \geq 1\}$. We also need to know how the sequences starts out; that is, the distribution of $X_1$ must be given. We call the distribution of $X_1$ for a Markov sequence the initial distribution. Often in applications, $X_1$ has a degenerate distribution.

**Theorem 3.6.1.**  Let $\{X_i, i \geq 1\}$ be a Markov sequence with initial distribution $P_{X_1}$ and transition probability $(p(\cdot, \cdot)$. Then

$$P[X_1 \leq x_1, X_2 \leq x_2, \ldots, X_n \leq x_n]$$
$$= \int_{-\infty}^{x_1} \cdots \int_{-\infty}^{x_n} p(y_{n-1}, dy_n) p(y_{n-2}, dy_{n-1}) \cdots p(y_1, dy_2)\, dP_{X_1}(y_1)$$

for all $(x_1, x_2, \ldots, x_n) \in R_n$ and $n \geq 1$.

**Proof.**  We prove the result for the case $n = 3$ only. This gives the main idea of the proof. Fix $(x_1, x_2, x_3) \in R_3$.

$$P[X_1 \leq x_1, X_2 \leq x_2, X_3 \leq x_3]$$
$$= E\{E[I(X_3 \leq x_3)I(X_2 \leq x_2)I(X_1 \leq x_1) \mid X_2, X_1]\}$$
$$= E\{I(X_2 \leq x_2)I(X_1 \leq x_1)E[I(X_3 \leq x_3) \mid X_2]\}$$
$$= E\{E[I(X_2 \leq x_2)I(X_1 \leq x_1)E(I(X_3 \leq x_3) \mid X_2) \mid X_1]\}$$
$$= E\{I(X_1 \leq x_1)E[I(X_2 \leq x_2)E(I(X_3 \leq x_3) \mid X_2) \mid X_1]\}.$$

Now by Lemma 3.6.2,

$$E(I(X_3 \leq x_3) \mid X_2) = \int_{-\infty}^{x_3} p(X_2, dy_3) \qquad \text{a.s.,}$$

$$E[I(X_2 \leq x_2)E(I(X_3 \leq x_3) \mid X_2) \mid X_1])$$
$$= \int_{-\infty}^{x_2} \int_{-\infty}^{x_3} p(y_2, dy_3)p(X_1, dy_2) \qquad \text{a.s.,}$$

and

$$E\{I(X_1 \le x_1)E[I(X_2 \le x_2)E(I(X_3 \le x_3) \mid X_2) \mid X_1]\}$$

$$= \int_{-\infty}^{x_1} \int_{-\infty}^{x_2} \int_{-\infty}^{x_3} p(y_2, dy_3)p(y_1, dy_2) \, dF_{X_1}(y_1) \qquad \text{a.s.} \quad \blacksquare$$

**Exercise 3.6.2.** Three people play a game. Label the people 1, 2, and 3. Let $X_i = j$ if person $j$ loses the $i$th round. Each round consists of two people competing. The winner then competes with the idle person in the next round while the loser sits out a round. Assume persons 1 and 2 compete in the first round. Let $r_{j|k} = P$ [$j$ wins a round given that $j$ and $k$ compete]. Let

$$r_{1|2} = \tfrac{3}{4}, \qquad r_{1|3} = \tfrac{1}{3}, \qquad r_{2|3} = \tfrac{1}{2}.$$

Is $\{X_i, i \ge 1\}$ a Markov sequence? Does it have stationary transition probabilities? What is its initial distribution? What is its transition function? What is the distribution of $X_2$? What is the joint distribution of $X_1$ and $X_2$? Let $Y_i = j$ if person $j$ wins the $i$th round. Is $\{Y_i, i \ge 1\}$ a Markov sequence? $\blacksquare$

Einstein's mathematical model for the physical phenomenon Brownian motion is one of the most important and most widely studied probability models in which the Markov assumption is made. Brownian motion results when particles suspended in a fluid undergo displacement resulting from molecular bombardment by the molecules of the fluid. Fix a particular direction in the fluid. Let $X_t$ be the displacement at time $t$ in the fixed direction of a particle undergoing Brownian motion. The usual model is obtained by considering the family of random variables $\{X_t, t \ge 0\}$. Here, in order to obtain another example of a Markov sequence we discretise the time axis (thereby destroying much of the mathematical interest in the model) and look at $\{X_i, i = 0, 1, \ldots\}$. If the fluid is fairly viscous (therefore rapidly damping out the velocity of a bombarded particle) it seems a reasonable approximation to reality to assume that $\{X_i, i = 0, 1, \ldots\}$ is a Markov sequence. If the physical process is homogeneous in time, it seems reasonable to assume stationary transition probabilities. If the fluid medium is spatially homogeneous, it seems reasonable to assume that $E[X_i \mid X_{i-1}] = X_{i-1}$ a.s. and $\text{var}(X_i \mid X_{i-1}) = \alpha$ for all $i \ge 1$. $X_i - X_{i-1}$ is the result of many independent bombardments in the time interval $(i - 1, i]$. Recalling the central limit theorem, it seems reasonable to assume that the distribution of $X_i - X_{i-1}$ is normal for each $i \ge 1$. Because of the high viscosity, it seems reasonable to assume that $X_i - X_{i-1}$ is independent of $X_{i-1}$ for each $i \geq 1$.

Condensing these conclusions into a definition, discrete time Brownian motion is taken to be Markov sequence $\{X_i, i \geq 0\}$ with transition probability $p(\cdot, \cdot)$ given by

$$p(x, A) = (1/(2\pi\alpha)^{1/2}) \int_A \exp[-(y - x)^2/(2\alpha)]\, dy$$

for all $x \in R_1$ and Borel sets $A$. Here $\alpha > 0$ is a physical constant. We assume that the particle has displacement 0 at time 0; that is, $P[X_0 = 0] = 1$, thus specifying the initial distribution.

**Exercise 3.6.3.**    Find the distribution of $X_2$. Find the distribution of $X_n$ for arbitrary $n \geq 1$. Find the joint distribution of $X_1, X_2$. Find the distribution of the sequence $\{X_i - X_{i-1}, i \geq 1\}$.  ∎

The study of continuous time Brownian motion will prove useful in the proof of the Hartman–Wintner law of the iterated logarithm in Chapter 5.

Another example of a Markov sequence is provided by the branching process described in Example 2.8.1. The Polya urn scheme described in Example 2.5.4 provides still another example. Birth and death processes and queuing processes provide two more important classes of Markov sequences (see Feller [1966, 1968a]). Indeed much of the field of applied probability is concerned with Markov sequences. This ends the introductory discussion of Markov sequences.

For our present purposes, it is important to know when a Markov sequence is a stationary sequence. The answer is pleasantly simple.

**Theorem 3.6.2.**    Let $\{X_i, i \geq 1\}$ be a Markov sequence. Then $\{X_i, i \geq 1\}$ is stationary if and only if $X_1$ and $X_2$ have the same distribution.

**Proof.**    If $\{X_i, i \geq 1\}$ is stationary, then trivially $X_1$ and $X_2$ have the same distribution.

To go the other way, assume $X_1$ and $X_2$ have the same distribution. It suffices to show that $(X_1, X_2, \ldots, X_n)$ and $(X_2, X_3, \ldots, X_{n+1})$ have the same distribution for each $n \geq 1$ since this implies that $(X_1, X_2, \ldots)$ and $(X_2, X_3, \ldots)$ have the same distribution which implies stationarity. Fix $n \geq 1$ and $(x_1, x_2, \ldots, x_n) \in R_n$. By Theorem 3.6.1,

$$P[X_1 \leq x_1, X_2 \leq x_2, \ldots, X_n \leq x_n]$$

$$= \int_{-\infty}^{x_1} \cdots \int_{-\infty}^{x_n} p(y_{n-1}, dy_n) \cdots p(y_1, dy_2)\, dP_{X_1}(y_1). \qquad (3.6.1)$$

$$P[X_2 \leq x_1, X_3 \leq x_2, \ldots, X_{n+1} \leq x_n]$$

$$= P[X_1 < \infty, X_2 \leq x_1, \ldots, X_{n+1} \leq x_n]$$

$$= \int_{-\infty}^{\infty} \int_{-\infty}^{x_1} \cdots \int_{-\infty}^{x_n} p(y_n, dy_{n+1}) \cdots p(y_1, dy_2) \, dP_{X_1}(y_1)$$

$$= \int_{-\infty}^{x_1} \cdots \int_{-\infty}^{x_n} p(y_n, dy_{n+1}) \cdots p(y_2, dy_3) \, dP_{X_2}(y_2)$$

since

$$dP_{X_2}(y_2) = \int_{-\infty}^{\infty} p(y_1, dy_2) \, dP_{X_1}(y_1).$$

Thus by Eq. (3.6.1) and the fact $P_{X_1} = P_{X_2}$,

$$P[X_2 \leq x_1, X_3 \leq x_2, \ldots, X_{n+1} \leq x_n]$$

$$= \int_{-\infty}^{x_1} \cdots \int_{-\infty}^{x_n} p(y_n, dy_{n+1}) \cdots p(y_2, dy_3) \, dP_{X_1}(y_2)$$

$$= P[X_1 \leq x_1, X_2 \leq x_2, \ldots, X_n \leq x_n]. \quad \blacksquare$$

Given a random variable $X$, $F_X$ denotes the distribution function of $X$.

**Corollary 3.6.1.**   Let $\{X_i, i \geq 1\}$ be a Markov sequence. Then $\{X_i, i \geq 1\}$ is stationary if and only if

$$F_{X_1}(x_1) = \int_{-\infty}^{\infty} \int_{-\infty}^{x_1} p(y_1, dy_2) \, dP_{X_1}(y_1) \qquad (3.6.2)$$

for all $x_1 \in R_1$.

*Proof.*   $X_1$ and $X_2$ have the same distribution if and only if $\{X_i, i \geq 1\}$ is stationary; that is, $F_{X_1}(x) = F_{X_2}(x)$ for all $x \in R_1$ if and only if $\{X_i, i \geq 1\}$ is stationary. But

$$F_{X_2}(x) = \int_{-\infty}^{\infty} \int_{-\infty}^{x} p(y_1, dy_2) \, dP_{X_1}(y_1)$$

for all $x \in R_1$.   $\blacksquare$

**Definition 3.6.6.**   Let $\{X_i, i \geq 1\}$ be a Markov sequence with transition probability $p(\cdot, \cdot)$. If $F$ is a distribution function satisfying

$$F(x) = \int_{-\infty}^{\infty} \int_{-\infty}^{x} p(y_1, dy_2) \, dF(y_1)$$

for all $x \in R_1$, $F$ is said to be a stationary initial distribution.   $\blacksquare$

Note that whether a stationary initial distribution exists or not depends solely on $p(\cdot, \cdot)$. If $X_1$ has a stationary initial distribution as its distribution the Markov sequence is stationary. Of course, a stationary initial distribution may exist and yet the Markov sequence be nonstationary because $X_1$ does not have a stationary initial distribution as its distribution. It turns out that some Markov sequences can be made stationary by changing the distribution of $X_1$, while others are nonstationary for every choice of distribution for $X_1$.

The following important application of the pointwise ergodic theorem is obvious.

**Theorem 3.6.3.** Let $\{X_i, i \geq 1\}$ be a Markov sequence with $X_1$ having a stationary initial distribution as its distribution. Let $f$ be a measurable function from $R_1$ to $R_1$ such that $E |f(X_1)| < \infty$. Then

$$\sum_{i=1}^{n} f(X_i)/n \qquad \text{converges a.s.}$$

**Proof.** $\{X_i, i \geq 1\}$ is stationary since $X_1$ has a stationary initial distribution. Thus $\{f(X_i), i \geq 1\}$ is stationary. The desired result follows by the pointwise ergodic theorem for stationary sequences. ∎

Theorem 3.6.3 leaves two important questions unanswered. Given a stationary Markov sequence $\{X_i, i \geq 1\}$, when can we conclude that $\sum_{i=1}^{n} X_i/n$ converges almost surely to a constant? That is, what conditions for a stationary Markov sequence imply that it is ergodic? Secondly, can we sometimes conclude that $\sum_{i=1}^{n} X_i/n$ converges almost surely when $\{X_i, i \geq 1\}$ is Markovian, but not necessarily stationary? Both of these questions hinge on a study of the existence of stationary initial distributions. The existence of stationary initial distributions is analyzed in Doob's text [1953, Chapter 5] where a rather general sufficient condition known as Doeblin's condition together with aperiodicity is shown to guarantee the existence of at least one stationary initial distribution.

**Definition 3.6.7.** (i) A Markov sequence $\{X_i, i \geq 1\}$ with state space $S$ is said to satisfy Doeblin's condition if there exists a probability measure $\mu$ defined on the Borel sets $\mathscr{B}_1(S)$ of the state space $S$, an integer $n \geq 1$, and a number $\varepsilon > 0$ such that

$$p^{(n)}(x, A) \leq 1 - \varepsilon$$

for all $x \in S$ and $A \in \mathscr{C}_1(S)$ such that $\mu(A) \leq \varepsilon$. Here $p^{(n)}(\cdot, \cdot)$ is a regular version of $P[X_{n+1} \in \cdot \mid X_1 = \cdot]$. $p^{(n)}(\cdot, \cdot)$ is called the $n$th order transition probability. ∎

There are a variety of situations where Doeblin's condition is satisfied. For example, suppose $S$ is finite. Without loss of generality, let $S = \{1, 2, \ldots, n\}$ for some positive integer $n$. Here all subsets of $S$ are Borel sets. Let $\{p_{ij}, 1 \leq i \leq n, 1 \leq j \leq n\}$ satisfy $\sum_{j=1}^{n} p_{ij} = 1$ for all $1 \leq i \leq n$ and $p_{ij} \geq 0$ for all $1 \leq i \leq n$, $1 \leq j \leq n$. Let $p(i, A) = \sum_{j \in A} p_{ij}$ define the transition probability $p(\cdot, \cdot)$ of a Markov sequence $\{X_i, i \geq 1\}$ defined on $S$. When a Markov sequence has a countable state space it is called a Markov chain. When, in addition, the state space is finite, the Markov sequence is called a finite dimensional Markov chain. Thus $\{X_i, i \geq 1\}$ is a finite dimensional Markov chain. Is Doeblin's condition satisfied? For $A \subset S$, let $\mu(A) = $ (number of integers in $A$)$/n$ define $\mu$. Let $\varepsilon = (n + 1)^{-1}$. Trivially, Doeblin's condition is satisfied.

**Example 3.6.1.** (i) Let $\{Y_i, i \geq 1\}$ be independent with $P[Y_i = 1] = P[Y_i = -1] = \frac{1}{3}$ and $P[Y_i = 0] = \frac{1}{3}$ for all $i \geq 1$. Let $T_n = \sum_{i=1}^{n} Y_i$. Note that $\{T_n, n \geq 1\}$ is Markovian with stationary transition probabilities. Is Doeblin's condition satisfied? Suppose $\mu$ is a probability measure defined on $S = \{0, \pm 1, \ldots\}$. Fix $\varepsilon > 0$. There clearly exists $A = \{n, n + 1, \ldots\}$ such that $\mu(A) \leq \varepsilon$.

$$p^{(1)}(n + 1, \{n, n + 1, n + 2\}) = 1,$$

$$p^{(2)}(n + 2, \{n, n + 1, n + 2, n + 3, n + 4\}) = 1, \ldots .$$

Thus Doeblin's condition fails. It is fairly easy to see that a stationary initial distribution does not exist.

(ii) Let $\{X_i, i \geq 1\}$ be a Markov sequence with state space $S = \{0, 1, \ldots\}$. Let $p_{ij} = P[X_2 = j \mid X_1 = i]$. Let $p_{0j} = 2^{-j}$ for $j \geq 1$, $p_{00} = 0$, and $p_{i0} = 1$ for $i \geq 1$ and $p_{ij} = 0$ otherwise. Is Doeblin's condition satisfied? Let $\mu(0) = 1$ and $\mu(i) = 0$ for $i \neq 0$ define $\mu$. Let $n = 1$. Let $\varepsilon = \frac{1}{4}$. Choose $A \subset S$ such that $\mu(A) \leq \varepsilon$. $0 \notin A$. Thus

$$p^{(1)}(x, A) \leq \tfrac{1}{2} \leq 1 - \tfrac{1}{4}$$

for all $x \in S$. Hence Doeblin's condition is satisfied. It is easy to see that a stationary initial distribution does exist. (Find it.) ∎

**Theorem 3.6.4.**   Let $\{X_i, i \geq 1\}$ be a Markov sequence satisfying Doeblin's condition. Then there exists at least one stationary initial distribution.

*Proof.*   Omitted (see Doob [1953, p. 214]). The class of all stationary initial distributions is characterized by Doob [1953] also. ∎

Theorem 3.6.4 obviously gives a method of establishing the existence of stationary initial distributions. For more information, see Doob [1953].

**Definition 3.6.8.**   Let $\{X_i, i \geq 1\}$ be a Markov sequence with state space $S$ and transition probability $p(\cdot, \cdot)$. Then $C \in \mathscr{C}_1(S)$ is Markov invariant if $p(x, C) = 1$ for all $x \in C$. If there do not exist two disjoint Markov invariant sets, $\{X_i, i \geq 1\}$ is said to be Markov ergodic or indecomposable. ∎

The idea of Markov ergodicity is that the state space of a Markov sequence cannot be decomposed into two subsets closed in the sense that the Markov sequence cannot ever leave such a subset after once entering it.

**Exercise 3.6.4.**   Let $\{X_1, i \geq 1\}$ be a Markov sequence with state space $S = \{1, 2, 3, 4\}$. Let $p_{ij} = P[X_2 = j \mid X_1 = i]$ be given by

$$[p_{ij}] = \begin{bmatrix} \frac{1}{3} & \frac{2}{3} & 0 & 0 \\ \frac{3}{4} & \frac{1}{4} & 0 & 0 \\ \frac{1}{4} & \frac{1}{4} & \frac{1}{4} & \frac{1}{4} \\ 0 & 0 & 0 & 1 \end{bmatrix}.$$

Find all Markov invariant sets. Is the sequence Markov ergodic? Find two stationary initial distributions. Are there more? For any Markov sequence with finite state space, what matrix property characterizes Markov ergodicity? ∎

We have defined two kinds of ergodicity, namely ergodicity for Markov sequences and ergodicity for stationary sequences. Do these two definitions coincide for a stationary Markov sequence? In order to answer this question, we need two preparatory lemmas. The first extends the notion of the Markov assumption.

**Lemma 3.6.3.** If $\{X_i, i \geq 1\}$ is a Markov sequence, then

$$P[(X_{n+1}, X_{n+2}, \ldots) \in C_\infty \mid X_n, X_{n-1}, \ldots, X_1]$$
$$= P[(X_{n+1}, X_{n+2}, \ldots) \in C_\infty \mid X_n]$$
$$= P[(X_2, X_3, \ldots) \in C_\infty \mid X_1] \qquad \text{a.s.}$$

for all $C_\infty \in \mathscr{C}_\infty$ and $n \geq 1$.

**Exercise 3.6.5.** Prove Lemma 3.6.3. Hint: Prove the result first for measurable finite dimensional product cylinder sets of $R_\infty$. ∎

The invariant events of a stationary sequence which is also Markovian have a particularly simple and useful representation.

**Lemma 3.6.4.** Let $\{X_i, i \geq 1\}$ be a stationary Markov process. Let $A$ be an almost invariant event of this stationary sequence. Then

$$A = [X_1 \in C_1]$$

for some $C_1 \in \mathscr{C}_1$. (Recall by our convention that $A = [X_1 \in C_1]$ means $P[A \bigtriangleup (X_1 \in C_1)] = 0$.)

**Remark.** Note the use of martingale theory in the proof.

**Proof.** By hypothesis,

$$A = [(X_n, X_{n+1}, \ldots) \in C_\infty]$$

for some $C_\infty \in \mathscr{C}_\infty$ and all $n \geq 1$. Let

$$Z_n = E[I(A) \mid X_1, X_2, \ldots, X_n].$$

By Theorem 2.8.6 (i),

$$Z_n \to E[I(A) \mid X_1, X_2, \ldots] \qquad \text{a.s.}$$

Fix $n \geq 1$. Since $A$ is a tail event, it follows by Lemma 3.6.3 that $Z_n = E[I(A) \mid X_n]$ a.s. Thus $E[I(A) \mid X_1, X_2, \ldots]$ is a tail random variable. Let $\mathscr{G}$ be the tail $\sigma$ field.

$$E[I(A) \mid X_1, X_2, \ldots] = E\{E[I(A) \mid X_1, X_2, \ldots] \mid \mathscr{G}\}$$
$$= E[I(A) \mid \mathscr{G}] \qquad \text{a.s.}$$

since $\mathscr{G} \subset \mathscr{B}\{X_i, i \geq 1\}$. But $A \in \mathscr{G}$ implies that $E[I(A) \mid \mathscr{G}] = I(A)$ a.s. Hence

$$E[I(A) \mid X_n] \to I(A) \qquad \text{a.s.} \tag{3.6.3}$$

Choose $g$ measurable such that $E[I(A) \mid X_1] = g(X_1)$. $E[I(A) \mid X_1] = P[(X_2, \ldots) \in C_\infty \mid X_1]$. By Lemma 3.6.3 and the almost sure invariance of $A$, $g(X_1) = E[I(A) \mid X_n] = E[I(A) \mid X_1]$ a.s. for each $n \geq 1$. Using Eq. (3.6.3), it thus follows that

$$P[A \mid X_n] = I(A) \qquad \text{a.s.}$$

for all $n \geq 1$. In particular

$$P[A \mid X_1] = I(A) \qquad \text{a.s.}$$

Thus there exists $[X_1 \in C_1] \in \mathscr{B}(X_1)$ such that

$$P[(X_1 \in C_1) \triangle A] = 0;$$

that is, $[X_1 \in C_1] = A$ as desired. ∎

The following corollary helps explain the apparently surprising conclusion of Lemma 3.6.4.

**Corollary 3.6.2.** Let $\{X_i, i \geq 1\}$ be a stationary Markov sequence. Then $A$ is an almost invariant event of this stationary sequence if and only if

$$A = \{X_i \in C_1 \text{ for all } i \geq 1\}$$

for some $C_1 \in \mathscr{C}_1$.

**Proof.** Suppose $A = \{X_i \in C_1 \text{ for all } i \geq 1\}$. $A \subset \{X_i \in C_1 \text{ for all } i \geq 2\}$. Thus, by Lemma 3.5.7 (ii), $A$ is almost invariant.

Suppose, on the other hand, that $A$ is almost invariant. By Lemma 3.6.4, $A = [X_1 \in C_1]$ for some $C_1 \in \mathscr{C}_1$. By definition of almost invariance,

$$A = [X_i \in C_1] \qquad \text{for each } i \geq 1.$$

Hence $A = [X_i \in C_1 \text{ for all } i \geq 1]$. ∎

**Theorem. 3.6.5**   Let $\{X_i, i \geq 1\}$ be a stationary Markov sequence.

(i)   If $\{X_i, i \geq 1\}$ is Markov ergodic, it is ergodic as a stationary sequence.

(ii)   If there exists a Markov invariant set $C_1$ such that $0 < P[X_1 \in C_1] < 1$, then the sequence is not ergodic as a stationary sequence.

**Remark.**   Let $\{X_i, i \geq 1\}$ be stationary Markov. $\{X_i, i \geq 1\}$ not Markov ergodic implies that there exist two disjoint Markov invariant sets $C_1$ and $C_2$. However it does not follow that $1 > P[X_1 \in C_1] > 0$ or $1 > P[X_1 \in C_2] > 0$. Hence it does not follow by Theorem 3.6.5 (ii) that ergodicity implies Markov ergodicity for stationary sequences.

**Exercise 3.6.6.**   Construct a simple example where $\{X_i, i \geq 1\}$ is a stationary Markov sequence which is not Markov ergodic but is ergodic as a stationary sequence.   ∎

**Proof of Theorem 3.6.5.**   (i)   Assume $\{X_i, i \geq 1\}$ is Markov ergodic. Let $S$ be the state space. Let $A$ be an invariant event of the stationary sequence $\{X_i, i \geq 1\}$. By Lemma 3.6.4, $A = [X_1 \in C]$ for some $C \in \mathscr{C}_1$. We wish to show that

$$P[X_1 \in C] = 0 \qquad \text{or} \qquad P[X_1 \in C] = 1.$$

If we could show that $C$ and its complement both assumed to be nonempty implies that they are both Markov invariant, this would complete the proof. For, by the hypothesis of Markov ergodicity, there cannot exist two disjoint Markov invariant sets. Hence either $C = S$ or $C = \varnothing$, implying that either $P[X_1 \in C] = 1$ or $P[X_1 \in C] = 0$.

$$[X_1 \in C] = [X_2 \in C]$$

by the invariance of $A$. Thus

$$P_{X_1}(C) = P[X_1 \in C, X_2 \in C] = \int_C p(x, C) \, dP_{X_1}(x). \qquad (3.6.4)$$

Hence,

$$P_{X_1}[C \cap (x \mid p(x, C) \neq 1)] = 0. \qquad (3.6.5)$$

Thus $C$ is "almost" Markov invariant. To produce a Markov invariant set we remove a subset of $P_{X_1}$ measure 0 from $C$, thereby creating a Markov invariant set. Let $C_0$ be $C$ and $C_i$ be $C_{i-1} \cap [x \mid p(x, C_{i-1}) = 1]$ for $i \geq 1$. Let $C_\infty$ be $\bigcap_{i=0}^\infty C_i$. $x \in C_\infty$ implies $p(x, C_i) = 1$ for $i \geq 0$. Hence, since

$p(x, \cdot)$ is a probability, $\lim_i p(x, C_i) = p(x, C_\infty)$ for all $x \in C_\infty$. Thus, $C_\infty$, if nonempty, is Markov invariant. Using Eq. (3.6.5), $P_{X_1}(C_0 \triangle C_1) = 0$. Hence $[X_1 \in C_0] = [X_1 \in C_1]$. Thus, using Eq. (3.6.4) with $C_1$ substituted for $C$, it follows that Eq. (3.6.5) holds with $C_1$ substituted for $C$. Thus by induction, $[X_1 \in C_k] = [X_1 \in C]$ for all $k \geq 1$. It follows that $P_{X_1}(C) = P_{X_1}(C_\infty)$.

Let $D$ be the complement of $C$. $[X_1 \in D] = [X_2 \in D]$. Thus $P_{X_1}[D \cap (x \mid p(D, x) \neq 1)] = 0$. Let $D_0$ be $D$ and $D_i$ be $D_{i-1} \cap [x \mid p(D_{i-1}, x) = 1]$ for $i \geq 1$. Let $D_\infty$ be $\bigcap_{i=0}^{\infty} D_i$. As above, $D_\infty$, if nonempty, is Markov invariant and

$$P_{X_1}(D) = P_{X_1}(D_\infty).$$

$C_\infty$ and $D_\infty$ are disjoint and both are Markov invariant if both are nonempty. But by hypothesis, $\{X_i, i \geq 1\}$ is Markov ergodic. Thus $C_\infty = S$ or $C_\infty = \emptyset$. Thus

$$1 = P_{X_1}(S) = P_{X_1}(C_\infty) = P_{X_1}(C)$$

or

$$0 = P_{X_1}(\emptyset) = P_{X_1}(C_\infty) = P_{X_1}(C),$$

as desired.

(ii) Assume there exists a Markov invariant set $C_1$ such that $0 < P[X_1 \in C_1] < 1$.

$$P[(X_1 \in C_1) \cap (X_2 \notin C_1)] = 0.$$

By Lemma 3.5.7 (ii) $[X_i \in C_1]$ is an almost invariant event. Thus $\{X_i, i \geq 1\}$ is not ergodic as a stationary sequence. ∎

**Theorem 3.6.6.**   (Strong law of large numbers for Markov sequences). If $\{X_i, i \geq 1\}$ is a stationary Markov sequence which is Markov ergodic with $E \mid X_1 \mid < \infty$ then $\sum_{i=1}^{n} X_i/n \to EX_1$ a.s.

*Proof.*   This follows immediately from the pointwise ergodic theorem and Theorem 3.6.5. ∎

*Exercise 3.6.7.*   Construct a stationary Markov sequence with $\sum_{i=1}^{n} X_i/n$ converging almost surely, but to a nonconstant random variable. ∎

For completeness, we mention the following theorem which treats the case of a nonstationary Markov sequence.

**Theorem 3.6.7.** Let $\{X_i, i \geq 1\}$ be a Markov sequence having at least one stationary initial distribution $F$. Let $f$ be a Borel function $f: R_1 \rightarrow R_1$ such that

$$\int_S |f(x)| \, dG(x) < \infty$$

for all stationary initial distributions $G$. Then

(i) $\sum_{i=1}^{n} f(X_i)/n$ converges a.s.

(ii) If, in addition, $\{X_i, i \geq 1\}$ is Markov ergodic, then $F$ is the only stationary initial distribution and

$$\sum_{i=1}^{n} f(X_i)/n \rightarrow \int_S f(x) \, dF(x) \qquad \text{a.s.}$$

**Proof.** The proof is very involved and is omitted (see Doob [1953, p. 220]). ∎

## 3.7. Stability Results for $\{S_n, n \geq 1\}$ Assuming Only Moment Restrictions

In Section 2.4, certain moment conditions on the $S_n$ without any dependence assumptions on $\{X_i, i \geq 1\}$ were shown to imply the almost sure convergence of $S_n$. In Section 3.7 an analogous study of the stability of $S_n$ is undertaken.

For each $a \geq 0$ and $n \geq 1$, let $F_{a,n}$ be the joint distribution function of $X_{a+1}, X_{a+2}, \ldots, X_{a+n}$ and let $S_{a,n} = \sum_{i=a+1}^{a+n} X_i$. We work with nonnegative functionals defined on the collection of joint distribution functions. The basic assumption will always be that

$$E |S_{a,n}|^v \leq g^{v/2}(F_{a,n}) \tag{3.7.1}$$

for all $a \geq 0$, $n \geq 1$, some $v \geq 2$, and some functional $g$. Providing $g$ satisfies certain mild requirements, Eq. (3.7.1) implies two maximal inequalities, one holding for $v = 2$ and the second holding for $v > 2$. These maximal inequalities, without any additional assumptions, are shown to imply interesting stability results.

To then apply these results to specific classes of stochastic sequences $\{X_i, i \geq 1\}$, it is only necessary to find dependence assumptions and moment conditions on the $X_i$ which together imply Eq. (3.7.1) for reasonable choices of $g$ and $v$.

It is interesting to note that an inequality of the form of Eq. (3.7.1) always holds, namely

$$E \mid S_{a,n} \mid^{\nu} \leq \left[ \sum_{i=a+1}^{a+n} (E \mid X_i \mid^{\nu})^{1/\nu} \right]^{\nu}$$

for all $a \geq 0$, $n \geq 1$, and $\nu \geq 1$.

This follows by the Minkowski inequality for integrals. If $E \mid X_i \mid^{\nu} \leq K$ for all $i \geq 1$, some $\nu \geq 2$, and some $K < \infty$, then

$$E \mid S_{a,n} \mid^{\nu} \leq Kn^{\nu}$$

for all $a \geq 0$ and $n \geq 1$. Thus $E \mid X_i \mid^{\nu} \leq K$ for all $i \geq 1$, some $\nu \geq 2$, and some $K < \infty$ implies that Eq. (3.7.1) holds with

$$g(F_{a,n}) = K^{2/\nu} n^2.$$

Dependence assumptions on $\{X_i, i \geq 1\}$ of course improve this inequality. For example, $EX_i^2 \leq K < \infty$ for all $i \geq 1$ and $\{X_i, i \geq 1\}$ orthogonal imply $ES_{a,n}^2 \leq Kn$ for all $a \geq 0$, $n \geq 1$; that is, $g(F_{a,n}) = Kn$ for $\nu = 2$. We shall see that Eq. (3.7.1) often holds for some $\nu \geq 2$ and $g(F_{a,n}) = Kn$ and that this particular inequality produces interesting stability results.

Let

$$M_{a,n} = \max_{1 \leq k \leq n} \left| \sum_{i=a+1}^{a+k} X_i \right|$$

for each $a \geq 0$ and $n \geq 1$. We already have a maximal inequality from Chapter 2, namely:

**Theorem 2.4.1.**   Suppose $g$ is a functional defined on the joint distribution functions of $\{X_i, i \geq 1\}$ such that

$$g(F_{a,k}) + g(F_{a+k,m}) \leq g(F_{a,k+m}) \tag{3.7.2}$$

for all $1 \leq k < k + m$ and $a \geq 0$ and

$$ES_{a,n}^2 \leq g(F_{a,n}) \tag{3.7.3}$$

for all $n \geq 1$ and $a \geq 0$.

Then

$$EM_{a,n}^2 \leq (\log 2n/\log 2)^2 g(F_{a,n}) \tag{3.7.4}$$

for all $n \geq 1$ and $a \geq 0$.  ∎

Theorem 2.4.1 was used to establish Theorem 2.4.2 and Corollary 2.4.1, these being convergence results for $S_n$. We can of course convert Theorem 2.4.1 and Corollary 2.4.1 to stability results by means of the Kronecker lemma.

**Theorem 3.7.1.** Let $\{a_n, n \geq 1\}$ be constants such that $a_n \uparrow \infty$ as $n \to \infty$. Suppose Eq. (3.7.2) holds. Suppose

$$E\left(\sum_{i=a+1}^{a+n} X_i/a_i\right)^2 \leq g(F_{a,n})$$

for all $n \geq 1$ and $a \geq 0$. Suppose, in addition, that

$$h(F_{a,k}) + h(F_{a+k,m}) \leq h(F_{a,k+m})$$

for all $1 \leq k < k + m$ and $a \geq 0$,

$$h(F_{a,n}) \leq K < \infty$$

for all $n \geq 1$ and $a \geq 0$, and

$$g(F_{a,n}) \leq Kh(F_{a,n})/\log^2 (a + 1)$$

for all $n \geq 1$ and $a > 0$. Then

$$S_n/a_n \to 0 \qquad \text{a.s.}$$

**Proof.** Immediate from the Kronecker lemma and Theorem 2.4.2. ∎

**Exercise 3.7.1.** Let $\{X_i, i \geq 1\}$ satisfy $EX_i^2 \leq K < \infty$ for all $i \geq 1$ and some $K < \infty$. Prove that

$$S_n/(n \log^{2+\varepsilon} n) \to 0 \qquad \text{a.s.}$$

for each $\varepsilon > 0$. Hint: Use Theorem 3.7.1. ∎

**Theorem 3.7.2.** Let $\{a_i, i \geq 1\}$ be constants such that $a_n \uparrow \infty$ as $n \to \infty$. Suppose there exist constants $\{\varrho_i, i \geq 0\}$ such that $0 \leq \varrho_i \leq 1$ and

$$EX_iX_j \leq \varrho_{j-i}(EX_i^2 EX_j^2)^{1/2}$$

for all $j \geq i > 0$,

$$\sum_{i=0}^{\infty} \varrho_i < \infty,$$

and

$$\sum_{i=1}^{\infty} \log^2 i \, EX_i^2/a_i^2 < \infty.$$

Then

$$\sum_{i=1}^{n} X_i/a_n \to 0 \qquad \text{a.s.}$$

**Proof.**  Immediate from Corollary 2.4.1 and the Kronecker lemma.  ∎

**Example 3.7.1.**  We apply Theorem 3.7.2 to an example in queuing theory. Let $\{N(t), t \geq 0\}$ be a Poisson process with $EN(t) = vt$ for all $t \geq 0$. $\{N(t), t \geq 0\}$ governs the arrivals to the queue. Arrivals begin being served immediately (an infinite server queue). Let $\{Y_i, i \geq 1\}$ be a sequence of independent identically distributed exponential random variables with mean $1/\mu$. Suppose $\{Y_i, i \geq 1\}$ is independent of $\{N(t), t > 0\}$. $Y_i$ is the amount of time to service the $i$th arrival. For each $t \geq 0$, let $X(t)$ be the number of people in the queue at time $t$. For each $i \geq 1$ let $\tau_i$ be the time of the $i$th arrival, that is $\tau_i$ is the instant of time $t$ such that $N(t^-) = i - 1$, $N(t) = i$. Let

$$w_0(s, y) = \begin{cases} 1 & \text{if} \quad 0 \leq s \leq y \\ 0 & \text{if} \quad s < 0 \text{ or } s > y. \end{cases}$$

Then $X(t)$ can be represented

$$X(t) = \sum_{n=1}^{N(t)} w_0(t - \tau_n, Y_n) \qquad \text{for each} \quad t \geq 0.$$

Because of this representation it is fairly easy to show that for each $t \geq 0$, $X(t)$ is a Poisson random variable with mean $(v/\mu)(1 - e^{-\mu t})$ and that $X(s)$ and $X(t)$ for each $0 \leq s < t$ has covariance

$$(v/\mu)(e^{-\mu(t-s)} - e^{-\mu t}).$$

Let $X_i = X(i) - EX(i)$ for integers $i \geq 1$ and consider $\{X_i, i \geq 1\}$. $EX_iX_j = (v/\mu)(e^{-\mu(j-i)} - e^{-\mu j})$ for all $j > i \geq 1$. $EX_i^2 = (v/\mu)(1 - e^{-\mu i})$ for all $i \geq 1$. Elementary computations show that to apply Theorem 3.7.2 we may take $\varrho_{j-i} = Ce^{-\mu(j-i)}$ for some $C < \infty$, and all $j \geq i$. Then $\sum_{i=0}^{\infty} \varrho_i < \infty$ as required. Let $a_i = i^{1/2} \log^2 i$ for all $i \geq 2$ and $a_1 = 1$. Then

$$\sum_{i=1}^{\infty} \log^2 i EX_i^2/a_i^2 \leq (v/\mu) \sum_{i=1}^{\infty} \log^2 i/(i \log^4 i) < \infty$$

as required. Thus, by Theorem 3.7.2,

$$\sum_{i=1}^{n} X_i/(n^{1/2} \log^2 n) \to 0 \qquad \text{a.s.}$$

In particular

$$\sum_{i=1}^{n} [X(i) - EX(i)]/n \to 0 \qquad \text{a.s.}$$

$EX(i) \to v/\mu$ as $n \to \infty$. Hence

$$\sum_{i=1}^{n} X(i)/n \to v/\mu \qquad \text{a.s. as} \quad n \to \infty.$$

That is, the long time average of the number of individuals in the queue approaches $v/\mu$. ∎

Theorem 2.4.1 can be used to derive a strong law of large numbers:

**Theorem 3.7.3.** [Serfling, 1970b]. Suppose $g$ is a functional defined on the joint distribution functions of $\{X_i, i \geq 1\}$ such that

$$g(F_{a,k}) + g(F_{a+k,m}) \leq g(F_{a,k+m}) \tag{3.7.5}$$

for all $1 \leq k < k + m$ and $a \geq 0$,

$$ES_{a,n}^2 \leq g(F_{a,n}) \tag{3.7.6}$$

for all $u \geq 1$ and $a \geq 0$.
Suppose in addition that

$$g(F_{a,n}) \leq Kn^2 (\log n \log_2 n)^{-2} \tag{3.7.7}$$

for all $a \geq 0$, $n \geq 1$. Then

$$S_n/n \to 0 \qquad \text{a.s.}$$

**Remark.** In order to get some feel for Eq. (3.7.7) recall that $EX_i^2 \leq K$ for all $i \geq 1$ and some $K < \infty$ implies

$$ES_{a,n}^2 \leq Kn^2$$

for all $n \geq 1$. Hence uniformly bounded second moments almost implies Eq. (3.7.7). Also note the connection with Exercise 3.7.1.

***Proof of Theorem 3.7.3.*** We proceed by the method of sub-sequences. Throughout, $K$ will denote a positive constant, not necessarily the same constant at each appearance. Fix $n \geq 1$ and $\varepsilon > 0$. By the Chebyshev inequality,

$$P[|\,S_n\,| > \varepsilon n] \leq g(F_{a,n})/(\varepsilon n)^2$$
$$\leq K(\log n \log_2 n)^{-2}.$$

Letting $n_k = [\exp k^{1/2}]$ define an integer subsequence $\{n_k, k \geq 1\}$ we obtain

$$P[|\,S_{n_k}\,| > \varepsilon n_k] \leq K(k \log^2 k)^{-1}$$

for all $k \geq 1$. Thus, using the Borel–Cantelli lemma,

$$S_{n_k}/n_k \to 0 \qquad \text{a.s.}$$

as $k \to \infty$. For ease of notation, let

$$\zeta_k = M_{n_k, n_{k+1}-n_k} \qquad \text{for} \quad k \geq 1.$$

Fix $k \geq 1$. By Theorem 2.4.1,

$$E(\zeta_k{}^2) \leq K[\log 2(n_{k+1} - n_k)]^2 g(F_{n_k, n_{k+1}-n_k}).$$

By Eq. (3.7.7)

$$E(\zeta_k{}^2) \leq K(n_{k+1} - n_k)^2/[\log_2 (n_{k+1} - n_k)]^2.$$

Thus

$$P[\zeta_k > \varepsilon n_k] \leq K\left(\frac{n_{k+1} - n_k}{n_k}\right)^2 [\log_2 (n_{k+1} - n_k)]^2.$$

Elementary analysis shows that

$$\left(\frac{n_{k+1} - n_k}{n_k}\right)^2 [\log_2(n_{k+1} - n_k)]^2 \leq K(k \log^2 k)^{-1}.$$

Thus

$$\zeta_k/k \to 0 \qquad \text{a.s.}$$

by the Borel–Cantelli lemma. For $n_k < n \leq n_{k+1}$,

$$|\,S_n\,|/n \leq |\,S_{n_k}\,|/n_k + \zeta_k/n_k.$$

Thus $S_n/n \to 0$ a.s. as $n \to \infty$, establishing the desired result. ∎

Clearly stability results analogous to Theorem 3.7.3 could be stated for norming constants other than $a_n = n$. Let $a_n = n$ in the statement of Theorem 3.7.2. This result is not a corollary of Theorem 3.7.3. However, results similar to this result do follow from Theorem 3.7.3, as is shown below.

**Definition 3.7.1.** $\{X_i, i \geq 1\}$ is said to be weakly stationary if

$$EX_i^2 = EX_j^2 < \infty \qquad \text{and} \qquad E[X_1 X_{1+j}] = E[X_i X_{i+j}]$$

for all $i \geq 1$ and $j \geq 1$. ▮

The name "weak stationarity" suggests that stationarity should imply weak stationarity. This is true when second moments are finite. Weak stationarity is not enough to imply that $S_n/n \to 0$ a.s., but it is almost enough.

**Theorem 3.7.4.** Let $\{X_i, i \geq 1\}$ be weakly stationary with $EX_iX_{i+j} = r_j$ for all $i \geq 1$, $j \geq 0$. If

$$\sum_{j=1}^{n} | r_j | \leq Kn(\log n \log_2 n)^{-2} \tag{3.7.8}$$

for each $n \geq 1$ and some $K < \infty$, then $S_n/n \to 0$ a.s.

***Remark.*** Since $E | X_iX_j | \leq EX_1^2 \equiv r_0^2$ for all $i \geq 1$, $j \geq 1$ for a weakly stationary sequence, it is always true that

$$\sum_{j=1}^{n} | r_j | \leq Kn$$

for each $n \geq 1$. Thus Eq. (3.7.8) is a rather mild restriction.

***Proof of Theorem 3.7.4.*** We apply Theorem 3.7.3. Let

$$g(F_{a,n}) = 2n \sum_{i=0}^{n} | r_i |$$

for $a \geq 0$ and $n \geq 1$. By hypothesis, Eq. (3.7.7) holds.

$$\begin{aligned} ES_{a,n}^2 &= \sum_{i=a+1}^{a+n} EX_i^2 + 2 \sum_{i=a+1}^{a+n-1} \sum_{j=i+1}^{a+n} EX_iX_j \\ &\leq nEX_1^2 + 2n \sum_{i=1}^{n} | r_i | \leq 2n \sum_{i=0}^{n} | r_i |. \end{aligned}$$

Thus, Eq. (3.7.6) holds.

$$g(F_{a,k}) + g(F_{a+k,m}) = 2k \sum_{i=0}^{k} |r_i| + 2m \sum_{i=0}^{m} |r_i|$$
$$\leq 2(k+m) \sum_{i=0}^{k+m} |r_i| = g(F_{a+k,m})$$

for $k \geq 0$ and $m \geq 1$. Thus Eq. (3.7.5) holds. Hence $S_n/n \to 0$ a.s. by Theorem 3.7.3.  ∎

Recall that $m$-dependent random variables were mentioned briefly in Section 3.3 as an example of $*$-mixing random variables. $\{X_i, i \geq 1\}$ is an $m$-dependent sequence if $\{X_1, \ldots, X_n\}$ and $\{X_{n+j+1}, X_{n+j+2}, \ldots\}$ are independent classes of random variables for each $n \geq 1$ and $j \geq m$. (That is, $A \in \mathscr{B}(X_i, i \leq n)$ and $B \in \mathscr{B}(X_i, i \geq m+n+1)$ are independent for each $n \geq 1$.)

**Definition 3.7.2.**  $\{X_i, i \geq 1\}$ is said to be a pairwise $m$-dependent sequence if $X_i$ and $X_j$ are independent for all $i$ and $j$ such that $|i - j| \geq m+1$.  ∎

**Corollary 3.7.1.**  Let $\{X_i, i \geq 1\}$ be weakly stationary and pairwise $m$-dependent for some $m < \infty$. Let $EX_i = 0$ for $i \geq 1$. Then

$$S_n/n \to 0 \qquad \text{a.s.}$$

**Proof.**   Immediate from Theorem 3.7.2 or from Theorem 3.7.4.  ∎

Serfling [1970b] also investigates the role of moments higher than two in the study of stability. The following maximal inequality is central to this investigation.

**Theorem 3.7.5.**   Let $g$ be a nondecreasing function with

$$2g(n) \leq g(2n) \qquad\qquad (3.7.9)$$

for all $n \geq 1$ and $g(n)/g(n+1) \to 1$ as $n \to \infty$. Suppose

$$E|S_{a,n}|^\nu \leq g^{\nu/2}(n) \qquad\qquad (3.7.10)$$

for all $a \geq 0$, $n \geq 1$, and some $\nu > 2$. Then

$$EM_{a,n}^\nu \leq K_\nu g^{\nu/2}(n) \qquad\qquad (3.7.11)$$

for all $a \geq 0$, $n \geq 1$, and some $K_\nu < \infty$.

**Remark.** It is important to note that the bound on $EM_{a,n}^\nu$ is of the same order of magnitude as the bound on $E\mid S_{a,n}\mid^\nu$. This contrasts sharply with Theorem 2.4.1.

**Proof of Theorem 3.7.5.** Choose $\delta$ such that $2/\nu < \delta < 1$. Choose $n_0$ such that

$$n \geq n_0 \quad \text{implies} \quad g(n)/g(n-1) \leq 2^{1-\delta}.$$

Let

$$K_0 = \max_{n \leq n_0} \sup_{a \geq 0} \frac{EM_{a,n}^\nu}{g^{\nu/2}(n)}.$$

For fixed $n$,

$$EM_{a,n}^\nu \leq E(\mid S_{a,1}\mid^\nu + \mid S_{a,2}\mid^\nu + \cdots + \mid S_{a,n}\mid^\nu) \leq ng^{\nu/2}(n) < \infty$$

by Eq. (3.7.10) and the fact that $g$ is nondecreasing. Thus $K_0 < \infty$. Fix $K_1$ to be determined later. Let $K_\nu = \max(K_0, K_1)$. We have established that

$$E\mid M_{a,n}\mid^\nu \leq K_\nu g^{\nu/2}(n)$$

for all $n \leq n_0$ and $a \geq 0$. An induction completes the proof. Fix $N > n_0$. Assume Eq. (3.7.11) for all $n < N$ and all $a \geq 0$. We show (with $K_1$ chosen independently of $N$) that Eq. (3.7.11) holds for $n \leq N$ and $a \geq 0$. Let $m = [(N+1)/2]$. For any $1 \leq j \leq k$,

$$\mid S_{a,j}\mid^\nu \leq \mid M_{a,k}\mid^\nu. \tag{3.7.12}$$

Choose an integer $k$ and $0 < \varepsilon \leq 1$ such that $\nu = k + \varepsilon$. For $m < n \leq N$,

$$\mid S_{a,n}\mid^\nu = \mid S_{a,m} + S_{a+m,n-m}\mid^\nu$$
$$\leq \mid S_{a,m} + S_{a+m,n-m}\mid^k(\mid S_{a,m}\mid^\varepsilon + \mid S_{a+m,n-m}\mid^\varepsilon).$$

Using the binomial expansion and (3.7.12), for $m < n \leq N$,

$$\mid S_{a,n}\mid^\nu \leq \left[\sum_{j=0}^k \binom{k}{j}\mid S_{a,m}\mid^j M_{a+m,N-m}^{k-j}\right](\mid S_{a,m}\mid^\varepsilon + M_{a+m,N-m}^\varepsilon)$$
$$\leq \sum_{j=0}^{k-1}\binom{k}{j}\mid S_{a,m}\mid^{j+\varepsilon} M_{a+m,N-m}^{k-j} + \mid S_{a,m}\mid^\nu + M_{a+m,N-m}^\nu$$
$$+ \sum_{j=1}^k \binom{k}{j}\mid S_{a,m}\mid^j M_{a+m,N-m}^{k-j+\varepsilon}.$$

Recalling Eq. (3.7.12),

$$| S_{a,n} |^{\nu} \leq \sum_{j=0}^{k-1} \binom{k}{j} | S_{a,m} |^{j+\varepsilon} M_{a+m,N-m}^{k-j} + M_{a,m}^{\nu} + M_{a+m,N-m}^{\nu}$$

$$+ \sum_{j=1}^{k} \binom{k}{j} | S_{a,m} |^{j} M_{a+m,N-m}^{k-j+\varepsilon}. \qquad (3.7.13)$$

Note that each of the terms of the two summations consists of the products of two factors whose exponents add to $\nu$. With this in mind, consider $r > 0$ and $s > 0$ such that $r + s = \nu$. By the Holder inequality

$$E | S_{a,m} |^{r} M_{a+m,N-m}^{s} \leq (E | S_{a,m} |^{r+s})^{r/(r+s)} (E M_{a+m,N-m}^{r+s})^{s/(r+s)}.$$

Hence, by Eq. (3.7.10) and the induction hypothesis,

$$E | S_{a,m} |^{r} M_{a+m,N-m}^{s} \leq g^{r/2}(m)[K_{\nu}^{s/\nu} g^{s/2}(m)]$$
$$= K_{\nu}^{s/\nu} g^{\nu/2}(m).$$

Combining this with Eq. (3.7.13) and again using the induction hypothesis yields

$$E | M_{a,N} |^{\nu} \leq 2K_{\nu} g^{\nu/2}(m) + g^{\nu/2}(m) \left[ \sum_{j=0}^{k-1} \binom{k}{j} K_{\nu}^{(k-j)/\nu} + \sum_{j=1}^{k} \binom{k}{j} K_{\nu}^{(k-j+\varepsilon)/\nu} \right]$$

$$= K_{\nu} g^{\nu/2}(m) \left[ 2 + \sum_{j=0}^{k-1} \binom{k}{j} K_{\nu}^{-(j+\varepsilon)/\nu} + \sum_{j=1}^{k} \binom{k}{j} K_{\nu}^{-j/\nu} \right]$$

$$\equiv K_{\nu} g^{\nu/2}(m)[2 + f(K_{\nu})].$$

Note that $f(K) \to 0$ as $K \to \infty$. Recalling that $\nu\delta > 2$, choose $K_1$ such that

$$2 + f(K) \leq 2^{\nu\delta/2}$$

for all $K \geq K_1$. Note $K_1$ has been chosen independent of $N$. Recalling that $K_{\nu} = \max(K_0, K_1)$, we have

$$EM_{a,N}^{\nu} \leq K_{\nu} g^{\nu/2}(m) 2^{\nu\delta/2}$$
$$\leq K_{\nu} 2^{\nu(\delta-1)/2} g^{\nu/2}(2m)$$

by Eq. (3.7.9).

$$K_{\nu} 2^{\nu(\delta-1)/2} g^{\nu/2}(2m) = K_{\nu} 2^{\nu(\delta-1)/2} \frac{g^{\nu/2}(2m)}{g^{\nu/2}(2m-1)} g^{\nu/2}(2m-1)$$

$$\leq K_{\nu} 2^{\nu(\delta-1)/2} 2^{(1-\delta)\nu/2} g^{\nu/2}(N)$$

$$= K_{\nu} g^{\nu/2}(N)$$

since $2m \geq n_0$, $2m - 1 \leq N$, and $g$ increasing.  ∎

Billingsley [1968, p. 94] establishes a result in the spirit of Theorem 3.7.5. He translates bounds on $P[|S_{a,n}| > \lambda]$ into bounds on $P[\max_{k \leq n} |S_{a,k}| > \lambda]$, thus avoiding consideration of moments. Many of the convergence results of this section could have been derived using Billingsley's result. It is an easy matter to obtain stability results from Theorem 3.7.5.

**Theorem 3.7.6.** Let $g$ be a nondecreasing function with

$$2g(n) \leq g(2n) \quad \text{for all} \quad n \geq 1 \quad \text{and} \quad g(n)/g(n+1) \to 1$$

as $n \to \infty$. Suppose

$$E \mid S_{a,n} \mid^\nu \leq g^{\nu/2}(n)$$

for all $a \geq 0$, $n \geq 1$, and some $\nu > 2$. Then

$$\mid S_n \mid /[g^{1/2}(n)(\log n)^{1/\nu}(\log_2 n)^{2/\nu}] \to 0 \qquad \text{a.s.}$$

**Proof.** We employ the method of subsequences. Let

$$a(n) = g^{1/2}(n)(\log n)^{1/\nu}(\log_2 n)^{2/\nu}.$$

Fix $\varepsilon > 0$.

$$P[|S_{2^k}| > \varepsilon a(2^k)] \leq E \mid S_{2^k} \mid^\nu \varepsilon^{-\nu} a^{-\nu}(2^k)$$
$$\leq \varepsilon^{-\nu}(\log 2^k)^{-1}(\log_2 2^k)^{-2}$$

for each $k \geq 1$ since $E \mid S_{a,n} \mid^\nu \leq g^{\nu/2}(n)$ for all $n \geq 1$ by hypothesis. Thus

$$\sum_{k=1}^{\infty} P[|S_{2^k}| > \varepsilon a(2^k)] < \infty.$$

Hence

$$S_{2^k}/a(2^k) \to 0 \qquad \text{a.s.}$$

as $k \to \infty$. By Theorem 3.7.5,

$$P[|M_{2^k,2^k}| > \varepsilon a(2^k)] \leq EM_{2^k,2^k}^\nu \varepsilon^{-\nu} a^{-\nu}(2^k)$$
$$\leq K_\nu \varepsilon^{-\nu}(\log 2^k)^{-1}(\log_2 2^k)^{-2}$$

for all $k \geq 1$. Hence

$$M_{2^k,2^k}/a(2^k) \to 0 \qquad \text{a.s.}$$

Fix $n \geq 1$ and let $k$ satisfy $2^k < n \leq 2^{k+1}$.

$$| S_n | / a(n) \leq | S_n - S_{2^k} | / a(2^k) + | S_{2^k} | / a(2^k)$$
$$\leq M_{2^k, 2^k} / a(2^k) + | S_{2^k} | / a(2^k).$$

Thus

$$S_n / a(n) \to 0 \qquad \text{a.s.}$$

as $n \to \infty$. ∎

Serfling [1970b] proved a slightly weaker version of Theorem 3.7.6, namely, that $\limsup | S_n | / a(n) \leq K$ a.s. for some $K < \infty$. One should note that the conclusion of Theorem 3.7.6 improves as $v \to \infty$.

**Corollary 3.7.2.** (Strong law of large numbers). Let $g$ be a nondecreasing function with

$$2g(n) \leq g(2n)$$

for all $n \geq 1$ and $g(n)/g(n + 1) \to 1$ as $n \to \infty$. Suppose

$$E | S_{a,n} |^v \leq g^{v/2}(n)$$

for all $a \geq 0$, $n \geq 1$, and some $v > 2$.

Suppose further that

$$g(n) \leq K n^2 (\log n)^{-2/v} (\log_2 n)^{-4/v} \tag{3.7.14}$$

for all $n \geq 1$ and some $K < \infty$. Then $S_n/n \to 0$ a.s.

**Proof.** Without loss of generality, assume equality in Eq. (3.7.14). Then the result is immediate from Theorem 3.7.6. ∎

Corollary 3.7.2 should be compared with the analogous second moment result, Theorem 3.7.3.

$$E | S_{a,n} |^v \leq K n^{v/2} \qquad \text{for all} \quad a \geq 0,$$

all $n \geq 1$, some $v > 2$, and $K < \infty$ holds in many situations as we shall see. This motivates the following theorem.

**Theorem 3.7.7.** Let

$$E | S_{a,n} |^v \leq K n^{v/2} \tag{3.7.15}$$

for all $a \geq 0$, all $n \geq 1$, some $\nu > 2$, and some $K < \infty$. Then

$$S_n / [n^{1/2}(\log n)^{1/\nu}(\log_2 n)^{2/\nu}] \to 0 \qquad \text{a.s.}$$

**Proof.** Immediate from Theorem 3.7.6 with $g(n) = K^{2/\nu}n$. ∎

When does Eq. (3.7.15) hold? Because it is related to this question, we return to the concept of mixing random variables. (Recall Definition 3.3.1.) Intuitively $\{X_i, i \geq 1\}$ is a mixing sequence if the $X_i$ with indices far apart are almost independent. This was formalized in the definition of *-mixing. Other formalizations appear in the literature:

**Definition 3.7.3.** Let

$$\mathscr{B}_{ij} = \mathscr{B}(X_k, i \leq k \leq j)$$

for all $1 \leq i \leq j \leq \infty$.

(i) $\{X_i, i \geq 1\}$ is said to be $\phi$-mixing if there exists an integer $M$ and a function $\phi$ for which $\phi(m) \to 0$ as $m \to \infty$ and $A \in \mathscr{B}_{1n}$, $B \in \mathscr{B}_{m+n,\infty}$ implies

$$| P(A \cap B) - P(A)P(B) | \leq \phi(m)PA \qquad (3.7.16)$$

for all $m \geq M$ and all $n \geq 1$.

If instead of Eq. (3.7.16), we have

$$| P(A \cap B) - P(A)P(B) | \leq \phi(m)P(A)P(B) \qquad (3.7.17)$$

for all $m \geq M$ and all $n \geq 1$, then $\{X_i, i \geq 1\}$ is said to be strong $\phi$-mixing. If instead of Eq. (3.7.16) we have

$$| P(A \cap B) - P(A)P(B) | \leq \phi(m) \qquad (3.7.18)$$

for all $m \geq 1$ and all $n \geq 1$, then $\{X_i, i \geq 1\}$ is said to be weak $\phi$-mixing (not standard terminology). ∎

**Lemma 3.7.1.** $\{X_i, i \geq 1\}$ $m$-dependent for some $m \geq 0$ implies $\{X_i, i \geq 1\}$ strong $\phi$-mixing implies $\{X_i, i \geq 1\}$ $\phi$-mixing implies $\{X_i, i \geq 1\}$ weak $\phi$-mixing. Also $\{X_i, i \geq 1\}$ strong $\phi$-mixing implies $\{X_i, i \geq 1\}$ *-mixing.

**Proof.** Immediate. ∎

Now back to the question of when Eq. (3.7.15) holds:

**Theorem 3.7.8.**  Let $\{X_i, i \geq 1\}$ satisfy $E \mid X_i \mid^\nu \leq C$ for all $i \geq 1$, some $\nu > 2$, and some $C < \infty$. Then

$$E \mid S_{a,n} \mid^\nu \leq K n^{\nu/2} \qquad (3.7.19)$$

for all $a \geq 0$, all $n \geq 1$, and some $K < \infty$ holds if either

(i)  $\{X_i, i \geq 1\}$ is a martingale difference sequence;

(ii)  $\{X_i, i \geq 1\}$ is stationary and $\phi$-mixing with $EX_1 = 0$ and

$$E\left(\sum_{i=1}^n X_i\right)^2 \leq Kn$$

for some $K < \infty$ and all $n \geq 1$;

(iii)  $\{X_i, i \geq 1\}$ is a stationary aperiodic Markov sequence which is Markov ergodic and satisfies Doeblin's condition; or

(iv)  [Dharmadhikari and Jogdeo, 1969] $\nu$ is an even integer and

$$E(X_{i_1}^{j_1} X_{i_2}^{j_2} \cdots X_{i_m}^{j_m}) = 0$$

for all positive integers $1 \leq i_1 \leq i_2 < \cdots < i_m$, all $m \leq \nu$, and all positive integer exponents $(j_1, j_2, \ldots, j_m)$ such that $\min(j_1, j_2, \ldots, j_m) = 1$ and $j_1 + j_2 + \cdots + j_m = \nu$.

**Proof.**  The proofs of (ii) and (iii) are somewhat involved and hence omitted. (ii) is proved by Ibragimov [1962] and (iii) is proved in Doob's text [1953, p. 225]. An indication of the proof of (ii) is given by the proof of lemma 5.4.8. Actually (iii) is a special case of (ii).

(i)  This was essentially done before. Fix $n \geq 1$. According to Eq. (3.3.14),

$$E\left|\sum_{i=1}^n X_i\right|^\nu \leq C n^{\nu/2-1} \sum_{i=1}^n E \mid X_i \mid^\nu.$$

But

$$\sum_{i=1}^n E \mid X_i \mid^\nu \leq Kn, \text{ establishing (i)}.$$

(iv)  Suppose $\nu$ even.

$$ES_n^\nu = \sum_{m=1}^\nu \sum{}' \sum{}'' \frac{\nu!}{j_1! j_2! \cdots j_m!} E\,(X_{i_1}^{j_1} \cdots X_{i_m}^{j_m}),$$

where $\sum'$ is over all $1 \leq i_1 < i_2 < \cdots < i_m \leq n$ and $\sum''$ is over all $(j_1, j_2, \ldots, j_m)$ such that $j_1 + j_2 + \cdots + j_m = \nu$ and $\min(j_1, j_2, \ldots, j_m) \geq 1$.

If $m > \nu/2$, then $\min(j_1, \ldots, j_m) = 1$ and by hypothesis

$$E(X_{i_1}^{j_1} \cdots X_{i_m}^{j_m}) = 0.$$

By Holder's inequality and hypothesis

$$E \mid X_{i_1}^{j_1} \cdots X_{i_m}^{j_m} \mid \leq (E \mid X_{i_1} \mid^{\nu})^{j_1/\nu} \cdots (E \mid X_{i_m} \mid^{\nu})^{j_m/\nu} \leq C.$$

Thus

$$ES_n^{\nu} \leq C \sum_{m=1}^{\nu/2} \binom{n}{m} \sum{}'' \frac{\nu!}{j_1! j_2! \cdots j_m!}$$

$$\leq C \sum_{m=1}^{\nu/2} \binom{n}{m} m^{\nu}$$

$$\leq C \sum_{m=1}^{\nu/2} n^m m^{\nu}$$

$$\leq C n^{\nu/2} \sum_{m=1}^{\nu/2} m^{\nu}.$$

Thus, taking $K = C \sum_{m=1}^{\nu/2} m^{\nu}$, the result is proved for even integers $\nu \geq 4$. ∎

**Remark.** In (iv), $K$ is more carefully estimated.

One special dependence structure where the hypothesis of (iv) and, hence, Eq. (3.7.19) holds is worth mentioning. Let $\{X_i, i \geq 1\}$ be a stochastic sequence with $E \mid X_i \mid^{\nu} \leq C < \infty$ for all $i \geq 1$ and some even integer $\nu > 2$. Suppose that $(r_1 X_1, \ldots, r_n X_n)$ has the same distribution as $(X_1, X_2, \ldots, X_n)$ for all $n \geq 1$ and all $s_i$ equal $\pm 1$. This is clearly an assumption about symmetry. (Recall Exercise 2.8.3.) According to Berman [1965], $\{X_i, i \geq 1\}$ is said to be sign invariant. Clearly the hypotheses of (iv) hold and hence Eq. (3.7.19) holds.

**Corollary 3.7.3.** Let $\{X_i, i \geq 1\}$ satisfy $E \mid X_i \mid^{\nu} \leq C$ for all $i \geq 1$, some $\nu > 2$, and some $C < \infty$. Then if either (i), (ii), (iii), or (iv) of Theorem 3.7.8 holds,

$$S_n/[n^{1/2}(\log n)^{1/\nu}(\log_2 n)^{2/\nu}] \to 0 \qquad \text{a.s.}$$

**Proof.** Immediate from Theorems 3.7.7 and 3.7.8. ∎

**Example 3.7.2.** We consider a probabilistic model of an individual in a learning situation. On each trial the individual is presented with a stimulus and either responds correctly or incorrectly. Let $X_n = 1$ if the response is correct on the $n$th trial and $X_n = 0$ if the response is incorrect on the $n$th trial. Suppose $\{X_n, n \geq 0\}$ is a Markov chain with stationary transition probabilities having transition matrix

$$P = \begin{bmatrix} 1 - \alpha & \alpha \\ p & 1 - p \end{bmatrix}.$$

Here the state space $S = \{0, 1\}$, 0 denoting an incorrect response and 1 a correct response. To avoid trivialities suppose $0 < \alpha < 1$ and $0 < p < 1$. Elementary matrix analysis shows that

$$P^m = \frac{1}{\alpha + p} \begin{bmatrix} p & \alpha \\ p & \alpha \end{bmatrix} + \frac{(1 - \alpha - p)^m}{\alpha + p} \begin{bmatrix} \alpha & -\alpha \\ -p & p \end{bmatrix}.$$

Let
$$p_0^{(0)} = P[X_0 = 0].$$

Let
$$p^{(0)} = (p_0^{(0)}, 1 - p_0^{(0)}).$$

Let
$$p_0^{(n)} = P[X_n = 0]$$

and
$$p^{(n)} = (p_0^{(n)}, 1 - p_0^{(n)}).$$

Then
$$p^{(n)} = p^{(0)} P^n$$

is easily shown, this relationship holding for an arbitrary Markov chain. Let $\mathscr{B}_{i,j} = \mathscr{B}(X_i, X_{i+1}, \ldots, X_j)$ for $0 \leq i \leq j \leq \infty$. Let $A \in \mathscr{B}_{0,n}$ and $B \in \mathscr{B}_{n+m,\infty}$, supposing $A \notin \mathscr{B}_{0,n-1}$ and $B \notin \mathscr{B}_{n+m+1,\infty}$. Then $A = A' \cap (X_n = a_n)$ where $a_n = 0$ or $a_n = 1$ and $A' \in \mathscr{B}_{0,n-1}$ and $B = B' \cap (X_{n+m} = a_{n+m})$ where $a_{n+m} = 0$ or $a_{n+m} = 1$ and $B' \in \mathscr{B}_{n+m+1,\infty}$

$$|P(A \cap B) - P(A)P(B)|$$
$$= |P(A)[P(B \mid A) - P(B)]|$$
$$= |P(A)[P(B \mid X_n = a_n) - P(B)]|$$
$$= |P(A)[P(B' \mid X_{n+m} = a_{n+m})P(X_{n+m} = a_{n+m} \mid X_n = a_n)$$
$$- P(B' \mid X_{n+m} = a_{n+m})P(X_{n+m} = a_{n+m})]|$$
$$\leq P(A) |P(X_{n+m} = a_{n+m} \mid X_n = a_n) - P(X_{n+m} = a_{n+m})|.$$

Computation shows that

$$| P[X_{n+m} = a_{n+m} \mid X_n = a_n] - P[X_{n+m} = a_{n+m}] | \le c(1 - \alpha - p)^m,$$

where $c$ is independent of $a_n$ and $a_{n+m}$.

Thus, $| P(A \cap B) - P(A)P(B) | \le P(A)c(1 - \alpha - p)^m$ for all $A \in \mathscr{B}_{0,n}$, $B \in \mathscr{B}_{n+m,\infty}$. Then, recalling that $0 < \alpha + p < 2$ by assumption, $\{X_n, n \ge 0\}$ is $\phi$-mixing. It is easy to see that

$$p^{(0)} = \begin{pmatrix} \dfrac{p}{\alpha + p} \\[2mm] \dfrac{\alpha}{\alpha + p} \end{pmatrix}$$

makes $\{X_n, n \ge 0\}$ stationary. Suppose this choice for $p^{(0)}$. $E(\sum_{i=0}^{n-1} X_i)^2 \le n$ for $n \ge 1$. $EX_0 = \alpha/(\alpha + p)$. Thus, it follows from Theorems 3.7.8 (ii) and 3.7.7 that

$$\frac{\sum_{i=1}^{n} X_i - n\alpha(\alpha + p)^{-1}}{n^{1/2}(\log n)^{1/\nu}(\log_2 n)^{2/\nu}} \to 0 \qquad \text{a.s.} \qquad (3.7.20)$$

for each $\nu > 2$. In particular $\sum_{i=1}^{n} X_i/n \to \alpha(\alpha + p)^{-1}$ a.s., a result which also follows from stationarity, $E \mid X_1 \mid < \infty$ and Markov ergodicity. Of course, Theorem 3.7.8 (iii) could have been used instead of Theorem 3.7.8 (ii) to establish Eq. (3.7.20). ∎

Example 3.7.2 is a special case of a general result. Let $\{X_n, n \ge 1\}$ be a Markov sequence with finite state space $S$. Suppose moreover that this Markov chain is aperiodic and that for each $i, j \in S$, there exists $n \ge 1$ such that

$$P[X_n = i \mid X_1 = j] > 0.$$

Then it can be shown that $\{X_n, n \ge 1\}$ is $\phi$-mixing (see Billingsley [1968, pp. 167–168]). Indeed $\phi(n) = K\varrho^n$ can be shown for some $0 < \varrho < 1$.

We close Section 3.7 by mentioning a striking result done to Komlós [1967].

**Theorem 3.7.9.** Let $\{X_i, i \ge 1\}$ satisfy $E \mid X_i \mid \le M < \infty$ for all $i \ge 1$. Then there exists an increasing subsequence of positive integers $\{n_k, k \ge 1\}$ and a random variable $X$ such that

$$\sum_{k=1}^{n} X_{n_k}/n \to X \qquad \text{a.s.}$$

*Proof.*   Omitted. See Komlós [1967]. ∎

*Exercise 3.7.2.*   Show that the strong law of large numbers for independent identically distributed random variables is a corollary of Theorem 3.7.9. ∎

# Stability of Weighted Sums
# of Random Variables

Throughout, let $\{a_{nk}, k \geq 1, n \geq 1\}$ denote a matrix of real numbers and

$$T_{nm} = \sum_{k=1}^{m} a_{nk}X_k$$

for each $m \geq 1$, $n \geq 1$. For each $n \geq 1$ let $T_n$ denote the limit in probability of $T_{nm}$ as $m \to \infty$. (In each situation studied, the hypotheses will imply that $T_{nm}$ converges in probability as $m \to \infty$.) Let

$$C_n = \sum_{k=1}^{\infty} a_{nk}^2$$

for each $n \geq 1$. Our purpose in Chapter 4 is to find sufficient conditions on $\{a_{nk}, k \geq 1, n \geq 1\}$ and $\{X_k, k \geq 1\}$ such that $T_n - c_n \to 0$ a.s. for appropriate centering constants $\{c_n, n > 1\}$. Many such results were presented in Chapter 3 for specific matrices such as

$$a_{nk} = \begin{cases} n^{-\alpha} & \text{for some } \alpha > 0 \text{ and all } k \leq n \\ 0 & \text{for all } k > n. \end{cases}$$

In Chapter 4 assumptions made concerning the $a_{nk}$ are much less specific. $C_n$ converging to zero at a specific rate as $n \to \infty$ is a typical assumption. Many of the results assume the $X_k$ are independent identically distributed; however, results are also stated for independent but not necessarily identically distributed random variables, martingale differences, strongly multiplicative random variables, and stationary random variables.

## 4.1.  Independent Identically Distributed $X_k$

Let $\{a_k, k \geq 1\}$ be a sequence of positive real numbers,

$$A_n = \sum_{k=1}^{n} a_k$$

for all $n > 1$, $\{X_k, k \geq 1\}$ be a sequence of independent identically distributed random variables, and

$$T_n = \sum_{k=1}^{n} a_k X_k / A_n$$

for all $n \geq 1$. Thus

$$a_{nk} = \begin{cases} a_k/A_n & \text{for} \quad k \leq n \\ 0 & \text{for} \quad k > n \end{cases}$$

for all $n \geq 1$. $T_n$ is a weighted average of the first $n$ $X_k$. Under these assumptions, Jamison et al. [1965] obtain rather precise results.

If $\sum_{k=1}^{\infty} a_k < \infty$, then $T_n - c_n \to 0$ a.s. is impossible unless the $X_k$ are degenerate. Hence we assume $\sum_{k=1}^{\infty} a_k = \infty$. In this case, $T_n - c_n$ either converges almost surely to zero for some choice of $\{c_n, n \geq 1\}$ or else diverges almost surely for every choice of $\{c_n, n \geq 1\}$. This follows from the Kolmogorov 0–1 law since $[T_n - c_n$ converges] is a tail event. When $T_n - c_n$ converges almost surely to 0 for some choice of $\{c_n, n \geq 1\}$ we say that $T_n$ is stable. If $T_n - c_n$ converges almost surely to 0 for a particular choice of $\{c_n, n \geq 1\}$ we say that $T_n$ is stable when centered at $c_n$.

The first result to be stated is perhaps best viewed as a generalization of Kolmogorov's strong law of large numbers for independent identically distributed random variables. In the notation of this section the strong law says that $E \mid X_1 \mid < \infty$ implies that $T_n$ is stable when $a_n = 1$ for all $n \geq 1$. Let $N(n)$ be the number of subscripts $k$ such that $A_k/a_k \leq n$ thus defining $N(\cdot)$. The rate of growth of $N(\cdot)$ will be critical in establishing stability results.

***Exercise 4.1.1.***  Prove that $N(n) = \infty$ for some $n > 0$ implies that $T_n$ is not stable. Hint: Show that $T_n$ stable implies $A_k/a_k \to \infty$ as $k \to \infty$ by noting that

$$\sum_{i=1}^{n} \frac{a_i X_i}{A_n} - \sum_{i=1}^{n-1} \frac{a_i X_i}{A_{n-1}} = \frac{a_n}{A_n} \left[ X_n - \sum_{i=1}^{n-1} \frac{a_i X_i}{A_{n-1}} \right]. \quad \blacksquare$$

We assume throughout that $A_k/a_k \to \infty$. This together with the assumption that $A_k \to \infty$ holds if and only if $\max_{k \leq n} a_k/A_n \to 0$. $\text{Max}_{k \leq n} a_k/A_n$ merely says that each individual random variable $X_i$ contributes little to the value of $T_n$ when $n$ is large.

**Theorem 4.1.1.** [Jamison *et al.*, 1965]. If $E \mid X_1 \mid \, < \infty$ and

$$N(n)/n \leq K \qquad\qquad (4.1.1)$$

for all $n \geq 1$ and some $K < \infty$, then $T_n$ is stable when centered at $EX_1$.

If Eq. (4.1.1) does not hold, then there exists $\{X_i, i \geq 1\}$ such that $E \mid X_1 \mid \, < \infty$ and $T_n$ is not stable.

*Remark.* The following exercise is helpful in the interpretation of Eq. (4.1.1):

*Exercise 4.1.2.* Show that $a_k = 1$ for all $k \geq 1$ implies that Eq. (4.1.1) holds. (Hence the Kolmogorov strong law is included in the statement of Theorem 4.1.1.) Show that $a_1 = a_2 = 1$, $a_k = A_{k-1}/(\log k - 1)$ for all $k \geq 3$ implies that Eq. (4.1.1) fails. Roughly Eq. (4.1.1) implies that there cannot be too many large $a_k$, the largeness of an $a_k$ being relative to the magnitude of $A_k$. ∎

*Proof of Theorem 4.1.1.* Assume Eq. (4.1.1). We proceed essentially as in the proof of Kolmogorov's strong law (Theorem 3.2.2). Without loss of generality, assume $EX_1 = 0$.

$$\sum_{k=1}^{\infty} P\left[\mid X_k \mid \, \geq \frac{A_k}{a_k}\right] \leq \sum_{k=1}^{\infty} [N(k) - N(k-1)]P[\mid X_1 \mid \, \geq k - 1].$$

Summing the last sum by parts, it follows that

$$\sum_{k=1}^{\infty} P\left[\mid X_k \mid \, \geq \frac{A_k}{a_k}\right] \leq \sum_{k=1}^{\infty} k \, \frac{N(k)}{k} \, [P(\mid X_1 \mid \, \geq k - 1) - P(\mid X_1 \mid \, \geq k)]$$

$$\leq K \sum_{k=1}^{\infty} kP[k - 1 \leq \mid X_1 \mid \, < k] < \infty,$$

using Eq. (4.1.1) and the fact that $E \mid X_1 \mid \, < \infty$. Thus

$$P[\mid X_k \mid \, > A_k/a_k \text{ i.o.}] = 0$$

by the Borel–Cantelli lemma. Hence, recalling that $A_n \to \infty$, it suffices to show that

$$\sum_{k=1}^{n} a_k X_k I(\mid X_k \mid \, \leq A_k/a_k)/A_n \to 0 \qquad \text{a.s.}$$

$$E[X_k I(\mid X_k \mid \, \leq A_k/a_k)] = E[X_1 I(\mid X_1 \mid \, \leq A_k/a_k)] \to 0$$

as $k \to \infty$ follows from the Lebesgue dominated convergence theorem and the fact that $A_k/a_k \to \infty$. Let $a_{nk} = a_k/A_n$ for $k \leq n$; $= 0$ for $k > n$. $a_{nk} \to 0$

as $n \to \infty$, and $\sum_{k=1}^{\infty} |a_{nk}| = 1$ for all $n \geq 1$. Let

$$x_k = E[X_k I(|X_k| \leq A_k/a_k)].$$

By the Toeplitz lemma (Lemma 3.2.3 (i))

$$\sum_{k=1}^{n} a_{nk} x_k = \sum_{k=1}^{n} a_k E[X_k I(|X_k| \leq A_k/a_k)]/A_n \to 0$$

as $n \to \infty$. Thus it suffices to show that

$$\sum_{k=1}^{n} a_k X_k I\left(|X_k| \leq \frac{A_k}{a_k}\right) - a_k E\left[X_k I\left(|X_k| \leq \frac{A_k}{a_k}\right)\right]\bigg/ A_n \to 0 \qquad \text{a.s.}$$

By the Kronecker lemma, it suffices to show that

$$\sum_{k=1}^{\infty} \left\{a_k X_k I\left(|X_k| \leq \frac{A_k}{a_k}\right) - a_k E\left[X_k I\left(|X_k| \leq \frac{A_k}{a_k}\right)\right]\right\}\bigg/ A_k \qquad \text{converges a.s.}$$

Thus it suffices to show that

$$\sum_{k=1}^{\infty} \text{var}[a_k X_k I(|X_k| \leq A_k/a_k)]/A_k^2 < \infty$$

and hence that

$$\sum_{k=1}^{\infty} (a_k^2/A_k^2) E[X_k^2 I(|X_k| \leq A_k/a_k)] < \infty.$$

$$\sum_{k=1}^{\infty} (a_k^2/A_k^2) E[X_k^2 I(|X_k| \leq A_k/a_k)]$$

$$\leq \sum_{k=1}^{\infty} \{[N(k) - N(k-1)]/(k-1)^2\} E[X_1^2 I(|X_1| \leq k)]$$

$$= \sum_{k=1}^{\infty} \{[N(k) - N(k-1)]/(k-1)^2\} \sum_{j=1}^{k} E[X_1^2 I(j-1 < |X_1| \leq j)]$$

$$= \sum_{j=1}^{\infty} E[X_1^2 I(j-1 < |X_1| \leq j)] \sum_{k=j}^{\infty} [N(k) - N(k-1)]/(k-1)^2$$

$$= \sum_{j=1}^{\infty} E[X_1^2 I(j-1 < |X_1| \leq j)]$$

$$\times \left\{-N(j-1)/(j-1)^2 + \sum_{k=j}^{\infty} [(k-1)^{-2} - k^{-2}]N(k)\right\}$$

$$\leq K \sum_{j=1}^{\infty} E[X_1^2 I(j-1 < |X_1| \leq j)] \sum_{k=j}^{\infty} (2k-1)(k-1)^{-2}k^{-1}$$

$$\leq K' \sum_{j=1}^{\infty} E[X_1^2 I(j-1 < |X_1| \leq j)]/j$$

for some $K' < \infty$.

$$K' \sum_{j=1}^{\infty} E[X_1^2 I(j-1 < |X_1| \le j)]/j$$

$$\le K' \sum_{j=1}^{\infty} E[|X_1| I(j-1 < |X_1| \le j)] < \infty$$

since $E|X_1| < \infty$.

Thus, $T_n$ is stable when centered at $EX_1$, as desired.

Assume Eq. (4.1.1) fails for $\{a_k, k \ge 1\}$. Hence there exists positive integers $n_k$ such that $N(n_k)/n_k \to \infty$ as $k \to \infty$. This permits us to choose $p_k \ge 0$ such that

$$\sum_{k=1}^{\infty} p_k = 1, \quad \sum_{k=1}^{\infty} p_k n_k < \infty, \quad \text{and} \quad \sum_{k=1}^{\infty} p_k N(n_k) = \infty.$$

Let

$$P[X_1 = n_k] = P[X_1 = -n_k] = p_k/2$$

for all $k \ge 1$ and assume the $X_k$ are independent identically distributed. Then $E|X_1| < \infty$ and the $X_k$ are symmetric. Let $\{i(j), j \ge 1\}$ be a re-ordering of the positive integers such that $A_{i(j)}/a_{i(j)}$ increases in $j$. This choice of $\{i(j), j \ge 1\}$ is possible because $A_i/a_i \to \infty$ as $i \to \infty$.

$$\infty = \sum_{k=1}^{\infty} p_k N(n_k) = EN(|X_1|)$$

$$= \sum_{i=1}^{\infty} i P[N(|X_1|) = i]$$

$$= \sum_{i=1}^{\infty} P[N(|X_1|) \ge i] = \sum_{j=1}^{\infty} P[N(|X_1|) \ge i(j)].$$

$$[N(|X_1|) \ge i(j)] = [\text{there exist at least}$$

$$i(j) \text{ integers } k \text{ such that } A_k/a_k \le |X_1|] = [A_{i(j)}/a_{i(j)} \le |X_1|].$$

Hence

$$\sum_{i=1}^{\infty} P[|X_i| \ge A_i/a_i] = \sum_{j=1}^{\infty} P[|X_j| \ge A_{i(j)}/a_{i(j)}]$$

$$= \sum_{j=1}^{\infty} P[|X_1| \ge A_{i(j)}/a_{i(j)}]$$

$$= \sum_{j=1}^{\infty} P[N(|X_1|) \ge i(j)] = \infty.$$

Hence

$$P[a_i \mid X_i \mid /A_i \geq 1 \text{ i.o.}] = 1.$$

Hence

$$P\left[\sum_{i=1}^{n} a_i X_i / A_n \to 0\right] = 0.$$

Since the $X_i$ are symmetric, $T_n$ diverges almost surely for every choice of $\{c_i, i \geq 1\}$; that is, $T_n$ is not stable. ∎

It is interesting to note that there exists $\{a_k, k \geq 1\}$ such that $\mid a_k \mid \leq M$ for all $k \geq 1$ and some $M < \infty$ and yet

$$\limsup N(n)/n = \infty$$

(see Jamison *et al.* [1965, p. 43]). What then, if anything, can we say about bounded sequences of weights?

**Theorem 4.1.2.** [Jamison *et al.*, 1965]. Let $\{a_k, k \geq 1\}$ be a bounded sequence of weights. If

$$E(\mid X_1 \mid \log^+ \mid X_1 \mid) < \infty,$$

then $T_n$ is stable when centered at $EX_1$.

***Exercise 4.1.3.*** Prove Theorem 4.1.2. Hint: Prove that

$$\limsup N(n)/(n \log n) < \infty$$

and then proceed similarly to the proof of Theorem 4.1.1. ∎

***Exercise 4.1.4.*** Prove that

$$\sum_{i=1}^{\infty} E(X_i^2/(a_i^2 + X_i^2)) < \infty$$

implies that $T_n$ is stable. If in addition, $EX_1 = 0$, does it follow that $T_n \to 0$ a.s.? ∎

Chow and Teicher [1971] have recently done some work on the stability of $T_n$. In particular they show that

$$\liminf_{x \to \infty} xP[\mid X_1 \mid > x] > 0 \qquad (4.1.2)$$

implies that $T_n$ is not stable for any choice of $\{a_k, k \geq 1\}$. For proper interpretation of this, note that

$$\liminf_{x \to \infty} xP[|\,X_1\,| > x] > 0$$

implies $E\,|\,X_1\,| = \infty$.

The Cauchy distribution satisfies Eq. (4.1.2), as does the random variable of the St. Petersburg paradox ($P[X = 2^k] = 2^{-k}$ for $k \geq 1$). Chow and Teicher also present a result closely related to Theorem 4.1.2.

It is perhaps interesting to compare the stochastic theory of summability for $\{X_k, k \geq 1\}$, a sequence of independent identically distributed random variables, with the classical theory of summability for $\{x_k, k \geq 1\}$, a sequence of constants. Here we have been considering summability matrices defined by

$$a_{nk} = \begin{cases} a_k/A_n & \text{for } k \leq n \\ 0 & \text{for } k > n \end{cases}$$

for $n \geq 1$. An important question of the classical theory was to characterize matrices $\{a_{nk}\}$ for which $x_k \to x$ implies $t_n = \sum_{k=1}^{\infty} a_{nk}x_k \to x$ for all choices of $\{x_k\}$. Such summability matrices are called *regular*. The answer is pleasantly simple. $\{a_{nk}\}$ is regular if and only if it is a Toeplitz matrix (see Lemma 3.2.3 (i)). In the classical case, the $x_k$ arise as successive partial sums of an infinite series. Note that the matrices considered in Section 4.1 in the stochastic case are trivially Toeplitz since $\sum_{k=1}^{\infty} |\,a_{nk}\,| = \sum_{k=1}^{\infty} a_{nk} = 1$ and $a_{nk} = a_k/A_n \to 0$ as $n \to \infty$ since $A_n \to \infty$ is assumed. Clearly not all regular matrices of the form $a_{nk} = a_k/A_n$ satisfy the additional requirement imposed in the stochastic case; that is, that $A_n/a_n \to \infty$ as $n \to \infty$.

One motivation for the classical summability theory was to give meaning to the "sum" of divergent series like $1 - 1 + 1 - 1 \cdots$. Converting to the sequence of partial sums, one wishes to define a "limit" for the $x_k$ sequence $1, 0, 1, 0, \ldots$. The stochastic theory can similarly be viewed as a method of defining a "limit" for divergent sequences like $X_1, X_2, \ldots$ where the $X_k$ are independent with $P[X_k = 0] = P[X_k = 1] = \frac{1}{2}$ for each $k \geq 1$. Hardy [1967] can be referred to for an exhaustive treatment of the classical theory.

***Example 4.1.1.*** Consider $\{a_{nk}\}$ defined by

$$a_{nk} = \begin{cases} k^{-1}\left(\sum_{i=1}^{n} i^{-1}\right)^{-1} & \text{if } k \leq n \\ 0 & \text{if } k > n. \end{cases}$$

This is the regular logarithmic method of summation of Riesz. Clearly $A_n/a_n \to \infty$ and thus the stochastic theory given above applies. Since $N(n)/n \leq K$ for all $n \geq 1$, by Theorem 4.1.1, if $EX_1 = 0$, then

$$\sum_{k=1}^{\infty} a_{nk} X_k \to 0 \qquad \text{a.s.} \quad \blacksquare$$

When $a_{nk} = a_k/A_n$ for $k \leq n$; $= 0$ for $k > n$,

$$A_j T_j - A_i T_i = \sum_{k=i+1}^{j} a_k X_k$$

for all $j > i$. Thus, the sequence $\{T_n, n \geq 1\}$ is strongly dependent. In the remainder of Section 4.1 we look at the question of when $T_n$ is stable when the only restrictions on $\{a_{nk}\}$ are ones of magnitude. In this case, we will no longer necessarily have the strong dependence between the $T_n$ mentioned above. As a result, the Kronecker lemma approach, which has served us so well, cannot possibly work. It is even possible for the $T_n$ to be independent (for example, if each column of the $\{a_{nk}\}$ matrix has at most one nonzero term). Thus,

$$\sum_{n=1}^{\infty} [P \mid T_n - c_n \mid > \varepsilon] < \infty$$

for some $\{c_n, n \geq 1\}$ and all $\varepsilon > 0$ is a sufficient condition for the stability of $T_n$ which in some cases (independence of the $T_n$) is also necessary.

**Definition 4.1.1.** A sequence of random variables $\{U_n, n \geq 1\}$ is said to converge completely to a constant $c$ if

$$\sum_{n=1}^{\infty} P[\mid U_n - c \mid > \varepsilon] < \infty \qquad \text{for each} \quad \varepsilon > 0. \quad \blacksquare$$

This definition is due to Hsu and Robbins [1947]. Note that complete convergence of $U_n$ to $c$ is a property of the marginal distributions of the $U_n$. We will study the complete convergence of $T_n$ to zero as a means of establishing that $T_n$ is stable. This approach is conceptually simpler than other approaches we have used which take into account the simultaneous behavior of the $T_n$ for different subscripts, for example, the method of subsequences and the upcrossings approach. When using the complete convergence approach, accurate estimation of probabilities of the form $P[\mid T_n \mid > \varepsilon]$ is crucial. A clever truncation technique due to Erdös [1949] is useful in this respect. In order to illuminate Erdös's method clearly we present a special case of some work due to Stout [1968]. First, two simple lemmas are needed.

**Lemma 4.1.1.**  If $Y \leq 1$ a.s. then

$$E[\exp Y] \leq \exp(EY + EY^2).$$

**Proof.**

$$\exp(Y) = 1 + Y + (Y^2/2)(1 + Y/3 + Y^2/(3 \cdot 4) + \cdots)$$
$$\leq 1 + Y + (Y^2/2)(1 + 1/3 + 1/(3 \cdot 4) + \cdots).$$

Hence

$$E[\exp Y] \leq 1 + EY + EY^2 \leq \exp(EY + EY^2). \quad \blacksquare$$

**Lemma 4.1.2.**  (One of Bonferroni's inequalities).  Let $\{A_i, 1 \leq i \leq m\}$ be a sequence of events. Let

$$q_N = \sum_{1 \leq i_1 < i_2 < \cdots < i_N \leq m} P[A_{i_1} \cap A_{i_2} \cap \ldots \cap A_{i_N}].$$

Prove that

$$P[\text{at least } N \ A_i \text{ occur}] \leq q_N.$$

**Proof.**  Exercise. (For more combinatorial inequalities of this type, see Feller [1968a, Chapter IV].)  $\blacksquare$

**Theorem 4.1.3.**  Let $E \mid X_1 \mid^{2/\alpha} < \infty$ for some $0 < \alpha \leq 1$, $EX_1 = 0$,

$$\mid a_{nk} \mid \leq Kn^{-\alpha} \quad \text{for} \quad k \leq n \quad \text{and some} \quad K < \infty,$$
$$a_{nk} = 0 \quad \text{for} \quad k > n,$$

and

$$C_n = o(\log^{-1} n)$$

(recall $C_n = \sum_{k=1}^{\infty} a_{nk}^2$). Then, $T_n$ converges completely to zero.

**Proof.**  Let $\varepsilon > 0$ be given. Without loss of generality, assume $a_{nk} \geq 0$ for all $k \geq 1$, $n \geq 1$, and $EX_1^2 = 1$. Let

$$X'_{nk} = X_k I(a_{nk} X_k \leq n^{-\varrho})$$

for $\varrho > 0$ to be specified below and let

$$T_n' = \sum_{k=1}^{n} a_{nk} X_{nk}'.$$

Let

$$X_{nk}'' = X_k I(a_{nk} X_k > \varepsilon/N)$$

for an integer $N$ to be specified below and let

$$T_n'' = \sum_{k=1}^{n} a_{nk} X_{nk}''.$$

Let

$$X_{nk}''' = X_k - X_k' - X_k'' = X_k I(n^{-\varrho} < a_{nk} X_k \leq \varepsilon/N)$$

and

$$T_n''' = \sum_{k=1}^{n} a_{nk} X_{nk}'''.$$

We shall prove that $\sum_{n=1}^{\infty} P[T_n > 3\varepsilon] < \infty$ by proving that

$$\sum_{n=1}^{\infty} P[T_n' > \varepsilon] < \infty, \qquad \sum_{n=1}^{\infty} P[T_n'' > \varepsilon],$$

and

$$\sum_{k=1}^{\infty} P[T_n''' > \varepsilon] < \infty,$$

noting that

$$[T_n > 3\varepsilon] \subset [T_n' > \varepsilon] \cup [T_n'' > \varepsilon] \cup [T_n''' > \varepsilon]$$

for all $n \geq 1$. Replacing the $X_k$ by $-X_k$ and repeating the argument will then establish $\sum_{n=1}^{\infty} P[T_n < -3\varepsilon] < \infty$ and hence that $\sum_{n=1}^{\infty} P[|T_n| > 3\varepsilon] < \infty$ for all $\varepsilon > 0$, which implies the desired result.

Fix $n \geq 1$. Let $u = \min[\varepsilon/(2C_n), n^\varrho]$.

$$E \exp(u T_n') = \prod_{k=1}^{n} E \exp(u a_{nk} X_{nk}').$$

Since $u a_{nk} X_{nk}' \leq 1$ it follows by Lemma 4.1.1 that

$$E \exp(u a_{nk} X_{nk}') \leq \exp[u a_{nk} E X_{nk}' + u^2 a_{nk}^2 E(X_{nk}')^2].$$

$EX'_{nk} \le EX_k = 0$ and $E(X'_{nk})^2 \le EX_k{}^2 = 1$. Thus

$$E \exp(ua_{nk}X'_{nk}) \le \exp(u^2 a^2_{nk})$$

and hence

$$E \exp(uT_n') \le \exp(u^2 C_n).$$

By the Chebyshev inequality

$$P[T_n' > \varepsilon] \le \exp(-\varepsilon u)E \exp(uT_n')$$
$$\le \exp(-\varepsilon u + u^2 C_n).$$

If $\varepsilon/(2C_n) > n^\varrho$, then by the definition of $u$

$$P[T_n' > \varepsilon] \le \exp(-\varepsilon n^\varrho/2).$$

If $\varepsilon/(2C_n) \le n^\varrho$, then by the definition of $u$

$$P[T_n' > \varepsilon] \le \exp(-\varepsilon^2/(4C_n)).$$

$$\sum_{n=1}^{\infty} \exp(-\varepsilon^2/(4C_n)) < \infty$$

since $C_n = o(\log^{-1} n)$.

$$\sum_{n=1}^{\infty} \exp(-\varepsilon n^\varrho/2) < \infty.$$

Thus

$$\sum_{n=1}^{\infty} P[T_n' > \varepsilon] < \infty.$$

$$[T_n'' > \varepsilon] \subset \bigcup_{k=1}^{n} [a_{nk}X_k > \varepsilon/N] \subset \bigcup_{k=1}^{n} [X_k > \varepsilon n^\alpha (KN)^{-1}]$$

since $0 \le a_{nk} \le Kn^{-\alpha}$ for $k \le n$. Thus

$$P[T_n'' > \varepsilon] \le \sum_{k=1}^{n} P[X_k > \varepsilon n^\alpha (KN)^{-1}]$$
$$= nP[X_1 > \varepsilon n^\alpha (KN)^{-1}]$$
$$\le nP\{|X_1|^{1/\alpha} > [\varepsilon(KN)^{-1}]^{1/\alpha} n\}.$$

$E|X_1|^{2/\alpha} < \infty$ implies

$$\sum_{n=1}^{\infty} nP[|X_1|^{1/\alpha} > Cn] < \infty$$

for all $C > 0$ by Exercise 3.2.1. Thus

$$\sum_{n=1}^{\infty} P[T_n'' > \varepsilon] \leq \sum_{n=1}^{\infty} nP\{|\,X_1\,|^{1/\alpha} > [\varepsilon(KN)^{-1}]^{1/\alpha}n\} < \infty.$$

$P[T_n''' > \varepsilon] \leq P$ [there exists at least $N$ indices $k$ such that

$$|\,X_k\,| > n^{-\varrho}/a_{nk}] \leq \binom{n}{N} P^N[|\,X_k\,|^{2/\alpha} > K^{2/\alpha}n^{(\alpha-\varrho)2/\alpha}]$$

follows by Lemma 4.1.2, by the independence of the $X_k$, and the fact that $|\,a_{nk}\,| \leq Kn^{-\alpha}$. (This step is the crucial step in the Erdös truncation scheme.)

$$\binom{n}{N} P^N[|\,X_k\,|^{2/\alpha} > n^{(\alpha-\varrho)2/\alpha}] \leq n^N n^{2N(\varrho-\alpha)/\alpha}(E\,|\,X_1\,|^{2/\alpha})^N$$
$$= n^{N(2\varrho/\alpha-1)}(E\,|\,X_1\,|^{2/\alpha})^N.$$

Thus, choosing $\varrho$ sufficiently small and $N$ sufficiently large,

$$\sum_{n=1}^{\infty} P[T_n''' > \varepsilon] \leq (E\,|\,X_1\,|^{2/\alpha})^N \sum_{n=1}^{\infty} n^{N(2\varrho/\alpha-1)} < \infty.$$

Thus

$$\sum_{n=1}^{\infty} P[|\,T_n\,| > 3\varepsilon] < \infty$$

for all $\varepsilon > 0$, as desired. ∎

In order to give some feeling for the hypotheses of Theorem 4.1.3, we state a corollary due to Erdös [1949].

**Corollary 4.1.1.**  Let $EX_1^4 < \infty$ and $EX_1 = 0$. Then

$$\sum_{k=1}^{n} X_k/[n^{1/2}(\log n)^{1/2+\delta}]$$

converges completely to zero for each $\delta > 0$.

*Proof.*  Apply Theorem 4.1.3 with $\alpha = \tfrac{1}{2}$ and

$$a_{nk} = \begin{cases} n^{-1/2}(\log n)^{-1/2-\delta} & \text{if } k \leq n \\ 0 & \text{if } k > n. \end{cases}$$

Note that $|\,a_{nk}\,| \leq n^{-\alpha}$ for $k \leq n$, $a_{nk} = 0$ for $k > n$, $E\,|\,X_1\,|^{2/\alpha} < \infty$, and

$$C_n = (\log n)^{-(1+2\delta)} = o(\log^{-1} n). \quad ∎$$

For a thorough discussion of the problem of complete and almost sure convergence for weighted sums $\sum_{k=1}^{\infty} a_{nk}X_k$, see Stout [1968]. Typical of the generality achievable is the following theorem.

**Theorem 4.1.4.**   Let

$$| a_{nk} | \leq Kn^{-\alpha}$$

for all $k \geq 1$, all $n \geq 1$, some $\alpha > 0$ and some $K < \infty$,

$$C_n \leq Kn^{\beta - \alpha}$$

for all $n \geq 1$ and some $\beta > -(1 + \alpha)$, and

$$E | X_1 |^{(1 + \alpha + \beta)/\alpha} < \infty.$$

(i)   If $(1 + \alpha + \beta)/\alpha > 2$,

$$\sum_{n=1}^{\infty} \exp(-u/C_n) < \infty$$

for all $u > 0$ and $EX_1 = 0$, then $T_n$ converges completely to zero.

(ii)   If $1 \leq (1 + \alpha + \beta)/\alpha < 2$,

$$\sum_{k=1}^{\infty} | a_{nk} |^{(1 + \alpha + \beta)/\alpha} \leq Kn^{-\gamma}$$

for all $n \geq 1$ and some $\gamma > 0$, and either

$$\sum_{k=1}^{\infty} | a_{nk} | \leq K < \infty \qquad \text{for all} \quad n \geq 1 \quad \text{with} \quad EX_1 = 0$$

or

$$\sum_{k=1}^{\infty} | a_{nk} | \to 0 \qquad \text{as} \quad n \to \infty,$$

then $T_n$ converges completely to zero.

(iii)   If $0 < (1 + \alpha + \beta)/\alpha < 1$,

$$\sum_{k=1}^{\infty} | a_{nk} |^{(1 + \alpha + \beta)/\alpha} \leq Kn^{-\gamma}$$

for all $n \geq 1$ and some $\gamma > 0$, and $a_{nk} = 0$ for $k > n^\zeta$ for all $n \geq 1$ and some $\zeta > 0$ where $\zeta < \gamma\alpha/(1 + \alpha + \beta)$, then $T_n$ converges completely to zero.

**Proof.**  Similar to Theorem 4.1.3, except for more computational complexity. Omitted. ∎

Theorem 4.1.4 includes many of the known results about complete convergence for weighted sums of independent identically distributed random variables. In particular, it includes results of Chow [1966], Pruitt [1966], and Franck and Hanson [1966].

**Exercise 4.1.5.**  [Chow and Lai, 1973].  Let $\{c_i, i \geq 0\}$ satisfy $\sum_{i=0}^{\infty} c_i^2 < \infty$ and let $T_n = \sum_{i=1}^{n} c_{n-i} X_i / n^{1/\varrho}$, where $E \mid X_1 \mid^\varrho < \infty$ for some $\varrho > 2$. Suppose $EX_1 = 0$. Prove that $T_n \to 0$ a.s., assuming the validity of Theorem 4.1.4. ∎

For many interesting results concerning $\{T_n, n \geq 1\}$ as defined in Exercise 4.1.5, see the paper by Chow and Lai. Such sums of weighted random variables occur naturally for the statistical quality control problem of detecting changes in the location of the distribution of a sequence of independent observations. For, it is intuitively clear that more recent observations (in time) should receive more weight in such statistical decision making.

It follows from the yet to be proved Hartman–Wintner law of the iterated logarithm (Theorem 3.2.9) that $EX_1^2 < \infty$ and $EX_1 = 0$ implies

$$\sum_{k=1}^{n} X_k / [n^{1/2} (\log n)^{1/2}] \to 0 \qquad \text{a.s.}$$

However computation shows that $\sum_{k=1}^{n} X_k / [n^{1/2} (\log n)^{1/2}]$ does not converge completely to zero even in the case that $X_1$ is normally distributed with mean zero. Thus, as one would certainly suspect, almost sure convergence can hold when complete convergence fails. When the matrix is subject only to magnitude restrictions, very few results establish almost sure convergence without establishing complete convergence. One exception is given next. It is unknown whether the hypotheses of Theorem 4.1.5 also imply complete convergence.

**Theorem 4.1.5.**  [Stout, 1968].  Let $C_n \leq K n^{-\alpha}$ for all $n \geq 1$, some $\alpha > 0$, and some $K < \infty$. Let $a_{nk}^2 \leq K k^{-1}$ for all $k \geq 1$ and $n \geq 1$. Suppose $EX_1^2 < \infty$ and $EX_1 = 0$. Then $T_n \to 0$ a.s.

**Proof.**  We modify the Erdös double truncation technique. Without loss of generality suppose $EX_1^2 = 1$ and $a_{nk} \geq 0$ for all $k \geq 1$, $n \geq 1$. Fix $\varepsilon > 0$. $T_n = \sum_{k=1}^{\infty} a_{nk} X_k$ is well defined since $\sum_{k=1}^{\infty} \text{var}(a_{nk} X_k) = C_n < \infty$.

Define $X'_{nk}$, $X''_{nk}$, $X'''_{nk}$, $T_n'$, $T_n''$, and $T_n'''$ exactly as in the proof of Theorem 4.1.3. $T_n'$, $T_n''$, and $T_n'''$ are well defined since $C_n < \infty$.

$$\sum_{n=1}^{\infty} P[T_n' > \varepsilon] < \infty$$

follows almost as in the proof of Theorem 4.1.3. To estimate $P[T_n'' > \varepsilon]$, we proceed much as in the proof of Theorem 4.1.3:

$$P[T_n''' > \varepsilon] \le P[\text{there exists at least } N \text{ indices } k \text{ such that } |X_k| > n^{-\varrho}/a_{nk}]$$

$$\le \left[ \sum_{k=1}^{\infty} P(|X_k| > n^{-\varrho}/a_{nk}) \right]^N$$

$$\le \left( \sum_{k=1}^{\infty} a_{nk}^2 n^{2\varrho} \right)^N \le K^N n^{(2\varrho-\alpha)N}$$

by hypothesis and Lemma 4.1.2 (letting $m \to \infty$). Choose $\varrho > 0$ sufficiently small and $N$ sufficiently large so that

$$\sum_{n=1}^{\infty} n^{(2\varrho-\alpha)N} < \infty$$

and hence $\sum_{n=1}^{\infty} P[T_n''' > \varepsilon] < \infty$.

We treat $T_n''$ differently than in the proof of Theorem 4.1.3. By the Holder inequality,

$$|T_n''|^2 \le C_n \sum_{k=1}^{\infty} (X''_{nk})^2 = C_n \sum_{k=1}^{\infty} X_k^2 I[X_k > \varepsilon(Na_{nk})^{-1}]$$

$$\le Kn^{-\alpha} \sum_{k=1}^{\infty} X_k^2 I[X_k^2 > \varepsilon^2 k(N^2 K)^{-1}],$$

using the hypotheses that

$$C_n \le Kn^{-\alpha} \quad \text{and} \quad a_{nk}^2 \le Kk^{-1}$$

for all $k \ge 1$. But $EX_1^2 < \infty$ implies

$$\sum_{k=1}^{\infty} P[X_k^2 > \varepsilon^2 k(N^2 K)^{-1}] < \infty.$$

Thus

$$X \equiv \sum_{k=1}^{\infty} X_k^2 I[X_k^2 > \varepsilon^2 k(N^2 k)^{-1}] < \infty \quad \text{a.s.}$$

by the Borel–Cantelli lemma. Thus

$$| T_n'' |^2 \le Kn^{-\alpha}X \qquad \text{a.s.}$$

Hence $T_n'' \to 0$ a.s. (Note that we have *not* proved that $T_n''$ converges completely to zero.)

Combining the results for $T_n'$, $T_n''$, and $T_n'''$, it follows that

$$T_n^+ \to 0 \qquad \text{a.s.}$$

Replacing $X_k$ by $-X_k$ implies that

$$T_n^- \to 0 \qquad \text{a.s.}$$

Thus $T_n \to 0$ a.s.  ∎

**Remark.**   [Chow, 1966].   It is easy to see that $T_n \to 0$ a.s. for all $a_{nk}$ satisfying the hypotheses of Theorem 4.1.5 implies that $EX_1^2 < \infty$ and $EX_1 = 0$. To see this, first let

$$a_{nk} = \begin{cases} n^{-1/2} & \text{if } k = n \\ 0 & \text{if } k \neq n. \end{cases}$$

$C_n = n^{-1}$. Thus

$$T_n = X_n/n^{1/2} \to 0 \qquad \text{a.s.}$$

Thus $\sum_{n=1}^{\infty} P[X_n^2 > n] < \infty$ and hence $EX_1^2 < \infty$. Now let

$$a_{nk} = \begin{cases} n^{-1} & \text{if } 1 \le k \le n \\ 0 & \text{if } k > n. \end{cases}$$

$C_n = n^{-1}$. Thus $T_n = \sum_{i=1}^{n} X_i/n \to 0$ a.s. which implies by Kolmogorov's strong law that $EX_1 = 0$. Thus Theorem 4.1.5 is sharp in the sense that the moment conditions on $X_1$ cannot be weakened.

Recall in the theory of Jamison *et al.* [1965] for weighted averages that the matrices considered were always Toeplitz. In contrast, the matrices considered in Theorems 4.1.3–4.1.5 are not necessarily Toeplitz. In particular the matrix of Corollary 4.1.1 is not Toeplitz since there $\sum_{k=1}^{\infty} | a_{nk} | = n^{1/2}(\log n)^{-1/2-\delta} \to \infty$ as $n \to \infty$. Thus, there are matrices which imply the almost sure convergence of $\sum_{k=1}^{\infty} a_{nk}X_k$ to zero when $EX_1 = 0$ and $EX_1^4 < \infty$ (indeed $E| X_1 |^{2+\delta} < \infty$ for some $\delta > 0$ will do) and yet for which there exist constants $x_k \to 0$ such that $\sum_{k=1}^{\infty} a_{nk}x_k \not\to 0$.

**Example 4.1.2.** The Abelian summability matrix (discretised) is given by

$$a_{nk} = \left(\frac{2^n - 1}{2^n}\right)^{k-1} \frac{1}{2^n}$$

for each $k \geq 1$, $n \geq 1$. It is clearly Toeplitz and, therefore, regular. Do any of the theorems on stochastic summability apply? We check the conditions of Theorem 4.1.5. For each $n \geq 1$,

$$a_{nk}^2 \leq k^{-1}$$

for each $k \geq 1$.

$$C_n = \sum_{k=1}^{\infty} a_{nk}^2 = (2^{n+1} - 1)^{-1} \leq n^{-1}$$

for each $n \geq 1$. Thus, by Theorem 4.1.5, $EX_1^2 < \infty$ and $EX_1 = 0$ imply

$$\sum_{k=1}^{\infty} a_{nk}X_k \to 0 \qquad \text{a.s.}$$

Clearly Theorem 4.1.4 can be made to apply also. We omit a discussion of this. ∎

**Exercise 4.1.6.**   [Varberg, 1966].   Let $T_n = \sum_{j,k=1}^{n} a_{jk}X_jX_k$ for $n \geq 1$. Suppose $EX_1 = 0$, $EX_1^2 = 1$, $EX_1^4 < \infty$, $\sum_{j,k=1}^{\infty} a_{jk}^2 < \infty$. Prove that $T_n - ET_n \to 0$ a.s. Hint: Show that $\{T_n - ET_n, n \geq 1\}$ is a martingale. ∎

## 4.2.   Other Dependence Structures

Theorem 4.1.3 and the more general Theorem 4.1.4 assumed that the $X_k$ were independent identically distributed. If the assumption of identical distributions is dropped, results analogous to Theorems 4.1.3 and 4.1.4 can be stated where the assumptions on the $\{a_{nk}\}$ matrix remain the same and the moment assumptions for the $X_k$ are mildly strengthened [Stout, 1968]. We will not discuss these results.

Komlos and Révész [1964] prove an interesting result for weighted averages of independent but not necessarily identically distributed random variables.

**Theorem 4.2.1.** Let $\{X_k, k \geq 1\}$ be a sequence of independent random variables with common mean $EX_1$ and finite variances $\{\sigma_k^2, k \geq 1\}$

such that $\sum_{k=1}^{\infty} \sigma_k^{-2} = \infty$. Then the weighted average

$$\sum_{k=1}^{n} X_k \sigma_k^{-2} \Big/ \sum_{k=1}^{n} \sigma_k^{-2} \to EX_1 \quad \text{a.s.}$$

**Proof.** Let

$$Y_k = \frac{X_k - EX_1}{\sigma_k^2 \sum_{j=1}^{k} \sigma_j^{-2}}$$

for $k \geq 1$. Then

$$\text{var } Y_k = \frac{1}{(\sigma_k \sum_{j=1}^{k} \sigma_j^{-2})^2} \leq \frac{1}{\sum_{j=1}^{k-1} \sigma_j^{-2}} - \frac{1}{\sum_{j=1}^{k} \sigma_j^{-2}}$$

for $k \geq 1$. Thus

$$\sum_{k=1}^{\infty} \text{var } Y_k < \infty$$

and hence $\sum_{k=1}^{\infty} Y_k$ converges a.s. By the Kronecker lemma the desired result follows. ∎

**Exercise 4.2.1.** Let $\{X_k, k \geq 1\}$ be a sequence of independent random variables with finite variances $\{\sigma_k^2, k \geq 1\}$ and zero means.

(i) Let $\{a_k, 1 \leq k \leq n\}$ be a sequence of real numbers such that $\sum_{k=1}^{n} a_k = 1$ for a positive integer $n \geq 1$. Prove that

$$\text{var}\left(\sum_{k=1}^{n} a_k X_k\right) \geq \left(\sum_{k=1}^{n} \sigma_k^{-2}\right)^{-1}$$

and that equality is obtained when

$$a_k = \sigma_k^{-2} \Big/ \sum_{j=1}^{n} \sigma_j^{-2} \quad \text{for} \quad k \leq n.$$

(ii) Assume $\sum_{k=1}^{\infty} \sigma_k^{-2} < \infty$ and that the $X_k$ are normally distributed. Let the coefficient matrix $\{a_{nk}\}$ be given by

$$a_{nk} = \begin{cases} a_k \Big/ \sum_{i=1}^{n} a_i & \text{if } k \leq n \\ 0 & \text{if } k > n \end{cases}$$

where $a_1 > 0$, $a_i \geq 0$ for $i \geq 1$, and $\sum_{i=1}^{\infty} a_i = \infty$ is assumed. Note $\sum_{k=1}^{n} a_{nk} X_k$ is normal with mean zero and variance $\sum_{k=1}^{n} a_{nk}^2 \sigma_k^2$. Prove that

$\sum_{k=1}^{n} a_{nk}X_k$ does not converge almost surely. Hint: Show that it suffices to prove that $\sum_{k=1}^{n} a_{nk}^2 \sigma_k^2$ does not converge to zero as $n \to \infty$. ∎

Note the significance of Exercise 4.2.1 (ii), namely, that when

$$\sum_{k=1}^{\infty} \sigma_k^{-2} < \infty,$$

it is possible that no weighted average for which $\sum_{k=1}^{\infty} a_k = \infty$ may converge almost surely. Note on the other hand by Theorem 4.2.1 that $\sum_{k=1}^{\infty} \sigma_k^{-2} = \infty$ is a sufficient condition for the almost sure convergence of the weighted average for at least one set of weights, those that minimize the variance of the weighted average.

**Example 4.2.1.** Theorem 4.2.1 has an interesting statistical interpretation. Suppose that $X_1, X_2, \ldots,$ is a sequence of independent measurements of an unknown quantity $\mu$. Suppose that each measurement is unbiased in the sense that $EX_i = \mu$ for $i \geq 1$. Is there a strongly consistent sequence of estimators of $\mu$; that is, does there exist

$$f_n(X_1, X_2, \ldots, X_n) \to \mu \qquad \text{a.s.?}$$

Theorem 4.2.1 says that the minimum variance estimators

$$\left\{ \sum_{k=1}^{n} X_k \sigma_k^{-2} \Big/ \sum_{k=1}^{n} \sigma_k^{-2}, n \geq 1 \right\}$$

is such a sequence provided the measurement imprecision (as given by the $\sigma_i^2$) does not grow too fast as $n \to \infty$. For example, $\sigma_i^2 \leq K < \infty$ for each $i \geq 1$ is more than sufficient. $\sigma_i^2 = i$ for each $i \geq 1$ is even sufficient. ∎

We now pass to the study of the stability of weighted sums $\{T_n, n \geq 1\}$ of martingale differences $\{X_k, k \geq 1\}$. As in the case of independent identically distributed random variables, the analysis is first conducted for $T_n$ of the form $T_n = \sum_{k=1}^{n} a_k X_k / B_n$, where $\{B_n, n \geq 1\}$ is a sequence of constants increasing to infinity, and then the analysis is conducted for $T_n$ of the form $T_n = \sum_{k=1}^{\infty} a_{nk}X_k$ with only magnitude restrictions placed on the $\{a_{nk}\}$ matrix.

Central to the first analysis mentioned above is the maximal inequality given in Theorem 4.2.2 below. We require some preliminary work.

According to Chow [1966], a random variable $X$ is *generalized Gaussian* if there exists a constant $\alpha > 0$ such that

$$E \exp(uX) \leq \exp(u^2\alpha^2/2)$$

for all $u \in R_1$. The minimum of such numbers $\alpha$ is denoted by $\alpha(X)$. Special cases of generalized Gaussian random variables include normal random variables with zero mean and random variables bounded almost surely by a constant with zero mean, two important classes of random variables. The term "generalized Gaussian" is, of course, motivated by the fact that normal random variables are often referred to as Gaussian random variables.

**Definition 4.2.1.** Let $X$ be a random variable and $\mathscr{G}$ a $\sigma$ field. $X$ is said to be locally generalized Gaussian with respect to $\mathscr{G}$ if there exists a constant $\alpha > 0$ such that

$$E[\exp(uX) \mid \mathscr{G}] \leq \exp(u^2\alpha^2/2) \qquad \text{a.s.}$$

for all $u \in R_1$. The minimum of such numbers is denoted by $\alpha(X)$. ∎

When $\{X_k, \mathscr{F}_k, k \geq 1\}$ is a martingale difference sequence and $\{a_k, k \geq 1\}$ a sequence of real numbers, Azuma [1967] studies the question of what norming constants $\{B_n, n \geq 1\}$ stabilize $\{\sum_{k=1}^{n} a_k X_k, n \geq 1\}$ under the additional hypotheses that $X_k$ is locally generalized Gaussian with respect to $\mathscr{F}_{k-1}$ with $\alpha(X_k) \leq 1$ for each $k \geq 1$, recalling that $\mathscr{F}_0 = \{\varnothing, \Omega\}$.

**Lemma 4.2.1.** Let $X$ be locally generalized Gaussian with respect to a $\sigma$ field $\mathscr{G}$ with constant $\alpha(X)$. Then $E[X \mid \mathscr{G}] = 0$ a.s. and $E[X^2 \mid \mathscr{G}] \leq \alpha^2(X)$ a.s.

*Remark.* Note by Lemma 4.2.1 that the conditional variance is a lower bound for $\alpha(X)$. When $X$ is normal with $EX = 0$ and $\mathscr{G} = \{\varnothing, \Omega\}$, this bound is sharp.

*Exercise 4.2.2.* (i) Prove Lemma 4.2.1: Hint: Expand the exponential function appearing in the definition of a locally generalized Gaussian random variable in a power series.

(ii) Let $E[X \mid \mathscr{G}] = 0$ a.s. and $|X| \leq C$ a.s. for a given $\sigma$ field $\mathscr{G}$ and constant $C$. Prove that $X$ is locally generalized Gaussian with $\alpha^2(X) \leq 2C^2$. (Actually, $\alpha^2(X) \leq C^2$ can be shown—see Lemma 4.2.2.) ∎

**Corollary 4.2.1.** Let $\{\mathscr{F}_k, k \geq 1\}$ be $\sigma$ fields adapted to $\{X_k, k \geq 1\}$. Suppose $X_k$ is locally generalized Gaussian with respect to $\mathscr{F}_{k-1}$ for each $k \geq 1$. Then $\{X_k, \mathscr{F}_k, k \geq 1\}$ is a martingale difference sequence.

*Proof.* Immediate from Lemma 4.2.1. ∎

In the remainder of this section, saying that $\{X_k, \mathscr{F}_k, k \geq 1\}$ is a locally generalized Gaussian stochastic sequence will mean that $\{X_k, \mathscr{F}_k, k \geq 1\}$ is a martingale difference sequence with $X_k$ locally generalized Gaussian with respect to $\mathscr{F}_{k-1}$. We can now state and prove Azuma's basic maximal inequality.

**Theorem 4.2.2.**   [Azuma, 1967].   Let $\{X_k, \mathscr{F}_k, k \geq 1\}$ be locally generalized Gaussian with $\alpha(X_k) \leq 1$ for each $k \geq 1$. Let $\{a_k, k \geq 1\}$ be a sequence of real numbers. Let $U_n = \sum_{k=1}^{n} a_k X_k$ for all $n \geq 1$. Then

$$E \exp(u \max_{k \leq n} \mid U_k \mid) \leq 8 \exp\left(u^2 \sum_{k=1}^{n} a_k^2/2\right) \tag{4.2.1}$$

for each $n \geq 1$ and all real $u$.

**Proof.**   We apply the submartingale maximal inequality of Corollary 3.3.2 after doing some preliminary computation.

$$\alpha(a_k X_k) = \mid a_k \mid \alpha(X_k) \leq \mid a_k \mid$$

for each $k \geq 1$ by hypothesis. Fix $u \in R_1$. Thus

$$\begin{aligned} E \exp(u U_k) &= E\{\exp(u U_{k-1}) E[\exp(u a_k X_k) \mid \mathscr{F}_{k-1}]\} \\ &\leq \exp(u^2 a_k^2/2) E \exp(u U_{k-1}) \qquad \text{a.s.} \end{aligned}$$

By backwards induction on $k$, we obtain

$$E \exp(u U_k) \leq \exp\left(u^2 \sum_{j=1}^{k} a_j^2/2\right) \tag{4.2.2}$$

for all $u \in R_1$.

Since $\{U_k, k \geq 1\}$ is a nonnegative submartingale, by Corollary 3.3.2,

$$E \max_{k \leq n} \mid U_k \mid^\alpha \leq \left(\frac{\alpha}{\alpha - 1}\right)^\alpha E \mid U_n \mid^\alpha$$

for each $n \geq 1$ and each $\alpha > 1$.

$$\begin{aligned} E \exp(u \max_{k \leq n} \mid U_k \mid) &\leq E[\exp(u \max_{k \leq n} \mid U_k \mid) + \exp(-u \max_{k \leq n} \mid U_k \mid)] \\ &= 2E\left[\sum_{j=0}^{\infty} u^{2j} \max_{k \leq n} \mid U_k \mid^{2j}/(2j)!\right] \\ &= 2 \sum_{j=0}^{\infty} u^{2j} E \max_{k \leq n} \mid U_k \mid^{2j}/(2j)!. \end{aligned}$$

But

$$E \max_{k \leq n} |U_k|^{2j} \leq \left(\frac{2j}{2j-1}\right)^{2j} E |U_n|^{2j} \leq 4E |U_n|^{2j}$$

for all $j \geq 1$, using the fact that $[\alpha/(\alpha - 1)]^\alpha \leq 4$ for $\alpha \geq 2$. Thus

$$2 \sum_{j=0}^{\infty} u^{2j} E \max_{k \leq n} |U_k|^{2j}/(2j)! \leq 8 \sum_{j=0}^{\infty} u^{2j} E |U_n|^{2j}/(2j)!$$

$$= 4[E \exp(uU_n) + E \exp(-uU_n)]$$

$$\leq 8 \exp\left(u^2 \sum_{j=1}^{n} a_j^2/2\right)$$

by Eq. (4.2.2), as desired. ∎

We now apply Theorem 4.2.2 to obtain a convergence result. Recall that $\log_2$ denotes log log.

**Theorem 4.2.3.** [Azuma, 1967]. Let $\{X_k, \mathscr{F}_k, k \geq 1\}$ be locally generalized Gaussian with $\alpha(X_k) \leq 1$ for each $k \geq 1$. Let $\{a_k, k \geq 1\}$ be a sequence of real numbers and $D_n = \sum_{k=1}^{n} a_k^2$ for all $n \geq 1$. Suppose $a_n^2/D_n \to 0$ and $D_n \to \infty$ as $n \to \infty$. Then

$$\limsup \left|\sum_{k=1}^{n} a_k X_k\right| \Big/ (2D_n \log_2 D_n)^{1/2} \leq 1 \qquad \text{a.s.} \qquad (4.2.3)$$

**Remarks.**   (i)   Note that

$$V_n \equiv \sum_{k=1}^{n} E[(a_k X_k)^2 \mid \mathscr{F}_{k-1}]$$

$$= \sum_{k=1}^{n} a_k^2 E[X_i^2 \mid \mathscr{F}_{k-1}] \leq \sum_{k=1}^{n} a_k^2 = D_n$$

by Lemma 4.2.1. The material of Chapter 5 on the law of the iterated logarithm will show that $(2V_n \log_2 V_n)^{1/2}$ would be the best possible denominator in Eq. (4.2.3) (instead of $(2D_n \log_2 D_n)^{1/2}$). Thus Azuma's result is reasonably sharp.

(ii)   Clearly $a_k = 1$ for each $k \geq 1$ satisfies the hypotheses of Theorem 4.2.3. Thus:

**Corollary 4.2.2.** Let $\{X_k, \mathscr{F}_k, k \geq 1\}$ be locally generalized Gaussian with $\alpha(X_k) \leq 1$ for each $k \geq 1$. Then

$$\limsup S_n/(2n \log_2 n)^{1/2} \leq 1 \qquad \text{a.s.}$$

**Proof of Theorem 4.2.3.**   Fix $\delta > 0$. Let $a(n) = (2D_n \log_2 D_n)^{1/2}$ for $n \geq 1$. It suffices to show that

$$P\left[\sum_{j=1}^{n} a_j X_j > (1 + \delta)a(n) \text{ i.o.}\right] = 0.$$

We prove this by choosing an increasing sequence of positive integers $\{n_k, k \geq 1\}$ such that

$$P\left[\max_{n \leq n_{k+1}} \sum_{j=1}^{n} a_j X_j > (1 + \delta)a(n_k) \text{ i.o.}\right] = 0.$$

Let (for the moment) $\{n_k, k \geq 1\}$ be any increasing sequence of positive integers. For any random variable $X$ and positive constants $a$ and $u$,

$$P[X > a] \leq \exp(-au)E \exp(uX).$$

Applying this to $X = \max_{n \leq n_{k+1}} |\sum_{j=1}^{n} a_j X_j|$, $a = (1 + \delta)a(n_k)$, and $u = (1 + \delta)a(n_k)/D_{n_{k+1}}$, and using Theorem 4.2.2, we obtain

$$P\left[\max_{n \leq n_{k+1}} \left|\sum_{j=1}^{n} a_j X_j\right| > (1 + \delta)a(n_k)\right] \leq 8 \exp\left[-(1 + \delta)^2 \frac{D_{n_k} \log_2 D_{n_k}}{D_{n_{k+1}}}\right]$$

for $k$ sufficiently large. For each $k \geq 1$ let $n_k$ be the smallest $n \geq 1$ such that $D_n > p^k$ where $p$ is chosen to satisfy $(1 + \delta)^2 p^{-1} > 1$. Using the hypothesis that $a_n^2/D_n \to 0$, it follows that $D_{n_k} \sim p^k$ and $D_{n_k}/D_{n_{k+1}} \to p^{-1}$. Thus, for $k$ sufficiently large,

$$8 \exp\left[-(1 + \delta)^2 \frac{D_{n_k} \log_2 D_{n_k}}{D_{n_{k+1}}}\right] \leq Ck^{-(1+\delta')}$$

for some $C < \infty$ and $\delta' > 0$. Thus

$$\sum_{k=1}^{\infty} P\left[\max_{n \leq n_{k+1}} \left|\sum_{j=1}^{n} a_j X_j\right| > (1 + \delta)a(n_k)\right] < \infty.$$

An application of the Borel–Cantelli lemma completes the proof.  ∎

Note that if $0 < B_n \to \infty$ such that $B_n/D_n \to \infty$, then supposing the hypotheses of Theorem 4.2.3 we have that $\sum_{k=1}^{n} a_k X_k/B_n \to 0$ a.s. In this manner Theorem 4.2.3 can be used to obtain stability results.

**Theorem 4.2.4.**   Let $\{X_k, \mathscr{F}_k, k \geq 1\}$ be locally generalized Gaussian with $\alpha(X_k) \leq 1$ for all $k \geq 1$. Let $\{a_k, k \geq 1\}$ be a sequence of positive real numbers with $A_n = \sum_{k=1}^{n} a_k$ for $n \geq 1$. Suppose

$$a_n/A_n = o(\log_2^{-1} A_n)$$

and $A_n \to \infty$. Then $T_n = \sum_{k=1}^{n} a_k X_k / A_n \to 0$ a.s.

**Proof.** Similar to that of Theorem 4.2.3. Omitted. ∎

**Remark.** Note the relationship between Theorem 4.2.4 and Theorems 4.1.1 and 4.1.2.

**Example 4.2.2.** Let $a_k = k^{-1}$ for each $k \geq 1$ and $\{X_k, \mathscr{F}_k, k \geq 1\}$ be locally generalized Gaussian with $\alpha(X_k) \leq 1$ for all $k \geq 1$. Since $A_n \sim \log n$, clearly $a_n/A_n = o(\log_2^{-1} A_n)$, and $A_n \to \infty$. Thus by Theorem 4.2.4,

$$\sum_{k=1}^{n} k^{-1} X_k / \log(n+1) \to 0 \qquad \text{a.s.}$$

This summability method defined by

$$a_{nk} = k^{-1}/\log(n+1) \qquad \text{for} \quad k \geq 1, n \geq 1$$

is essentially the regular logarithmic method of Riesz (see Example 4.1.1). ∎

Supposing that $\{X_i, i \geq 1\}$ is a martingale difference sequence, we now look at the question of when $T_n = \sum_{k=1}^{n} a_{nk} X_k$ is stable, assuming only magnitude restrictions on the $\{a_{nk}\}$ matrix.

**Lemma 4.2.2.** Let $\{X_k, \mathscr{F}_k, k \geq 1\}$ be locally generalized Gaussian with $\alpha(X_k) \leq 1$ for all $k \geq 1$. Suppose $C_n < \infty$ for each $n \geq 1$. Then

$$T_n = \sum_{k=1}^{\infty} a_{nk} X_k$$

is generalized Gaussian with

$$\alpha^2(T_n) \leq C_n$$

for each $n \geq 1$.

**Proof.** The proof is very similar to the initial segment of the proof of Theorem 4.2.3 and hence is omitted. ∎

**Theorem 4.2.5.** [Stout, 1968]. Let $\{X_k, \mathscr{F}_k, k \geq 1\}$ be locally generalized Gaussian with $\alpha(X_k) \leq 1$ for each $k \geq 1$. Suppose

$$C_n = o(\log^{-1} n).$$

Then $T_n = \sum_{k=1}^{\infty} a_{nk} X_k$ is stable at zero; indeed, $T_n$ converges completely to zero.

**Proof.**  Fix $\varepsilon > 0$ and $n \geq 1$. By Lemma 4.2.2,

$$E \exp(u T_n) \leq \exp(u^2 C_n / 2)$$

for all $u \in R_1$. Hence

$$
\begin{aligned}
P[|T_n| > \varepsilon] &= P[T_n < -\varepsilon] + P[T_n > \varepsilon] \\
&\leq \exp(-u\varepsilon)[E \exp(u T_n) + E \exp(-u T_n)] \\
&\leq \exp(-u\varepsilon + u^2 C_n / 2).
\end{aligned}
$$

Choosing $u = \varepsilon / C_n$ yields

$$\exp(-u\varepsilon + u^2 C_n / 2) = \exp[-\varepsilon^2 / (2 C_n)].$$

$C_n = o(\log^{-1} n)$ implies

$$\sum_{n=1}^{\infty} \exp[-\varepsilon^2 / (2 C_n)] < \infty.$$

Thus $\sum_{n=1}^{\infty} P[|T_n| > \varepsilon] < \infty$ for all $\varepsilon > 0$ as desired.  ∎

**Example 4.2.3.**  Let $\{a_{nk}\}$ be defined by

$$a_{nk} = (n - k + 1)^{-1} / \log(n + 1)$$

for each $k \geq 1$, $n \geq 1$. This is the harmonic method of summability. $\{a_{nk}\}$ is easily shown to be regular. Note incidentally that $a_{nk}$ is not of the form $a_{nk} = a_k / B_n$ for $k \leq n$, $= 0$ for $k > n$. Among regular methods of summability, the harmonic method is rather weak in that it brings about convergence of $\sum_{k=1}^{\infty} a_{nk} x_k$ for relatively few divergent sequences $\{x_k, k \geq 1\}$.

$$C_n \sim \log^{-2} n$$

is easily seen and thus

$$C_n = o(\log^{-1} n).$$

Thus by Theorem 4.2.5,

$$\left( \sum_{k=1}^{n} (n - k + 1)^{-1} X_k \right) \Big/ \log(n + 1) \to 0 \qquad \text{a.s.}$$

provided $\{X_k, \mathscr{F}_k, k \geq 1\}$ is locally generalized Gaussian with $\alpha(X_k) \leq 1$ for each $k \geq 1$.  ∎

Theorem 4.2.5 is sharp in the sense that replacing

$$C_n = o(\log^{-1} n)$$

by

$$C_n = O(\log^{-1} n)$$

renders the statement in Theorem 4.2.5 false. This is shown by an example of Erdös (see Hill [1951, p. 404]). Theorem 4.2.5 was proved in the special case that the $X_k$ are coin tossing random variables ($X_k$ independent with $P[X_k = 1] = P[X_k = -1] = \frac{1}{2}$ for all $k \geq 1$) by Hill [1951] and in the special case that the $X_k$ are independent generalized Gaussian with $\alpha(X_k) \leq 1$ for all $k \geq 1$ by Chow [1966].

One can relax the dependence assumptions even further on the $X_k$ and still obtain interesting stability results.

**Definition 4.2.2.** $\{X_k, k \geq 1\}$ is said to be a multiplicative sequence if

$$E[X_{i_1} X_{i_2} \cdots X_{i_n}] = 0$$

for all $n \geq 1$ and all $1 \leq i_1 < i_2 < \cdots < i_n$. ∎

Note that a multiplicative sequence is an orthogonal sequence.

**Theorem 4.2.6.** Let $\{X_k, \mathscr{F}_k, k \geq 1\}$ be a martingale difference sequence such that $E[X_{i_1} X_{i_2} \cdots X_{i_n}]$ is finite for all $n \geq 1$ and $1 \leq i_1 < i_2 < \cdots < i_n$. Then $\{X_k, k \geq 1\}$ is a multiplicative sequence.

**Proof.** Choose $n \geq 1$ and $1 \leq i_1 < i_2 < \cdots < i_n$.

$$
\begin{aligned}
E[X_{i_1} X_{i_2} \cdots X_{i_n}] &= E\{E[X_{i_1} X_{i_2} \cdots X_{i_n} \mid \mathscr{F}_{i_n-1}]\} \\
&= E\{X_{i_1} X_{i_2} \cdots X_{i_{n-1}} E[X_{i_n} \mid \mathscr{F}_{i_n-1}]\} = 0.
\end{aligned}
$$
∎

Thus assuming that a stochastic sequence is multiplicative strengthens the assumption of orthogonality and essentially weakens the assumption that the stochastic sequence is a martingale difference sequence. It is interesting to observe that certain lacunary trigonometric functions from a multiplicative sequence.

**Example 4.2.4.** Let $\{n_k, k \geq 1\}$ be a positive integer sequence such that $n_k/n_{k-1} \geq 3$ for $k \geq 2$. Let $\Omega = [0, \pi]$, $\mathscr{F}$ be the Borel sets of $\Omega$, and $2\pi P$ be Lebesgue measure. Then $\{\cos n_k x, k \geq 1\}$ and $\{\sin n_k x, k \geq 1\}$

can each be shown to be multiplicative stochastic sequences (see Zygmund [1959, pp. 208–209]). ∎

**Lemma 4.2.3.** (Azuma [1967] or Serfling [1969]). Let $\{X_k, k \geq 1\}$ be multiplicative with $|X_k| \leq 1$ a.s. for each $k \geq 1$ and $C_n < \infty$ for each $n \geq 1$. Then $T_n = \sum_{k=1}^{\infty} a_{nk} X_k$ is generalized Gaussian with $\alpha^2(T_n) \leq C_n$ for each $n \geq 1$.

*Proof.* Fix $u$ real and $n \geq 1$. $T_n$ is well defined since

$$\sum_{k=1}^{\infty} E a_{nk}^2 X_k^2 \leq C_n < \infty$$

and $\{a_{nk} X_k, k \geq 1\}$ is orthogonal, implying that $T_n$ exists as the $L_2$ limit of $T_{nK}$ as $K \to \infty$.

$$\exp(vx) \leq \cosh(v) + x \sinh(v)$$

is easily seen by the convexity of $f_v(x) = e^{vx}$, $-1 \leq x \leq 1$. Thus, for each integer $K \geq 1$,

$$E \exp(u T_{nK}) \leq E \prod_{k=1}^{K} [\cosh(u a_{nk}) + X_k \sinh(u a_{nk})]$$

$$= \prod_{k=1}^{K} \cosh(u a_{nk}) = \prod_{k=1}^{K} \sum_{m=0}^{\infty} u^{2m} a_{nk}^{2m}/(2m)!$$

$$\leq \prod_{k=1}^{K} \sum_{m=0}^{\infty} u^{2m} a_{nk}^{2m}/(2^m m!)$$

$$= \exp\left(u^2 \sum_{k=1}^{K} a_{nk}^2/2\right) \leq \exp(u^2 C_n/2).$$

Since $T_{nK} \to T_n$ in probability as $K \to \infty$, there exists a positive integer subsequence $\{K_j, j \geq 1\}$ such that $T_{nK_j} \to T_n$ a.s. as $j \to \infty$. By the Fatou lemma,

$$E \exp(u T_n) = E[\liminf_{j \to \infty} \exp(u T_{nK_j})]$$

$$\leq \liminf_{j \to \infty} E \exp(u T_{nK_j}) \leq \exp(u^2 C_n/2)$$

for all $u$ real. ∎

Lemma 4.2.3 enables us to generalize Theorem 4.2.5:

**Theorem 4.2.7.** [Azuma, 1967]. Let $\{X_k, k \geq 0\}$ be multiplicative with $|X_k| \leq 1$ a.s. for all $k \geq 1$ and $C_n < \infty$ for all $n \geq 1$. Then $C_n = o(\log^{-1} n)$ implies that $T_n = \sum_{k=1}^{\infty} a_{nk} X_k$ is stable at zero, indeed that $T_n$ converges completely to zero.

**Proof.** The proof is identical to that of Theorem 4.2.5 since both results follow from $T_n$ being generalized Gaussian with $\alpha^2(T_n) \leq C_n$ for $n \geq 1$. ∎

Recall that $\{X_k, k \geq 1\}$ stationary ergodic with $E|X_1| < \infty$ and

$$a_{nk} = \begin{cases} n^{-1} & \text{for} \quad k \leq n \\ 0 & \text{for} \quad k > n \end{cases}$$

for each $n \geq 1$ implies by the pointwise ergodic theorem that $T_n - EX_1 \to 0$ a.s. Throughout the remainder of Section 4.2, let $\{a_k, k \geq 1\}$ denote a sequence of positive real numbers and

$$a_{nk} = \begin{cases} a_k \Big/ \sum_{j=1}^{n} a_j & \text{if} \quad k \leq n \\ 0 & \text{if} \quad k > n \end{cases}$$

for each $n \geq 1$. Under somewhat restrictive hypotheses on the $a_k$, Baxter [1964] shows that $T_n - EX_1 \to 0$ a.s. The hypotheses of Baxter's theorem are satisfied for $a_k = 1$, $k \geq 1$. Hence Baxter's result can be viewed as an extension of the pointwise ergodic theorem to weighted averages of a stationary ergodic sequence. The heart of the proof of the pointwise ergodic theorem for weighted averages is a new maximal inequality analogous to the maximal ergodic theorem (Theorem 3.5.5).

Throughout the remainder of Section 4.2, $\{f_k, k \geq 1\}$ denotes a sequence of nonnegative constants such that $\sum_{k=1}^{\infty} f_k = 1$. It is assumed that the greatest common denominator of the indices $k$ for which $f_k > 0$ is one; that is, $\{f_k, k \geq 1\}$ has no "periodicities." Define $\{a_n, n \geq 1\}$ inductively by

$$a_1 = 1, \qquad a_2 = f_1 a_1, \ldots, a_n = \sum_{i=1}^{n-1} f_i a_{n-i}, \ldots .$$

Throughout the remainder of Section 4.2, $\{a_n, n \geq 1\}$ denotes this specially constructed sequence.

$\{a_n, n \geq 1\}$ has an interesting probabilistic interpretation that helps give some feel for its behavior. Let $\{Y_i, i \geq 1\}$ be a sequence of independent identically distributed random variables with $P[Y_1 = i] = f_i$ for each $i \geq 1$.

Let $U_n = \sum_{k=1}^{n} Y_k$ for $n \geq 1$. $\{U_n, n \geq 1\}$ has a physical interpretation worth mentioning. Consider a component in a system which operates a random number of times before failing and needing replaced. At that time a new component is inserted, renewing the system. The process is repeated indefinitely. Let $U_n$ be the time of the $n$th renewal. $\{U_n, n \geq 1\}$ is thus called the sequence of renewal times.

**Exercise 4.2.3.** Prove that the probability of a renewal at time $n - 1$ is $a_n$ for each $n \geq 2$. Hint: Condition on the time of the first renewal. ∎

Thus $\{a_n, n \geq 1\}$ can be thought of as a sequence of renewal probabilities. According to a famous result of probability theory, the sequence of renewal probabilities converges to one divided by the expected renewal time $EY_1$; that is

$$a_n \to \left( \sum_{k=1}^{\infty} kf_k \right)^{-1}$$

as $n \to \infty$. This rather intuitive result is somewhat messy to verify (see Feller [1968a, pp. 335–338]). The result shows that the behavior of $a_n$ is somewhat regular for large $n$. Indeed if $\sum_{k=1}^{\infty} kf_k < \infty$ (that is, $a_n \nrightarrow 0$),

$$a_n \sim a_{n+1}.$$

Thus given a stationary ergodic sequence $\{X_i, i \geq 1\}$ with $E|X_1| < \infty$ and $a_n \to (\sum_{k=1}^{\infty} kf_k)^{-1} \neq 0$, one might suspect on the basis of the pointwise ergodic theorem that

$$\sum_{k=1}^{n} a_k X_k \Big/ \sum_{k=1}^{n} a_k \to EX_1 \qquad \text{a.s.}$$

because of the asymptotic smoothness of the $a_n$. We shall see that this is true, even for the case $a_n \to 0$.

Baxter's maximal inequality depends on a numerical inequality:

**Lemma 4.2.4.** Let $\{x_n, n \geq 1\}$ be a sequence of real numbers. Let

$$y_1^{(n)} = x_n^+,$$

$$y_k^{(n)} = \left( x_{n-k+1} + \sum_{m=1}^{k-1} y_m^{(n)} f_{k-m} \right)^+$$

for $2 \leq k \leq n$ and all $n \geq 1$. Let $A_n = \{k \mid 1 \leq k \leq n, y_k^{(n)} > 0\}$ for $n \geq 1$.

Then

$$\sum_{k \in A_n} x_{n-k+1} \geq 0$$

for each $n \geq 1$.

**Proof.** Fix $n \geq 1$ and suppress superscripts. Let

$$B_{m,n} = \{k \mid m \leq k \leq n \text{ such that } y_k > 0\}$$

for all $1 \leq m \leq n$. Let $f_0 = 0$.

$$\sum_{k \in A_n} x_{n-k+1} = \sum_{k \in A_n} \left[ \left( x_{n-k+1} + \sum_{m=1}^{k-1} y_m f_{k-m} \right) - \sum_{m=1}^{k-1} y_m f_{k-m} \right]$$

$$= \sum_{k \in A_n} \left( y_k - \sum_{m=1}^{k-1} y_m f_{k-m} \right)$$

$$= \sum_{k \in A_n} y_k - \sum_{k \in A_n} \sum_{m \in A_n, m \leq k} y_m f_{k-m}$$

$$= \sum_{k \in A_n} y_k - \sum_{m \in A_n} y_m \sum_{k \in B_{m,n}} f_{k-m}$$

$$\geq \sum_{k \in A_n} y_k - \sum_{m \in A_n} y_m \sum_{j=1}^{n-m} f_j \geq 0. \quad \blacksquare$$

We can now prove Baxter's maximal ergodic theorem.

**Theorem 4.2.8.** Let $X$ be a random variable with $E \mid X \mid < \infty$ and $G$ a measure-preserving transformation. Let

$$U_k(\omega) = a_1 X(\omega) + a_2 X(G\omega) + \cdots + a_k X(G^{k-1}\omega)$$

define $U_k$ for $k \geq 1$. Then

$$\int_{[\sup_{k \geq 1} U_k > 0]} X \, dP \geq 0. \tag{4.2.4}$$

**Proof.** Fix $n \geq 1$. Let

$$X_k = XG^{k-1}$$

for $k \geq 1$, where $G^0$ is taken to be the identity transformation. Let

$$Y_1^{(n)} = X_n^{+}$$

and

$$Y_k^{(n)} = \left( X_{n-k+1} + \sum_{m=1}^{k-1} Y_m^{(n)} f_{k-m} \right)^+$$

for $2 \leq k \leq n$. Let

$$A_n = \{ k \mid 1 \leq k \leq n, \ Y_k^{(n)} > 0 \}.$$

($A_n$ is a random set.) Now

$$\sum_{k=1}^n \int_{[Y_k^{(k)} > 0]} X_1 \, dP = \sum_{k=1}^n \int_{[\omega \mid Y_k^{(k)}(G^{n-k}(\omega)) > 0]} X(G^{n-k}(\omega)) \, dP(\omega)$$

by Lemma 3.5.5. By the definition of the $Y_k^{(n)}$ and a simple induction $Y_k^{(n)}(\omega) = Y_k^{(k)}(G^{n-k}(\omega))$ for $1 \leq k \leq n$. Thus

$$\sum_{k=1}^n \int_{[Y_k^{(k)} > 0]} X_1 \, dP = \sum_{k=1}^n \int_{[Y_k^{(n)} > 0]} X_{n-k+1}(\omega) \, dP(\omega)$$

$$= \int_\Omega \sum_{k \in A_n} X_{n-k+1} \, dP.$$

By Lemma 4.2.4,

$$\int_\Omega \sum_{k \in A_n} X_{n-k+1} \, dP \geq 0.$$

Thus

$$\sum_{k=1}^n \int_{[Y_k^{(k)} > 0]} X_1 \, dP \geq 0. \qquad (4.2.5)$$

Let

$$Z_1^{(n)} = X_n$$

and

$$Z_k^{(n)} = X_{n-k+1} + \sum_{m=1}^{k-1} Z_m^{(n)} f_{k-m}$$

for $2 \leq k \leq n$. We show that $Y_m^{(n)} \geq Z_m^{(n)}$ for all $m \geq 1$ by induction. $Y_1^{(n)} \geq Z_1^{(n)}$. Suppose $Y_m^{(n)} \geq Z_m^{(n)}$ for all $1 \leq m \leq k - 1$.

$$Y_k^{(n)} = \left( X_{n-k+1} + \sum_{m=1}^{k-1} Y_m^{(n)} f_{k-m} \right)^+$$

$$\geq \left( X_{n-k+1} + \sum_{m=1}^{k-1} Z_m^{(n)} f_{k-m} \right)^+ \geq Z_k^{(n)},$$

completing the induction argument. A similar induction on $m$ shows that $Y_{m+1}^{(n+1)} \geq Y_m^{(n)}$ for each $1 \leq m \leq n$. Since $Y_{m+1}^{(n+1)} \geq Y_m^{(n)} \geq Z_m^{(n)}$ for $1 \leq m$ $\leq n$ and all $n \geq 1$,

$$Y_k^{(k)} \geq \sup_{1 \leq m \leq k} Z_m^{(m)} \qquad (4.2.6)$$

for all $k \geq 1$. Let

$$B_k = \{ \sup_{1 \leq m \leq k} Z_m^{(m)} > 0 \}$$

for each $k \geq 1$. Induction shows that

$$Z_k^{(n)} = \sum_{j=1}^{k} a_j X_{j+n-k}$$

for $1 \leq k \leq n$ and all $n \geq 1$ and hence that

$$Z_m^{(m)} = \sum_{j=1}^{m} a_j X_j = U_m$$

for all $m \geq 1$. Thus $B_k = [M_k > 0]$ where $M_k \equiv \max_{1 \leq m \leq k} U_m$. By definition of $B_k$, $X_1 = Z_1^{(1)} \leq 0$ on $B_k^c$ for all $k \geq 1$. Thus, by Eqs. (4.2.5) and (4.2.6),

$$\sum_{k=1}^{n} \int_{[M_k > 0]} X_1 \, dP = \sum_{k=1}^{n} \int_{B_k} X_1 \, dP \geq \sum_{k=1}^{n} \int_{Y_k^{(k)} > 0} X_1 \, dP \geq 0. \qquad (4.2.7)$$

This implies that

$$\sum_{k=1}^{n} \int_{[M_k > 0]} X_1 \, dP/n \geq 0$$

for all $n \geq 1$. $E \, | X_1 | < \infty$ implies

$$\int_{[M_k > 0]} X_1 \, dP \to \int_{[\sup_{m \geq 1} U_m > 0]} X_1 \, dP$$

as $k \to \infty$. Thus the Cesaro average

$$\sum_{k=1}^{n} \int_{[M_k > 0]} X_1 \, dP/n \to \int_{[\sup_{k \geq 1} U_k > 0]} X_1 \, dP.$$

Hence

$$\int_{[\sup_{k\geq 1} U_k > 0]} X_1 \, dP \geq 0$$

as desired.  ∎

It is now an easy matter to state and prove results analogous to the pointwise ergodic theorem and analogous to its application to stationary sequences.

**Theorem 4.2.9.**  [Baxter, 1964].  (i)  Let $X$ be a random variable with $E\,|\,X\,| < \infty$ and $G$ a measure preserving transformation. Let

$$U_n(\omega) = \sum_{k=1}^{n} a_k X G^{k-1}(\omega)$$

define $U_n$ for $n \geq 1$. Suppose

$$T_n = \sum_{k=1}^{n} a_k X G^{k-1} \Big/ \sum_{k=1}^{n} a_k$$

satisfies

$$T_n - T_n G \to 0 \qquad \text{a.s.} \tag{4.2.8}$$

Then $T_n \to E[X\,|\,\mathcal{I}]$ a.s., where $\mathcal{I}$ is the $\sigma$ field of invariant events of $G$. If, moreover, $G$ is ergodic, then $T_n \to EX$ a.s.

(ii)  Let $\{X_i, i \geq 1\}$ be a stationary sequence with $E\,|\,X_1\,| < \infty$. Suppose

$$\sum_{k=1}^{n} a_k (X_k - X_{k+1}) \Big/ \sum_{k=1}^{n} a_k \to 0 \qquad \text{a.s.} \tag{4.2.9}$$

Then $T_n = \sum_{k=1}^{n} a_k X_k / \sum_{k=1}^{n} a_k \to E[X\,|\,\mathcal{H}]$ a.s., where $\mathcal{H}$ is the $\sigma$ field of invariant events of $\{X_i, i \geq 1\}$. If, moreover, $\{X_i, i \geq 1\}$ is ergodic, then $T_n \to EX_1$ a.s.

**Proof.**  In the case of (i), Eq. (4.2.8) implies that lim sup $T_n / \sum_{i=1}^{n} a_i$ is almost surely an invariant random variable. In the case of (ii), Eq. (4.2.9) implies almost surely lim sup $T_n / \sum_{i=1}^{n} a_i$ is invariant. The rest of the proof is almost identical to the analogous proofs of Section 3.5 for the case $a_i = 1$ for all $i \geq 1$ and is omitted.  ∎

Of course, to be useful, it is necessary to find sufficient conditions for Eqs. (4.2.8) or (4.2.9). Papers of Baxter [1965] and Garsia and Sawyer

[1965], taking a more general approach that considers operators on function spaces, show that Eqs. (4.2.8) and (4.2.9) are always satisfied. We shall prove only a special case of this, namely that $a_n \sim a_{n+1}$ implies that Eqs. (4.2.8) and (4.2.9) are satisfied.

**Theorem 4.2.10.**  (i)  Suppose the hypotheses of Theorem 4.2.9 (i) with Eq. (4.2.8) replaced by $a_n \sim a_{n+1}$. Then $T_n$ converges a.s.

(ii)  Suppose the hypotheses of Theorem 4.2.9 (ii) with Eq. (4.2.9) replaced by $a_n \sim a_{n+1}$. Then $T_n$ converges a.s.

*Proof.*  We prove (i) only, (ii) following in the same manner. By Theorem 4.2.9 (i), it suffices to show that

$$T_n - T_n G \to 0 \qquad \text{a.s.}$$

It suffices to prove the result for $X \geq 0$ a.s. Fix $x > 0$. Applying Eq. (4.2.7) to $X - x$, it follows that

$$\sum_{k=1}^{n} \int_{[\max_{1 \leq j \leq k} T_j > x]} X \, dP \geq x \sum_{k=1}^{n} P[\max_{1 \leq j \leq k} T_j > x]$$

for all $n \geq 1$. Thus

$$\sum_{k=1}^{n} \int_{[\max_{1 \leq j \leq k} T_j > x]} X \, dP/n \geq x \sum_{k=1}^{n} P[\max_{1 \leq j \leq k} T_j > x]/n$$

for all $n \geq 1$. Letting $n \to \infty$ it follows that

$$\int_\Omega X \, dP \geq \int_{[\sup_{j \geq 1} T_j > x]} X \, dP \geq x P[\sup_{j \geq 1} T_j > x].$$

Since $\int_\Omega X \, dP < \infty$,

$$P[\sup_{j \geq 1} T_j = \infty] = 0. \qquad (4.2.10)$$

Let $X_k = XG^{k-1}$ for $k \geq 1$ define $\{X_k, k \geq 1\}$. Fix $\omega \in \Omega$ and $n \geq 1$ and let $a_0 = 0$.

$$| T_n(\omega) - T_n G(\omega) | \leq \left| \sum_{k=1}^{n} (a_k - a_{k-1}) X_k(\omega) \right| \bigg/ \sum_{k=1}^{n} a_k + a_n X_{n+1}(\omega) \bigg/ \sum_{k=1}^{n} a_k.$$

Fix $\varepsilon > 0$. By hypothesis

$$| (a_k - a_{k-1}) X_k(\omega) | < \varepsilon a_k X_k(\omega)$$

for $k$ sufficiently large. $\sum_{k=1}^{\infty} a_k = \infty$ follows easily from the definition of the $a_k$. Thus for all $n$ sufficiently large,

$$\frac{|\sum_{k=1}^{n} (a_k - a_{k-1})X_k(\omega)|}{\sum_{k=1}^{n} a_k} \leq \varepsilon \frac{1 + \sum_{k=1}^{n} a_k X_k(\omega)}{\sum_{k=1}^{n} a_k}$$

$$\leq \varepsilon(1 + \sup_{n \geq 1} T_n(\omega)) \qquad \text{a.s.}$$

Hence, recalling Eq. (4.2.10),

$$\left| \sum_{k=1}^{n} (a_k - a_{k-1})X_k \right| \Big/ \sum_{k=1}^{n} a_k \to 0 \qquad \text{a.s.}$$

To complete the proof, we must show that

$$a_n X_{n+1} \Big/ \sum_{k=1}^{n} a_k \to 0 \qquad \text{a.s.} \tag{4.2.11}$$

Since $a_{n+1} \sim a_n$, it suffices to show that $a_{n+1}X_{n+1}/\sum_{k=1}^{n} a_k \to 0$ a.s. Fix $\varepsilon > 0$. Let

$$Z_1 = X_1$$

and

$$Z_k = a_k X_k - \varepsilon \sum_{j=1}^{k-1} a_j$$

for $k \geq 2$ define $\{Z_k, k \geq 1\}$. We will show that $P[Z_k > 0 \text{ i.o.}] = 0$. In a similar way the result will follow for $\varepsilon < 0$ and $-Z_k$, implying Eq. (4.2.11) as desired. Fix $n \geq 2$.

Let

$$D_1 = D_2 = \cdots = D_{n-1} = -\varepsilon.$$

Let

$$Z_1^{(n)} = X_n$$

and

$$Z_k^{(n)} = D_{n-k+1} + \sum_{m=1}^{k-1} Z_m^{(n)} f_{k-m}$$

for $2 \leq k \leq n$. Let

$$Y_1^{(n)} = X_n^+$$

and

$$Y_k^{(n)} = \left( D_{n-k+1} + \sum_{m=1}^{k-1} Y_m^{(n)} f_{k-m} \right)^+$$

for $2 \leq k \leq n$. Applying Lemma 4.2.4 and integrating, it follows that

$$\sum_{k=1}^{n} \int_{[Y_k^{(n)} > 0]} D_{n-k+1} \, dP \geq 0.$$

Thus, by the definition of the $D_k$,

$$\int_{\Omega} X_n \, dP \geq \varepsilon \sum_{k=2}^{n} P[Y_k^{(n)} > 0]. \tag{4.2.12}$$

As in the proof of Theorem 4.2.8,

$$Z_k(G^{n-k}) = Z_k^{(n)} \leq Y_k^{(n)}$$

for $1 \leq k \leq n$ can be shown. Thus, since $G$ is measure preserving, using Eq. (4.2.12),

$$\infty > \int_{\Omega} X \, dP \geq \varepsilon \sum_{k=2}^{n} P[Z_k > 0].$$

Hence $\sum_{k=2}^{\infty} P[Z_k > 0] < \infty$ and $P[Z_k > 0 \text{ i.o.}] = 0$ as desired. ∎

**Example 4.2.5.** Let

$$f_i = (-1)^{i+1} \binom{\frac{1}{2}}{i}$$

for $i \geq 1$. It then follows that

$$a_{n+1} = \binom{2n}{n} 2^{-2n}$$

for each $n \geq 1$. Using Stirling's approximation,

$$a_{n+1} \sim (\pi n)^{-1/2}. \tag{4.2.13}$$

Thus by Theorem 4.2.10

$$T_n = \sum_{k=1}^{n} \binom{2k}{k} 2^{-2k} X_k \bigg/ \sum_{k=1}^{n} \binom{2k}{k} 2^{-2k}$$

converges a.s. Using Eq. (4.2.13), this implies

$$\sum_{k=1}^{n} (n/k)^{1/2} X_k / n$$

converges a.s. Indeed, a similar analysis shows that

$$\sum_{k=1}^{n} (n/k)^{\alpha} X_k / n$$

converges a.s. for each $0 \leq \alpha < 1$.
    In the case $\alpha = 1$,

$$T_n = \sum_{k=1}^{n} X_k / k.$$

It is easy to construct a stationary sequence for which $T_n = \sum_{k=1}^{n} X_k / k \rightarrow \infty$ a.s. For, let

$$P[(X_1, X_2, \ldots, X_n) = (0, 1, 0, \ldots)]$$
$$= P[(X_1, X_2, \ldots, X_n) = (1, 0, 1, \ldots)] = \tfrac{1}{2}$$

for each $n \geq 1$.    ∎

# Exponential Inequalities, the Law of the Iterated Logarithm, and Related Results

## 5.1. Introduction

Let $\{X_i, i \geq 1\}$ be a sequence of independent normally distributed random variables with $EX_i = 0$ and $EX_i^2 = \sigma^2 < \infty$ for each $i \geq 1$. The moment generating function of $X_1$ is given by

$$E \exp(uX_1) = \exp(u^2\sigma^2/2)$$

for all $u \in R_1$. Thus, each $X_i$ is generalized Gaussian with $\alpha^2(X_i) = \sigma^2$. (See Chapter 4, p. 237.) It thus follows from Theorem 4.2.3 and the remark following it that (recalling that $\log_2$ denotes log log)

$$\limsup |S_n|/[2n\sigma^2 \log_2(n\sigma^2)]^{1/2} \leq 1 \qquad \text{a.s.}$$

This upper class result is thus a rather simple consequence of the maximal exponential inequality Theorem 4.2.2. By the Kolmogorov 0–1 law, it follows that

$$\limsup S_n/[2n\sigma^2 \log_2(n\sigma^2)]^{1/2} = C \qquad \text{a.s.}$$

for some $-1 \leq C \leq 1$. We will shortly prove, independently of the material just discussed in Chapter 4, that $C = 1$, this result being the law of the iterated logarithm for normal random variables. Thus, besides the upper class result

$$\limsup S_n/[2n\sigma^2 \log_2 (n\sigma^2)]^{1/2} \leq 1 \qquad \text{a.s.,}$$

the lower class result

$$\limsup S_n/[2n\sigma^2 \log_2 (n\sigma^2)]^{1/2} \geq 1 \qquad \text{a.s.}$$

also holds. These results are, of course, a special case of the previously mentioned (but as yet unproved) Hartman–Wintner law of the iterated logarithm (Theorem 3.2.9), a result that will be proved in Chapter 5. Below we prove the law of the iterated logarithm for normal random variables mentioned above as a prototype of various laws of the iterated logarithm to follow. This approach has the advantage of illustrating the basic ideas of the standard method for proving a law of the iterated logarithm, with a minimum of computational complexity. The computational backbone for this result is provided by the simple exponential inequality given by Lemma 5.1.1.

**Lemma 5.1.1.**   [Feller, 1968a, p. 175].   Let $X$ be normal with $EX = 0$ and $EX^2 = 1$. Then

$$P[X > \varepsilon] < \exp[-\varepsilon^2/2] \qquad \text{for} \quad \varepsilon > 1. \qquad (5.1.1)$$

For any given $\gamma > 0$ there exists $\varepsilon(\gamma)$ such that $\varepsilon \geq \varepsilon(\gamma)$ implies

$$P[X > \varepsilon] > \exp[-(1 + \gamma)\varepsilon^2/2]. \qquad (5.1.2)$$

**Proof.**   Fix $\varepsilon > 0$. $X$ has density $(2\pi)^{-1/2} \exp(-x^2/2)$ for all $x \in R_1$. For any $x > 0$,

$$(1 - 3x^{-4}) \exp(-x^2/2) < \exp(-x^2/2) < (1 + x^{-2}) \exp(-x^2/2).$$

Multiplying by $(2\pi)^{-1/2}$ and integrating over $(\varepsilon, \infty)$, we obtain

$$(2\pi)^{-1/2}(\varepsilon^{-1} - \varepsilon^{-3}) \exp(-\varepsilon^2/2) < P[X > \varepsilon]$$
$$< (2\pi)^{-1/2}\varepsilon^{-1} \exp(-\varepsilon^2/2).$$

Equation (5.1.1) is now immediate. Equation (5.1.2) follows by choosing $\varepsilon(\gamma)$ sufficiently large.   ∎

**Theorem 5.1.1.**   Let $\{X_i, i \geq 1\}$ be independent normally distributed with $EX_i = 0$ and $EX_i^2 = \sigma^2 < \infty$ for each $i \geq 1$. Then

$$\lim \sup S_n/[2n\sigma^2 \log_2(n\sigma^2)]^{1/2} = 1 \qquad \text{a.s.}$$

**Proof.**   For notational convenience, let $a(n) = (2n \log_2 n)^{1/2}$ for $n \geq 1$. Replacing $X_i$ by $X_i/\sigma$ if necessary, assume $EX_i^2 = 1$ for each $i \geq 1$ without loss of generality. Although we already know by Theorem 4.2.3 that

$$\limsup S_n/a(n) \leq 1 \qquad \text{a.s.}, \tag{5.1.3}$$

we give another proof based on Lemma 5.1.1.

Fix $\delta > 0$. It suffices to show that

$$P[S_n > (1 + \delta)a(n) \text{ i.o.}] = 0. \tag{5.1.4}$$

Let $\{n_k, k \geq 1\}$ be any increasing sequence of positive integers. It suffices to show that

$$P[\max_{n \leq n_{k+1}} S_n > (1 + \delta)a(n_k) \text{ i.o.}] = 0. \tag{5.1.5}$$

Using the first Levy maximal inequality [Theorem 2.13.1 (i)] and Eq. (5.1.1) (choosing $k_1$ sufficiently large)

$$\sum_{k=k_1}^{\infty} P[\max_{n \leq n_{k+1}} S_n > (1 + \delta)a(n_k)]$$

$$\leq 2 \sum_{k=k_1}^{\infty} P[S_{n_{k+1}} > (1 + \delta)a(n_k)]$$

$$\leq 2 \sum_{k=k_1}^{\infty} \exp\left(-\frac{(1 + \delta)^2 n_k \log_2 n_k}{n_{k+1}}\right)$$

$$= 2 \sum_{k=k_1}^{\infty} (\log n_k)^{-(1+\delta)^2 n_k/n_{k+1}}.$$

Let $n_k = [p^k]$ where $(1 + \delta)^2 p^{-1} > 1$. The above series is then convergent. Thus by the Borel–Cantelli lemma, Eq. (5.1.5) and hence Eq. (5.1.4) holds. Thus Eq. (5.1.3) holds as desired.

Let

$$A_{n,\delta} = [S_n > (1 - \delta)a(n)]$$

for $n \geq 1$ and $\delta > 0$. Let $\{n_k, k \geq 1\}$ be an increasing sequence of positive integers to be specified later. In order to show that

$$\limsup S_n/a(n) \geq 1 \qquad \text{a.s.}, \tag{5.1.6}$$

it suffices to show that

$$P[A_{n_k,\delta} \text{ i.o. in } k] = 1$$

for each $\delta > 0$. It seems reasonable to try to apply the converse to the

Borel–Cantelli lemma. However, $\{A_{n_k,\delta}, k \geq 1\}$ is not a sequence of independent events. To surmount this difficulty, for each $\delta > 0$ we construct a sequence of independent events $\{B_{n_k,\delta}, k \geq 1\}$ such that

$$\sum_{k=1}^{\infty} P(B_{n_k,\delta}) = \infty$$

and such that

$$P[B_{n_k,\delta} \text{ i.o. in } k] = 1$$

for all $\delta > 0$ is easily seen to imply

$$P[A_{n_k,\delta} \text{ i.o. in } k] = 1$$

for all $\delta > 0$. Let

$$B_{n_k,\delta} = \{S_{n_{k+1}} - S_{n_k} > (1 - \delta)a(n_{k+1} - n_k)\}$$

for $k \geq 1$ and $\delta > 0$. Choose $n_{k+1} - n_k \to \infty$ as $k \to \infty$. Choose $\gamma > 0$. By Eq. (5.1.2), there exists $k_1$ such that

$$\sum_{k=k_1}^{\infty} P[B_{n_k,\delta}] \geq \sum_{k=k_1}^{\infty} \exp[-(1 + \gamma)(1 - \delta)^2 \log_2(n_{k+1} - n_k)]$$

$$= \sum_{k=k_1}^{\infty} \log(n_{k+1} - n_k)^{-(1+\gamma)(1-\delta)^2}.$$

Choosing $n_k = N^k$ for $k \geq 1$ and some integer $N > 1$ and $\gamma > 0$ such that $(1 + \gamma)(1 - \delta)^2 < 1$ makes the above series divergent. Thus, by the partial converse to the Borel–Cantelli lemma

$$P[B_{N^k,\delta} \text{ i.o. in } k] = 1$$

for all $\delta > 0$. By symmetry $X_i$ can be replaced by $-X_i$ for all $i \geq 1$ in Eq. (5.1.3) yielding

$$\liminf S_n/a(n) \geq -1 \qquad \text{a.s.}$$

Thus

$$P[S_{N^k} > -2\,a(N^k) \text{ for all } k \text{ sufficiently large}] = 1.$$

Let $C_k = [S_k > -2\,a(k)]$ for $k \geq 1$. It follows that

$$P[B_{N^k,\delta} \cap C_{N^k} \text{ i.o. in } k] = 1 \qquad (5.1.7)$$

for all $\delta > 0$.

$$B_{N^k,\delta} \cap C_{N^k} \subset [S_{N^{k+1}} > (1 - \delta)a(N^{k+1} - N^k) - 2a(N^k)]$$

$$= \left[ S_{N^{k+1}} > (1 - \delta)a(N^{k+1}) \left( \frac{a(N^{k+1} - N^k)}{a(N^{k+1})} - \frac{2a(N^k)}{(1 - \delta)a(N^{k+1})} \right) \right].$$

But

$$\frac{a(N^{k+1} - N^k)}{a(N^{k+1})} - \frac{2a(N^k)}{(1 - \delta)a(N^{k+1})} \sim \left( 1 - \frac{1}{N} \right)^{1/2} - \frac{2}{(1 - \delta)N^{1/2}}.$$

Thus,

$$[B_{N^k,\delta} \cap C_{N^k} \text{ i.o.}] \subset [S_{N^{k+1}} > (1 - \delta')a(N^{k+1}) \text{ i.o.}]$$

for fixed $\delta' > \delta$ and $N$ chosen sufficiently large. Thus, using Eq. (5.1.7),

$$P[A_{n,\delta'} \text{ i.o.}] = 1$$

for each $\delta' > 0$. ∎

Most of Chapter 5 is concerned with generalizing Theorem 5.1.1. With this in mind it is important to consider various implications of

$$\limsup S_n/[2n\sigma^2 \log_2(n\sigma^2)]^{1/2} = 1 \qquad \text{a.s.} \tag{5.1.8}$$

in order to better appreciate law of the iterated logarithm type results. First, Eq. (5.1.8) makes a remarkably precise statement about the magnitude of asymptotic fluctuations. It (and the symmetric analogue obtained by substituting $-X_i$ for $X_i$) implies, for arbitrary $\delta > 0$ that

$$P[S_n > (1 + \delta)[2n\sigma^2 \log_2 (n\sigma^2)]^{1/2} \text{ i.o.}] = 0$$

and

$$P[S_n < -(1 + \delta)[2n\sigma^2 \log_2 (n\sigma^2)]^{1/2} \text{ i.o.}] = 0$$

and yet

$$P[S_n > (1 - \delta)[2n\sigma^2 \log_2 (n\sigma^2)]^{1/2} \text{ i.o.}] = 1$$

and

$$P[S_n < -(1 - \delta)[2n\sigma^2 \log_2 (n\sigma^2)]^{1/2} \text{ i.o.}] = 1.$$

It is perhaps instructive to sketch $f(x) = \pm(1 + \delta)(2x\sigma^2 \log_2 x\sigma^2)^{1/2}$ for arbitrary real $\delta$ and fixed $\sigma^2$ in order to geometrically interpret these four probability statements. If $S_n$ represents a gambler's accumulated winnings at trial $n$, then the above four probability statements yield very precise information about the asymptotic fluctuations of the gambler's winnings.

Note that Eq. (5.1.8) implies that $S_n$ "crosses" zero infinitely often with probability one; that is, $P[(S_n \geq 0 \text{ eventually}) \cap (S_n \leq 0 \text{ eventually})] = 0$. It further implies that $\limsup S_n = \infty$ a.s. and $\liminf S_n = -\infty$ a.s. It also implies the strong law for $S_n$; that is, $S_n/n \to 0$ a.s. Indeed, it implies $S_n/n^{1/2+\varepsilon} \to 0$ a.s. for each $\varepsilon > 0$.

Consider $a \in [-1, 1]$. The law of the iterated logarithm does *not* imply that

$$P[S_n/(2n \log_2 n)^{1/2} \in (a - \varepsilon, a + \varepsilon) \text{ i.o.}] = 1$$

for every $\varepsilon > 0$ *unless* $a = -1$ or $a = 1$.

Generalizations of Theorem 5.1.1 in several directions are possible. Although discussing many of these, we concern ourselves mainly with two of these directions:

(i)   Let $\{X_i, i \geq 1\}$ be independent with zero means and finite variances but not necessarily normally distributed. Let $s_n{}^2 = \sum_{i=1}^{n} EX_i{}^2$. Under what conditions can we assert that

$$\limsup S_n/(2s_n{}^2 \log_2 s_n{}^2)^{1/2} = 1 \qquad \text{a.s.?}$$

This question culminates in the well known Kolmogorov law of the iterated logarithm [1929] and in the Hartman–Wintner law of the iterated logarithm.

(ii)   Let $\{X_i, i \geq 1\}$ be a stochastic sequence with $EX_i = 0$ and $EX_i{}^2 < \infty$ for each $i \geq 1$. Let $s_n{}^2 = \sum_{i=1}^{n} EX_i{}^2$. Under what dependence structures besides $\{X_i, i \geq 1\}$ an independent sequence can we assert that

$$\limsup S_n/(2s_n{}^2 \log_2 s_n{}^2)^{1/2} = 1 \qquad \text{a.s.}$$

or that a similar relationship holds? Pursuing this question, we obtain laws of the iterated logarithm for martingale differences, multiplicative random variables, and mixing random variables. In Section 5.2 we consider (i).

## 5.2.   The Kolmogorov–Prokhorov Exponential Inequalities with Applications

Throughout Section 5.2 $\{X_i, i \geq 1\}$ will denote a sequence of independent random variables with zero means and finite variances. Let $s_n{}^2 = \sum_{i=1}^{n} EX_i{}^2$ for all $n \geq 1$. Our ability to establish laws of the iterated logarithm for $S_n$ depends on our ability to establish inequalities similar to Eq. (5.1.1) when the $X_i$ are not assumed to be normal. Heuristically, one could expect the law of the iterated logarithm to hold for $\{X_i, i \geq 1\}$ such that for large $n$, $S_n/s_n$ is approximately normal with variance one. There are two approaches which develop this heuristic idea. Let $\Phi(x) = (2\pi)^{-1} \int_{-\infty}^{x} \exp(-y^2/2)\, dy$ for all real $x$.

One approach is to estimate

$$P[S_n/s_n > x] - [1 - \Phi(x)]$$

(or the ratio

$$P[S_n/s_n > x]/[1 - \Phi(x)])$$

for appropriate values of $x$. This amounts to finding a central limit theorem with error estimate. Exercise 5.2.1 illustrates this approach.

**Exercise 5.2.1.**    Let $S_0 = s_0^2 = 0$. Suppose

$$| P[(S_n - S_m)/(s_n^2 - s_m^2)^{1/2} > x] - [1 - \Phi(x)]|$$
$$\leq K[\log(s_n^2 - s_m^2)]^{-1-\varepsilon} \qquad (5.2.1)$$

for all $x > 0$, all $0 \leq m < n$ and some $K < \infty$ and $\varepsilon > 0$. Assume that the $X_i$ are symmetric. Suppose $EX_n^2/s_n^2 \to 0$ and $s_n^2 \to \infty$. Prove that

$$\limsup S_n/(2s_n^2 \log_2 s_n^2)^{1/2} = 1 \qquad \text{a.s.}$$

Hint: Follow the proof of Theorem 5.1.1, letting $n_k$ be the smallest $n$ such that $s_n^2 > \varrho^k$. ∎

This approach, which is quite productive, will be used in Section 5.4 to prove a law of the iterated logarithm for mixing random variables. The interested reader is also referred to Chapter 7 of Chung's text [1968] for an introductory treatment to this approach. A short and quite readable paper by Tomkins [1971b] also explores this approach as applied to sums of independent random variables. Work by Petrov [1966] is also closely related.

**Theorem 5.2.1.**    [Tomkins, 1971b].    Suppose $EX_n^2/s_n^2 \to 0$ and $s_n^2 \to \infty$. Let $u_n = (2 \log_2 s_n^2)^{1/2}$. Suppose

$$P[S_n/s_n > xu_n] \sim 1 - \Phi(xu_n)$$

and

$$P[S_n/s_n < -xu_n] \sim \Phi(-xu_n)$$

for all $x$ in some open interval containing 1. Then

$$\limsup S_n/(s_n u_n) = 1 \qquad \text{a.s.}$$

**Proof.**   Much like that of Theorem 5.1.1. Omitted. ∎

A second approach is to estimate

$$E \exp(uS_n/s_n)/\exp(u^2/2)$$

for appropriate values of $u$ and then to use this estimate to estimate the appropriate probabilities. We shall follow this second approach and establish the useful exponential inequalities in Theorem 5.2.2 below. Although the exponential inequalities necessary for the proof of Theorem 5.1.1 were easily obtained in Lemma 5.1.1, one should note that the derivation of the exponential inequalities given in Theorem 5.2.2 is genuinely delicate.

**Theorem 5.2.2.**   Let

$$|X_i| \le cs_n \qquad \text{a.s.}$$

for each $1 \le i \le n$ and $n \ge 1$. Suppose $\varepsilon > 0$ and $\gamma > 0$. Then for each $n \ge 1$,

(i)   $\varepsilon c \le 1$ implies that

$$P[S_n/s_n > \varepsilon] \le \exp[-(\varepsilon^2/2)(1 - \varepsilon c/2)],$$

(ii)

$$P[S_n/s_n \ge \varepsilon] \le \exp\left[-\frac{\varepsilon}{2c} \operatorname{arcsinh}\left(\frac{\varepsilon c}{2}\right)\right],$$

and

(iii)   there exist constants $\varepsilon(\gamma)$ and $\pi(\gamma)$ such that if $\varepsilon \ge \varepsilon(\gamma)$ and $\varepsilon c \le \pi(\gamma)$, then

$$P[S_n/s_n > \varepsilon] \ge \exp[-(\varepsilon^2/2)(1 + \gamma)].$$

**Remark.**   (i) and (iii) together with a weaker version of (ii) constitute what are referred to as the Kolmogorov exponential inequalities [1929]. (ii) is due to Prokhorov [1959a] and is very useful in studying the question of the characterization of the strong law of large numbers for independent random variables in terms of the magnitudes of the variances $\{EX_i^2, i \ge 1\}$, a topic we take up later in this section. ∎

**Proof.**   (i)   Fix $n \geq 1$. Suppose $\varepsilon c \leq 1$. For each $1 \leq i \leq n$,

$$E \exp(\varepsilon X_i / s_n) = 1 + \varepsilon^2 E X_i^2 / (2! s_n^2) + \varepsilon^3 E X_i^3 / (3! s_n^3) + \cdots$$

$$\leq 1 + \frac{\varepsilon^2 E X_i^2}{2 s_n^2} [1 + \varepsilon c / 3 + \varepsilon^2 c^2 / (3 \cdot 4) + \cdots]$$

$$\leq 1 + \frac{\varepsilon^2 E X_i^2}{2 s_n^2} [1 + \varepsilon c / 2] \leq \exp\left[\frac{\varepsilon^2 E X_i^2}{2 s_n^2} (1 + \varepsilon c / 2)\right]$$

since $1 + x \leq e^x$ for all $x$. By the independence of the $X_i$,

$$E \exp(\varepsilon S_n / s_n) \leq \exp[(\varepsilon^2 / 2)(1 + \varepsilon c / 2)].$$

Thus

$$P[S_n / s_n > \varepsilon] \leq \exp(-\varepsilon^2) E \exp(\varepsilon S_n / s_n)$$

$$\leq \exp[-(\varepsilon^2 / 2)(1 - \varepsilon c / 2)],$$

proving (i).

(ii)   Fix $n \geq 1$. Let $u > 0$. Since $e^x \geq ex$ for all $x \in R_1$,

$$E \exp(u X_i / s_n) \leq \exp[E \exp(u X_i / s_n) - 1]$$

for $1 \leq i \leq n$. Let

$$Q(x) = \sum_{i=1}^{n} P[X_i / s_n \leq x] / n$$

for all $x \in R_1$.

By the independence of the $X_i$ and the fact that $e^x - x - 1 \geq 0$ for all $x \in R_1$,

$$E \exp(u S_n / s_n) \leq \prod_{i=1}^{n} \exp[E \exp(u X_i / s_n) - 1]$$

$$= \exp\left[\sum_{i=1}^{n} E \exp(u X_i / s_n) - 1\right]$$

$$= \exp\left\{n \int_{-\infty}^{\infty} [\exp(ux) - 1 - ux] \, dQ(x)\right\}$$

$$\leq \exp\left\{2n \int_{-\infty}^{\infty} [\cosh(ux) - 1] \, dQ(x)\right\}.$$

By hypothesis, $Q$ assigns mass 1 to the interval $[-c, c]$ and

$$\int_{-\infty}^{\infty} x^2 \, dQ(x) = n^{-1}.$$

$$f(x) = \cosh(ux) - 1$$

is even and convex. Thus

$$\int_{-\infty}^{\infty} [\cosh(ux) - 1]\, dQ(x) \le \int_{-\infty}^{\infty} [\cosh(ux) - 1]\, dQ_1(x),$$

where $Q_1$ assigns mass to 0 and $c$ with as much mass as possible assigned to $c$ without violating

$$\int_{-\infty}^{\infty} x^2\, dQ_1(x) = n^{-1}.$$

Thus

$$Q_1\{c\} \le (nc^2)^{-1}.$$

This yields

$$\int_{-\infty}^{\infty} [\cosh(ux) - 1]\, dQ(x) \le [\cosh(uc) - 1](nc^2)^{-1}.$$

Combining,

$$E \exp(uS_n/s_n) \le \exp\left\{2n \int_{-\infty}^{\infty} [\cosh(ux) - 1]\, dQ(x)\right\}$$
$$\le \exp\{(2/c^2)[\cosh(uc) - 1]\}.$$

Thus

$$P[S_n/s_n > \varepsilon] \le \exp(-u\varepsilon) E \exp(uS_n/s_n)$$
$$\le \exp\{-u\varepsilon + (2/c^2)[\cosh(uc) - 1]\}.$$

Differentiation shows that

$$u = c^{-1} \operatorname{arcsinh}(\varepsilon c/2)$$

minimizes the right hand side of the above inequality. Setting $a = \operatorname{arcsinh}(\varepsilon c/2)$, it suffices to show

$$\cosh a - 1 \le (a \sinh a)/2.$$

But it is easily seen that

$$e^a(2 - a) + e^{-a}(2 + a) \le 4,$$

implying the above inequality.

(iii)  The proof (iii) is much more delicate than that of (i) or (ii). It suffices to prove the result for $\varepsilon$ sufficiently large and $\varepsilon c$ sufficiently small. Fix $u > 0$ and choose $c > 0$ such that $uc \le 1$, $u$ to be specified later.

Then, for $1 \leq i \leq n$,

$$E \exp(uX_i/s_n) = 1 + u^2 EX_i^2/(2!s_n^2) + u^3 EX_i^3/(3!s_n^3) + \cdots$$

$$\geq 1 + \frac{u^2 EX_i^2}{2s_n^2} (1 - uc/3 - u^2 c^2/(3 \cdot 4) \cdots)$$

$$\geq 1 + \frac{u^2 EX_i^2}{2s_n^2} (1 - uc/2) \geq \exp\left[\frac{u^2 EX_i^2}{2s_n^2} (1 - uc)\right].$$

The last inequality is computed from the fact that

$$1 + x \geq \exp(x - x^2)$$

for all $x > 0$. By the independence of the $X_i$,

$$E \exp(uS_n/s_n) \geq \exp[(u^2/2)(1 - uc)].$$

Fix $0 < \beta < 1$, to be specified later. Choose $c$ sufficiently small so that $uc \leq \beta^2/4$. Then

$$E \exp(uS_n/s_n) \geq \exp[(u^2/2)(1 - \beta^2/4)]. \qquad (5.2.2)$$

Let

$$Q(x) = P[S_n/s_n > x]$$

for all $x \in R_1$. Integration by parts yields

$$E \exp(uS_n/s_n) = -\int_{-\infty}^{\infty} \exp(ux)\, dQ(x)$$

$$= u \int_{-\infty}^{\infty} \exp(ux) Q(x)\, dx.$$

We now estimate $E \exp(uS_n/s_n)$ by splitting the last integral into parts; that is,

$$E \exp(uS_n/s_n) = \sum_{i=1}^{5} u \int_{A_i} \exp(ux) Q(x)\, dx,$$

where

$$A_1 = (-\infty, 0],$$

$$A_2 = (0, u(1 - \beta)),$$

$$A_3 = [u(1 - \beta), u(1 + \beta)],$$

$$A_4 = (u(1 + \beta), 8u],$$

$$A_5 = (8u, \infty).$$

$A_1$: $\int_{-\infty}^{0} u \exp(ux)Q(x)\,dx \leq \int_{-\infty}^{0} u \exp(ux)\,dx = 1$. $A_5$: Let $x \in A_5$. If $cx \leq 1$, (i) implies

$$Q(x) \leq \exp[-(x^2/2)(1 - xc/2)]$$
$$\leq \exp(-x^2/4) \leq \exp(-2ux)$$

since $x \geq 8u$. If $cx > 1$, by (ii),

$$Q(x) \leq \exp\left[-\frac{x}{2c}\,\text{arcsinh}(1/2)\right] \leq \exp(-2ux),$$

choosing

$$c \leq (4u)^{-1}\,\text{arcsinh}\,(1/2).$$

Thus

$$u\int_{A_5} \exp(ux)Q(x)\,dx \leq u\int_{8u}^{\infty} \exp(-ux)\,dx \leq 1.$$

$A_2$ and $A_4$: Choosing $c$ such that $c < (8u)^{-1}$ and then noting that $xc < 1$ for $x \in A_2 \cup A_4$, we have by (i) that

$$\exp(ux)Q(x) \leq \exp[ux - (x^2/2)(1 - xc/2)]$$
$$\leq \exp[ux - (x^2/2)(1 - 4uc)].$$

Let

$$g(x) = ux - (x^2/2)(1 - 4uc)$$

for all $x \in R_1$. $g$ is maximized at

$$x_0 = u/(1 - 4uc).$$

Choose $c$ sufficiently small so that $4uc(1 - \beta)^2 < \beta^2/2$ and $x_0 \in A_3$. Thus for $x \in A_2$,

$$g(x) \leq g[u(1 - \beta)]$$
$$= (u^2/2)(1 - \beta)(1 + \beta + 4uc - 4uc\beta)$$
$$< (u^2/2)(1 - \beta^2/2).$$

This yields

$$u\int_{0}^{u(1-\beta)} \exp(ux)Q(x)\,dx < u\int_{0}^{u(1-\beta)} \exp(g(x))\,dx$$
$$< u^2 \exp[(u^2/2)(1 - \beta^2/2)].$$

Similarly, using the fact that $x_0 \in A_3$,

$$u \int_{u(1+\beta)}^{8u} \exp(ux)Q(x)\,dx \le u \int_{u(1+\beta)}^{8u} \exp(g(x))\,dx$$

$$< 8u^2 \exp[(u^2/2)(1 - \beta^2/2)].$$

Combining,

$$u \int_{A_2 \cup A_4} \exp(ux)Q(x)\,dx < 9u^2 \exp[(u^2/2)(1 - \beta^2/2)]. \qquad (5.2.3)$$

Fix $\varepsilon(\gamma)$ to be chosen later and suppose $\varepsilon \ge \varepsilon(\gamma)$. Set $u = \varepsilon/(1 - \beta)$. Note that previous restrictions that $uc$ be small as a function of $\beta$ have now become the restriction that $\varepsilon c$ be small as a function of $\beta$. Applying Eq. (5.2.2) it follows that

$$9u^2 \exp[(u^2/2)(1 - \beta^2/2)] \le 9u^2 \exp[-u^2\beta^2/8]E \exp(uS_n/s_n). \qquad (5.2.4)$$

Choose $\varepsilon(\gamma)$ such that

$$u(1 - \beta) = \varepsilon \ge \varepsilon(\gamma)$$

implies

$$2 < (1/4)E \exp(uS_n/s_n) \qquad (5.2.5)$$

and

$$u \int_{A_2 \cup A_4} \exp(ux)Q(x)\,dx < \tfrac{1}{4}E \exp(uS_n/s_n).$$

This is possible by Eqs. (5.2.3) and (5.2.4), noting that

$$9u^2 \exp\left(\frac{-u^2\beta^2}{8}\right) \to 0 \qquad \text{as} \quad u \to \infty.$$

Recalling that

$$u \int_{A_1 \cup A_5} \exp(ux)Q(x)\,dx \le 2,$$

and using Eq. (5.2.5) it follows that

$$u \int_{A_1 \cup A_5} \exp(ux)Q(x)\,dx < (1/4)E \exp(uS_n/s_n).$$

Thus

$$E \exp(uS_n/s_n) = \sum_{i=1}^{5} u \int_{A_i} \exp(ux)Q(x)\, dx$$

$$< u \int_{A_3} \exp(ux)Q(x)\, dx + \tfrac{1}{2}E \exp(uS_n/s_n).$$

This implies

$$u \int_{u(1-\beta)}^{u(1+\beta)} \exp(ux)Q(x)\, dx > \tfrac{1}{2}E \exp(uS_n/s_n)$$

$$\geq \tfrac{1}{2} \exp[(u^2/2)(1 - \beta^2/4)]$$

by Eq. (5.2.2). But

$$u \int_{u(1-\beta)}^{u(1+\beta)} \exp(ux)Q(x)\, dx \leq 2u^2\beta \exp[u^2(1 + \beta)]Q(\varepsilon).$$

This implies

$$Q(\varepsilon) > \frac{1}{4u^2\beta} \exp(u^2\beta^2/8) \exp[-(u^2/2)(1 + 2\beta + \beta^2/2)].$$

$$\frac{1}{4u^2\beta} \exp(u^2\beta^2/8) \to \infty$$

as $u \to \infty$. Thus choosing $\varepsilon(\gamma)$ sufficiently large and substituting $u = \varepsilon/(1 - \beta)$,

$$Q(\varepsilon) > \exp\left[ - \frac{\varepsilon^2(1 + 2\beta + \beta^2/2)}{2(1 - \beta)^2} \right].$$

Choose $1 > \beta > 0$ such that

$$\frac{1 + 2\beta + \beta^2/2}{(1 - \beta)^2} \leq 1 + \gamma,$$

yielding

$$Q(\varepsilon) > \exp[-(\varepsilon^2/2)(1 + \gamma)]$$

for $\varepsilon \geq \varepsilon(\gamma)$ and $\varepsilon c$ sufficiently small as a function of $\gamma$, that is, for $\varepsilon c \leq \pi(\gamma)$, and for $\varepsilon \geq \varepsilon(\gamma)$, the desired result. ∎

**Remark.**  A careful check of the above proof is required if one is to be convinced that the parameters were not chosen in an inconsistent manner. $\gamma$ is the free parameter. $\gamma$ determines $\beta$. $\beta$ determines $\varepsilon(\gamma)$ and $\pi(\gamma)$.

Armed with the Kolmogorov exponential inequalities, we can now prove Kolmogorov's law of the iterated logarithm.

**Theorem 5.2.3.** Let $\{K_n, n \geq 1\}$ be positive constants such that $K_n \to 0$ as $n \to \infty$. Suppose $s_n^2 \to \infty$ and

$$| X_n | \leq K_n s_n (\log_2 s_n^2)^{-1/2} \qquad \text{a.s.}$$

for each $n \geq 1$. Then

$$\limsup S_n/(2s_n^2 \log_2 s_n^2)^{1/2} = 1 \qquad \text{a.s.}$$

**Proof.** We proceed much as in the proof of Theorem 5.1.1. Let

$$u_n = (2 \log_2 s_n^2)^{1/2}$$

for $n \geq 1$. First we show that

$$\limsup S_n/(s_n u_n) \leq 1 \qquad \text{a.s.}$$

Fix $\delta > 0$. Let $\{n_k, k \geq 1\}$ be an increasing sequence of positive integers.

$$P[S_n > (1 + \delta)s_n u_n \text{ i.o.}] \leq P[\max_{n \leq n_{k+1}} S_n > (1 + \delta)s_{n_k} u_{n_k} \text{ i.o.}].$$

By the first Lévy maximal inequality,

$$P\{\max_{n \leq n_{k+1}} [S_n - \mu(S_n - S_{n_{k+1}})] \geq \alpha\} \leq 2P[S_{n_{k+1}} \geq \alpha]$$

for all $\alpha > 0$. By Lemma 3.4.2,

$$\mu(S_n - S_{n_{k+1}}) \leq [2 \operatorname{var}(S_n - S_{n_{k+1}})]^{1/2} \leq 2^{1/2} s_{n_{k+1}}$$

for $n \leq n_{k+1}$. Thus

$$P[\max_{n \leq n_{k+1}} S_n \geq (1 + \delta)s_{n_k} u_{n_k}]$$

$$\leq P\{\max_{n \leq n_{k+1}} [S_n - \mu(S_n - S_{n_{k+1}})] + 2^{1/2} s_{n_{k+1}} > (1 + \delta)s_{n_k} u_{n_k}\}$$

$$\leq 2P[S_{n_{k+1}}/s_{n_{k+1}} \geq (1 + \delta)s_{n_k} u_{n_k}/s_{n_{k+1}} - 2^{1/2}].$$

By the Borel–Cantelli lemma, it suffices to show that

$$\sum_{k=1}^{\infty} P[S_{n_{k+1}}/s_{n_{k+1}} \geq (1 + \delta)s_{n_k} u_{n_k}/s_{n_{k+1}} - 2^{1/2}] < \infty.$$

Choose $\varrho > 1$ such that $(1 + \delta) > \varrho$. For each $k \geq 1$, let $n_k$ be the smallest integer $n$ such that $s_n > \varrho^k$. Since $s_n{}^2 \to \infty$, this can be done.

$$s_{n+1}^2/s_n{}^2 = 1 + \text{var}(X_{n+1})/s_n{}^2 \leq 1 + K_{n+1}^2 s_{n+1}^2/(s_n{}^2 \log_2 s_{n+1}^2)$$

as $n \to \infty$. It follows that

$$s_{n+1}^2/s_n{}^2 \sim 1, \qquad s_{n_k} \sim \varrho^k, \qquad \text{and that} \qquad s_{n_k}/s_{n_{k+1}} \sim \varrho^{-1}.$$

Noting that $1 + \delta > \varrho$, choose $0 < \delta' < \delta$ such that

$$(1 + \delta) \frac{s_{n_k} u_{n_k}}{s_{n_{k+1}} u_{n_{k+1}}} u_{n_{k+1}} - 2^{1/2} \geq (1 + \delta') u_{n_{k+1}}$$

for $k$ sufficiently large. Thus

$$P[S_{n_{k+1}}/s_{n_{k+1}} \geq (1 + \delta) s_{n_k} u_{n_k}/s_{n_{k+1}} - 2^{1/2}] \leq P[S_{n_{k+1}}/s_{n_{k+1}} \geq (1 + \delta') u_{n_{k+1}}]$$

for $k$ sufficiently large.

$$|X_i|/s_{n_{k+1}} \leq 2K_i s_i/(u_i s_{n_{k+1}}) \qquad \text{a.s.}$$

for $i \leq n_{k+1}$. Hence

$$c_{k+1} \equiv \max_{i \leq n_{k+1}} |X_i|/s_{n_{k+1}} \leq \max_{i \leq n_{k+1}} 2K_i s_i/(u_i s_{n_{k+1}}).$$

Thus, $c_{k+1}[(1 + \delta') u_{n_{k+1}}] \to 0$ as $k \to \infty$ since $K_i \to 0$ as $i \to \infty$ and $s_{n_k}/u_{n_k} \to \infty$ as $k \to \infty$. Thus by Theorem 5.2.2 (i), for any fixed $\eta > 0$,

$$P[S_{n_{k+1}}/s_{n_{k+1}} \geq (1 + \delta') u_{n_{k+1}}] \leq \exp[-(1 + \delta')^2 \log_2 s_{n_{k+1}}^2 (1 - \eta)]$$
$$= (\log s_{n_{k+1}}^2)^{-(1+\delta')^2(1-\eta)}.$$

for $k$ sufficiently large. Choose $\eta$ such that

$$(1 + \delta')^2 (1 - \eta) > 1.$$

Then, since $s_{n_{k+1}}^2 \sim \varrho^{2k}$,

$$\sum_{k=1}^{\infty} (\log s_{n_{k+1}}^2)^{-(1+\delta')^2(1-\eta)} < \infty.$$

Thus

$$P[S_n > (1 + \delta) s_n u_n \text{ i.o.}] = 0$$

for all $\delta > 0$ and hence

$$\limsup S_n/(s_n u_n) \leq 1 \qquad \text{a.s.}$$

as desired.

Fix $\delta > 0$ and let $\{n_k, k \geq 1\}$ be an increasing sequence of positive integers. To prove that

$$\limsup S_n/(s_n u_n) \geq 1 \qquad \text{a.s.,}$$

it suffices to show that

$$P[S_{n_k}/s_{n_k} > (1 - \delta)u_{n_k} \text{ i.o.}] = 1.$$

Let $n_k$ be the smallest integer $n$ such that $s_n > \varrho^k$, $\varrho > 1$ to be specified later. Let

$$B_{n_k, \delta} = \{S_{n_{k+1}} - S_{n_k} > (1 - \delta)[2(s_{n_{k+1}}^2 - s_{n_k}^2) \log_2 (s_{n_{k+1}}^2 - s_{n_k}^2)]^{1/2}\}$$

for each $k \geq 1$. We apply Theorem 5.2.2 (iii) with $\gamma$ chosen such that $(1 + \gamma)(1 - \delta)^2 < 1$. Note that

$$(1 - \delta)[2 \log_2 (s_{n_{k+1}}^2 - s_{n_k}^2)]^{1/2} \to \infty$$

and

$$\{(1 - \delta)[2 \log_2 (s_{n_{k+1}}^2 - s_{n_k}^2)]^{1/2}\}\{\max_{n_k < n \leq n_{k+1}} |X_n|/(s_{n_{k+1}}^2 - s_{n_k}^2)^{1/2}\}$$

$$\leq \{(1 - \delta)[2 \log_2(s_{n_{k+1}}^2 - s_{n_k}^2)]^{1/2}\}$$

$$\times \{\max_{n_k < n \leq n_{k+1}} K_n s_n \log_2^{-1/2} s_n^2/(s_{n_{k+1}}^2 - s_{n_k}^2)^{1/2}\} \to 0$$

as $k \to \infty$. Thus, by Theorem 5.2.2 (iii), for $k$ sufficiently large

$$P[B_{n_k, \delta}] \geq \exp[-(1 - \delta)^2 \log_2(s_{n_{k+1}}^2 - s_{n_k}^2)(1 + \gamma)] \sim Ck^{-(1+\gamma)(1-\delta)^2}$$

since

$$s_{n_{k+1}}^2 - s_{n_k}^2 = s_{n_{k+1}}^2(1 - s_{n_k}^2/s_{n_{k+1}}^2) \sim \varrho^{2(k+1)}(1 - \varrho^{-2}).$$

$$\sum_{k=1}^{\infty} k^{-(1+\gamma)(1-\delta)^2} = \infty$$

and hence by the independence of the $B_{n_k, \delta}$,

$$P[B_{n_k, \delta} \text{ i.o. in } k] = 1$$

for all $\delta > 0$. Lim inf $S_n/(2s_n^2 \log_2 s_n^2)^{1/2} \geq -1$ a.s. Thus

$$P[S_{n_k} > -2s_{n_k}u_{n_k} \text{ for all } k \text{ sufficiently large}] = 1.$$

Let

$$C_k = [S_k > -2s_k u_k]$$

for $k \geq 1$. It follows that

$$P[B_{n_k,\delta} \cap C_{n_k} \text{ i.o. in } k] = 1$$

for all $\delta > 0$.

$$B_{n_k,\delta} \cap C_{n_k} \subset \{S_{n_{k+1}} > (1-\delta)[2(s_{n_{k+1}}^2 - s_{n_k}^2) \log_2 (s_{n_{k+1}}^2 - s_{n_k}^2)]^{1/2} - 2s_{n_k}u_{n_k}\}.$$

$$(1-\delta)[2(s_{n_{k+1}}^2 - s_{n_k}^2) \log_2 (s_{n_{k+1}}^2 - s_{n_k}^2)]^{1/2} - 2s_{n_k}u_{n_k}$$
$$\sim [(1-\delta)(1-\varrho^{-2})^{1/2} - 2\varrho^{-1}]s_{n_{k+1}}u_{n_{k+1}}.$$

Choose $\delta' > \delta$ and then choose $\varrho$ sufficiently large so that

$$[(1-\delta)(1-\varrho^{-2})^{1/2} - 2\varrho^{-1}] > 1 - \delta'.$$

Thus

$$[B_{n_k,\delta} \cap C_{n_k} \text{ i.o.}] \subset [S_{n_{k+1}} > (1-\delta')s_{n_{k+1}}u_{n_{k+1}} \text{ i.o.}].$$

Since

$$P[B_{n_k,\delta} \cap C_{n_k} \text{ i.o.}] = 1$$

for all $\delta > 0$,

$$P[S_{n_{k+1}} > (1-\delta')s_{n_{k+1}}u_{n_{k+1}} \text{ i.o.}] = 1$$

for all $\delta' > 0$, this completing the proof. ∎

After the Herculean effort required to establish Theorem 5.2.3 one would hope that Theorem 5.2.3 is widely applicable and that it is in some sense about as good a result as possible. According to an example of Marcinkiewicz and Zygmund [1937], there exist $\{X_k, k \geq 1\}$ satisfying

$$|X_k|/s_k \leq K \log_2^{-1/2} s_k^2 \qquad \text{a.s.}$$

for all $k \geq 1$ and yet

$$\lim \sup S_n/(2s_n^2 \log_2 s_n^2)^{1/2} < 1 \qquad \text{a.s.}$$

Thus, our very detailed computations have produced a sharp result.

The following corollaries indicate how Theorem 5.2.3 may be applied to unbounded $X_i$ and thus indicate that the boundedness assumption is not as restrictive as it might first appear.

**Corollary 5.2.1.** Let $\{X_k, k \geq 1\}$ be identically distributed with $0 < E \mid X_1 \mid^{2+\delta} < \infty$ for some $\delta > 0$. Then

$$S_n/(2nEX_1^2 \log_2 nEX_1^2)^{1/2} = 1 \qquad \text{a.s.}$$

**Proof.** We truncate and then apply the preceding theorem. Without loss of generality, $EX_1^2 = 1$. Let

$$X_k' = X_k I(\mid X_k \mid \leq k^{1/2-\gamma})$$

for all $k \geq 1$ and some $\gamma > 0$ such that

$$(1/2 - \gamma)(2 + \delta) > 1.$$

Let $Y_k = X_k' - EX_k'$ for $k \geq 1$. Then

$$\mid Y_k \mid \leq 2k^{1/2-\gamma} = 2s_k^{1-2\gamma}.$$

$$EY_k^2 = EX_1^2 I(\mid X_1 \mid \leq k^{1/2-\gamma}) - E^2[X_1 I(\mid X_1 \mid \leq k^{1/2-\gamma})] \to EX_1^2$$

by the Lebesgue dominated convergence theorem. Moreover $\sum_{k=1}^n EY_k^2/n \to EX_1^2 = 1$. Thus $\sum_{k=1}^\infty EY_k^2 = \infty$. Let

$$(s_n')^2 = \sum_{k=1}^n \text{var } Y_k$$

for $n \geq 1$. $\{Y_k, k \geq 1\}$ thus satisfies the hypotheses of Theorem 5.2.3 with

$$K_k = \frac{2s_k \log^{1/2}(s_k')^2}{s_k'} s_k^{-2\gamma} \to 0.$$

Thus

$$\lim \sup \sum_{k=1}^n Y_k/[2(s_n')^2 \log_2 (s_n')^2]^{1/2} = 1 \qquad \text{a.s.}$$

Since $\sum_{k=1}^n EY_k^2/n \to EX_1^2 = 1$,

$$\lim \sup \sum_{k=1}^n Y_k/(2n \log_2 n)^{1/2} = 1 \qquad \text{a.s.}$$

$$\sum_{k=1}^\infty P[X_k' \neq X_k] = \sum_{k=1}^\infty P[\mid X_k \mid > k^{1/2-\gamma}]$$

$$= \sum_{k=1}^\infty P[\mid X_k \mid^{2+\delta} > k^{(1/2-\gamma)(2+\delta)}] < \infty$$

by Lemma 3.2.2 since $(1/2 - \gamma)(2 + \delta) > 1$ by the choice of $\gamma$. Hence

$$\limsup_{} \sum_{k=1}^{n} (X_k - EX_k')/(2n \log_2 n)^{1/2} = 1 \qquad \text{a.s.}$$

$$\left| \sum_{k=1}^{n} EX_k I(|X_k| \le k^{1/2-\gamma})/(2n \log_2 n)^{1/2} \right|$$
$$= \left| \sum_{k=1}^{n} EX_1 I(|X_1| > k^{1/2-\gamma})/(2n \log_2 n)^{1/2} \right|.$$

By a standard calculation,

$$\sum_{k=1}^{\infty} E|X_1| I(|X_1| > k^{1/2-\gamma})/(2k \log_2 k)^{1/2} < \infty.$$

Thus, using the Kronecker lemma,

$$\sum_{k=1}^{n} EX_k'/(2n \log_2 n)^{1/2} = \sum_{k=1}^{n} EX_k I(|X_k| \le k^{1/2-\gamma})/(2n \log_2 n)^{1/2} \to 0.$$

Hence

$$\limsup S_n/(2n \log_2 n)^{1/2} = 1 \qquad \text{a.s.}$$

as desired. ∎

**Corollary 5.2.2.** Let $\{K_i, i \ge 1\}$ be positive constants such that $K_i \to 0$. Suppose $s_n^2 \to \infty$. Then

$$\sum_{i=1}^{\infty} (K_i s_i)^{-2} \log_2 s_i^2 E[X_i^2 I(X_i^2 > s_i^2 K_i^2/\log_2 s_i^2)] < \infty$$

implies

$$\limsup S_n/(2s_n^2 \log_2 s_n^2)^{1/2} = 1 \qquad \text{a.s.}$$

**Exercise 5.2.2.** Prove Corollary 5.2.2.

**Corollary 5.2.3.** [Petrov, 1966]. Let

$$\liminf s_n^2/n > 0$$

and

$$E|X_i|^{2+\delta} \le K < \infty$$

for $i \ge 1$, some $\delta > 0$, and $K < \infty$. Then

$$\limsup S_n/(2s_n^2 \log_2 s_n^2)^{1/2} = 1 \qquad \text{a.s.}$$

**Proof.**  We apply Corollary 5.2.2.

$$E[X_i^2 I(X_i^2 > a)] \leq (E \mid X_i \mid^{2+\delta})^{2/(2+\delta)} (P[X_i^2 > a])^{\delta/(2+\delta)} \qquad (5.2.6)$$

for $i \geq 1$ by the Holder inequality. Let

$$K_i^2 = (\log_2 s_i^2)^{-1}$$

for $i \geq 1$. Fix $i_0$ such that $i \geq i_0$ implies $s_i^2/i > \beta$ for some $\beta > 0$. By Eq. (5.2.6) and the Chebyshev inequality,

$$\sum_{i=i_0}^{\infty} (K_i s_i)^{-2} (\log_2 s_i^2) E[X_i^2 I(X_i^2 > s_i^2 K_i^2/\log_2 s_i^2)]$$

$$\leq K^{2/(2+\delta)} \sum_{i=i_0}^{\infty} (\log_2 s_i^2)^2 s_i^{-2} (P[X_i^2 > s_i^2/(\log_2 s_i^2)^2])^{\delta/(2+\delta)}$$

$$\leq K^{2/(2+\delta)} \sum_{i=i_0}^{\infty} (\log s_i^2)^{(2+\delta)} s_i^{-(2+\delta)}$$

$$\leq K^{2/(2+\delta)} \sum_{i=i_0}^{\infty} (s_i)^{-(2+\alpha)}$$

for some $\delta > \alpha > 0$.

$$\sum_{i=i_0}^{\infty} s_i^{-(2+\alpha)} \leq \sum_{i=i_0}^{\infty} (\beta i)^{-(2+\alpha)/2} < \infty,$$

establishing the corollary.

The assumption that $E \mid X_i \mid^{2+\delta} \leq K < \infty$ can be weakened to

$$\limsup \sum_{i=1}^{n} EX_i^2 \mid \log \mid X_i \mid \mid^{1+\delta}/n < \infty$$

for some $\delta > 0$. ∎

The Kolmogorov–Prokhorov exponential inequalities are useful in seeking necessary and sufficient conditions for the strong law of large numbers in terms of the $EX_i^2$. This topic would have been included in Chapter 3 except for the fact that it requires knowledge of these exponential inequalities. Recall that necessary and sufficient conditions for the strong law were given in Chapter 3 in Theorems 3.4.1 and 3.4.2. However these conditions were not in terms of the second moments of $X_i$.

The question is not meaningful without restrictions since $S_n/n \to 0$ a.s. is possible with $EX_i^2 = \infty$ for all $i \geq 1$. $S_n/n \to C$ a.s. implies $X_n/n \to 0$ a.s.

implies $\sum_{n=1}^{\infty} P[|X_n| > n] < \infty$. Thus it seems reasonable to ask for necessary and sufficient conditions in terms of the second moments of the $X_i$ under the restriction that $|X_n| \leq n$ a.s. We will fail in the attempt to find such conditions, but we come close to success as the following theorem shows.

**Theorem 5.2.4.** [Prokhorov, 1959b]

(i)    Suppose

$$|X_i| \leq Ki/\log_2 i    \text{a.s.}$$

for all $i \geq 1$ and some constant $K$. Then

$$S_n/n \to 0    \text{a.s.}$$

if and only if

$$\sum_{n=1}^{\infty} \exp\left(-\varepsilon 2^{2n} \bigg/ \sum_{i=2^n+1}^{2^{n+1}} EX_i^2\right) < \infty$$

for all $\varepsilon > 0$.

(ii)    Let $\{a_i, i \geq 1\}$ be constants satisfying

$$a_i/(i \log_2^{-1} i) \to \infty.$$

Then there exist two sequences of independent symmetric random variables $\{Y_i, i \geq 1\}$ and $\{Z_i, i \geq 1\}$ such that

$$EY_i = EZ_i = 0, EY_i^2 = EZ_i^2, |Y_i| \leq a_i    \text{a.s.},$$

$$|Z_i| \leq a_i    \text{a.s.}$$

for each $i \geq 1$ and yet

$$\sum_{i=1}^{n} Y_i/n \to 0    \text{a.s.}$$

and

$$\sum_{i=1}^{n} Z_i/n    \text{diverges a.s.}$$

***Proof.***    (i)    Let

$$\sigma_n^2 = \sum_{i=2^n+1}^{2^{n+1}} EX_i^2.$$

Assume

$$\sum_{n=1}^{\infty} \exp(-\varepsilon 2^{2n}/\sigma_n^2) < \infty$$

for all $\varepsilon > 0$. For $2^n < i \leq 2^{n+1}$.

$$|X_i|/\sigma_n \leq Ki/(\sigma_n \log_2 i) \leq K2^{n+1}/(\sigma_n \log_2 2^{n+1}) \qquad \text{a.s.} \qquad (5.2.7)$$

Apply Theorem 5.2.2 (ii) with

$$c = K 2^{n+1}/(\sigma_n \log_2 2^{n+1})$$

to obtain

$$P\left[\left|\sum_{i=2^n+1}^{2^{n+1}} X_i\right|\Big/2^{n+1} \geq \varepsilon\right] \leq 2 \exp\left[\frac{-\varepsilon \log_2 2^{n+1}}{2K} \operatorname{arcsinh}\left(\frac{\varepsilon K 2^{2n+1}}{\sigma_n^2 \log_2 2^{n+1}}\right)\right].$$

We consider two cases. First, if

$$\operatorname{arcsinh}\left(\frac{\varepsilon K 2^{2n+1}}{\sigma_n^2 \log_2 2^{n+1}}\right) \geq 4K/\varepsilon,$$

$$P\left[\left|\sum_{i=2^n+1}^{2^{n+1}} X_i\right|\Big/2^{n+1} \geq \varepsilon\right] \leq 2 \exp(-2 \log_2 2^{n+1}) = 2[(\log 2)(n+1)]^{-2}.$$

Second, if

$$\operatorname{arcsinh}\left(\frac{\varepsilon K 2^{2n+1}}{\sigma_n^2 \log_2 2^{n+1}}\right) < \frac{4K}{\varepsilon},$$

then

$$\frac{\varepsilon \log_2 2^{n+1}}{2K} \operatorname{arcsinh}\left(\frac{\varepsilon K 2^{2n+1}}{\sigma_n^2 \log_2 2^{n+1}}\right) \geq \frac{C\varepsilon \log_2 2^{n+1}}{2K} \frac{\varepsilon K 2^{2n+1}}{\sigma_n^2 \log_2 2^{n+1}}$$

for some constant $C > 0$. This follows since given $a > 0$ there exists $C > 0$ such that $\operatorname{arcsinh} x \geq Cx$ for $0 < x < a$. Thus in the second case

$$P\left[\left|\sum_{i=2^n+1}^{2^{n+1}} X_i\right|\Big/2^{n+1} \geq \varepsilon\right] \leq 2 \exp(-C_0 2^{2n}/\sigma_n^2)$$

for some $C_0 > 0$.

$$\sum_{n=1}^{\infty} \{[\log 2](n+1)]^{-2} + \exp(-C_0 2^{2n}/\sigma_n^2)\} < \infty.$$

Thus by the Borel–Cantelli lemma,

$$\sum_{i=2^n+1}^{2^{n+1}} X_i/2^{n+1} \to 0 \qquad \text{a.s.}$$

By Theorem 3.4.2, $S_n/n \to 0$ a.s., proving half of (i).
Assume $S_n/n \to 0$ a.s. By Theorem 3.4.2,

$$T_n = \sum_{i=2^n+1}^{2^{n+1}} X_i/2^{n+1} \to 0 \qquad \text{a.s.}$$

Hence, as is shown in the proof that (ii) implies (i) for Theorem 3.4.2, $\sigma_n^2/2^{2n} \to 0$. Choose $\gamma > 0$ and let $\varepsilon(\gamma)$ and $\pi(\gamma)$ be as in Theorem 5.2.2 (iii). Fix $\varepsilon > 0$. Let $\varepsilon' = \varepsilon\delta$, where $\delta$ is chosen such that $8K\delta \leq \pi(\gamma)$. Let $A_1$ consist of all positive integers $n$ such that

$$\varepsilon' 2^{2n}/\sigma_n^2 < 2(\log_2 2^{n+1}) \tag{5.2.8}$$

and $A_2$ consist of the remaining positive integers. $n \in A_1$ implies

$$(\varepsilon' 2^{n+1}/\sigma_n)[K2^{n+1}/(\sigma_n \log_2 2^{n+1})] \leq 8K\delta.$$

Recall Eq. (5.2.7). There exists $N_0$ for which $n \geq N_0$ implies

$$\varepsilon' 2^{n+1}/\sigma_n \geq \varepsilon(\gamma).$$

$$\sum_{n=1}^{\infty} P[T_n > \varepsilon'] < \infty$$

since $T_n \to 0$ a.s. and the $T_n$ are independent. Thus by Theorem 5.2.2 (iii),

$$\infty > \sum_{n\geq N_0, n\in A_1} P[T_n > \varepsilon']$$

$$\geq \sum_{n\geq N_0, n\in A_1} \exp[-(\varepsilon')^2 2^{2n+2}(1+\gamma)/(2\sigma_n^2)]$$

$$\geq \sum_{n\geq N_0, n\in A_1} \exp(-\varepsilon 2^{2n}/\sigma_n^2)$$

by supposing without loss of generality $\varepsilon < [2\delta^2(1+\gamma)]^{-1}$.

By the complementary inequality to Eq. (5.2.8), choosing $\delta$ small,

$$\sum_{n \in A_2} \exp(-\varepsilon 2^{2n}/\sigma_n^2) \leq \sum_{n=1}^{\infty} \exp[-2 (\log_2 2^{n+1})] < \infty.$$

Thus

$$\sum_{n=1}^{\infty} \exp(-\varepsilon 2^{2n}/\sigma_n^2) < \infty$$

for all $\varepsilon > 0$, establishing (i).

For the construction necessary to verify (ii), see Prokhorov's paper [1959b]. ∎

Thus, the desired necessary and sufficient conditions on the second moments do not exist when only

$$| X_i | \leq i \qquad \text{a.s.}$$

is assumed. However, if the hypothesis is strengthened to

$$| X_i | \leq Ki/\log_2 i \qquad \text{a.s.},$$

then such conditions do exist. Moreover, the hypothesis $| X_i | \leq Ki/\log_2 i$ a.s. cannot be weakened.

**Remarks.** One can, as Prokhorov shows, construct an example similar to that of Theorem 5.2.4 (ii) such that $EY_i = EZ_i = 0, EY_i^2 = EZ_i^2$, $E | Y_i |^3 = E | Z_i |^3$, $EY_i^4 = EZ_i^4$, $| Y_i | \leq a_i$, $| Z_i | \leq a_i$ a.s. for each $i \geq 1$ and yet

$$\sum_{i=1}^{n} Y_i/n \to 0 \qquad \text{a.s.}$$

and

$$\sum_{i=1}^{n} Z_i/n \qquad \text{diverges a.s.}$$

One can ask the interesting question whether there exist necessary and sufficient conditions for the strong law in terms of the first $s$ absolute moments (assuming symmetry of the $X_i$) given that

$$\sum_{n=1}^{\infty} P[| X_i | > \varepsilon i] < \infty$$

for each $\varepsilon > 0$.

Here $s$ might have to be chosen very large. Nagaev [1972], in his recent paper giving necessary and sufficient conditions for the strong law—see

Section 3.4—gives an example which shows that such a result is impossible! Knowledge of $s$ absolute moments, even when $s = 10^9$ say, just does *not* give enough information!

## 5.3. An Almost Sure Invariance Principle with Applications

Throughout Section 5.3, $\{X_i, i \geq 1\}$ will denote a sequence of independent identically distributed random variables with zero means. An invariance principle due to Strassen [1964] is presented and then used to establish the Hartman–Wintner law of the iterated logarithm as well as other interesting results. Roughly speaking, $\{S_n, n \geq 1\}$ will be shown to fluctuate asymptotically very much like Brownian motion, this closeness in behavior to Brownian motion being *invariant* with respect to the distribution of $X_1$, provided $EX_1 = 0$ and $EX_1^2 = 1$ is assumed. This result in effect translates certain almost sure limit theorems for Brownian motion into almost sure limit theorems for $\{S_n, n \geq 1\}$.

Before looking at Strassen's work, we review briefly some of the theory of continuous parameter stochastic processes. Let $R^{[0,\infty)}$ be the collection of all real valued functions defined on $[0, \infty)$. Let $\mathscr{A}^{[0,\infty)}$ be the $\sigma$ field of subsets of $R^{[0,\infty)}$ generated by the measurable rectangles; that is, sets of the form

$$\prod_{i=1}^{n} A_{r_i} \times R^{[0,\infty)-\{r_1, r_2, \ldots, r_n\}}, \qquad 0 \leq r_1 < r_2 < \cdots < r_n < \infty,$$

$A_{r_i}$ Borel sets for $1 \leq i \leq n$, $n \geq 1$. Let $(\Omega, \mathscr{F}, P)$ be the underlying probability space. Then $X_{[0,\infty)} : \Omega \to R^{[0,\infty)}$ is called a random function if $X_{[0,\infty)}^{-1} (\mathscr{A}^{[0,\infty)}) \subset \mathscr{F}$. $[0, \infty)$ is called the parameter set. Often $[0, 1]$ or $R$ serve as parameter sets, the obvious changes being made in the above definition of a random function. A random function is merely a fancy way of describing a family of random variables as the following lemma shows.

**Lemma 5.3.1.** (i) Let $\{X(r), 0 \leq r < \infty\}$ be a family of random variables. Let $X(r, \omega)$ denote the value of the random variable $X(r)$ at $\omega$. Then $X_{[0,\infty)}$ defined by $X_{[0,\infty)}(\omega) = X(\cdot, \omega)$ for all $\omega \in \Omega$ is a random function. ($X(\cdot, \omega)$ denotes the function $r \to X(r, \omega)$ for fixed $\omega$.)

(ii) Let $X_{[0,\infty)}$ be a random function. Let $X(r, \omega)$ denote the value of the function $X_{[0,\infty)}(\omega)$ at $r$. Let $X(r) \equiv X(r, \cdot)$ define functions $X(r) : \Omega \to R$. Then $\{X(r), r \geq 0\}$ is a family of random variables.

**Proof.**  (i)  Consider a measurable rectangle

$$A = \prod_{i=1}^{n} A_{r_i} \times R^{[0,\infty)-\{r_1, r_2, \ldots, r_n\}}.$$

$$X_{[0,\infty)}^{-1}(A) = \bigcap_{i=1}^{n} (X_{r_i} \in A_{r_i}) \in \mathscr{F}.$$

Thus $X_{[0,\infty)}^{-1}(\mathscr{A}) \subset \mathscr{F}$, where $\mathscr{A}$ is the collection of measurable rectangles. Since $\mathscr{A}$ generates $\mathscr{A}^{[0,\infty)}$, the result is established. (ii) follows trivially. ∎

**Remark.**  $\{X(r), r \geq 0\}$ and $X_{[0,\infty)}$ will be used interchangeably, depending on convenience.

Let $P_X$ be defined by

$$P_X(A) = P[X_{[0,\infty)} \in A]$$

for all $A \in \mathscr{A}^{[0,\infty)}$. $P_X$ is called the distribution of the random function $X_{[0,\infty)}$. Usually, instead of $(\Omega, \mathscr{F}, P)$ and $X_{[0,\infty)}$ being given, we start with the joint distribution functions of $(X(r_1), X(r_2), \ldots, X(r_n))$ for all $0 \leq r_1 < r_2 < \cdots < r_n < \infty$, all $n \geq 1$, these possibly being motivated by corresponding physical assumptions (recall the discussion of discrete parameter Brownian motion). Then we are left with the hopefully accomplishable task of constructing $(\Omega, \mathscr{F}, P)$ and $X_{[0,\infty)}$. The foundational Kolmogorov extension theorem (see Loève [1963, p. 93]) guarantees that this is always possible with $\Omega = R^{[0,\infty)}, \mathscr{F} = \mathscr{A}^{[0,\infty)}$, and $X_{[0,\infty)} : R^{[0,\infty)} \to R^{[0,\infty)}$, the identity function.

**Definition 5.3.1.**  A random function $T_{[0,\infty)}$ is called Brownian motion if

(i)  $P[T(0) = 0] = 1$,

(ii)  $T(r + s) - T(s)$ is normal with mean 0 and variance $ar$ for some $a > 0$ and all $r > 0, s > 0$,

(iii)  $T(r_1), T(r_2) - T(r_1), \ldots, T(r_n) - T(r_{n-1}), \ldots$ are independent for all $0 < r_1 < r_2 < \cdots < r_n < \infty$ and $n \geq 1$, and

(iv)  $P[\omega \mid T(\cdot, \omega) \text{ continuous}] = 1$. ∎

**Remark.**  The continuity assumption introduces an important foundational question. It could be that all random functions satisfying (i)–(iii) (such a random function exists by the Kolmogorov extension theorem) also satisfy (iv) or that no random functions satisfying (i)–(iii) satisfy (iv). It turns out that (iv) is an independent assumption; that is, there exists a probability space and random function defined on it satisfying (i)–(iv) and there exists a probability space and random function defined on it satisfying (i)–(iii) and not satisfying (iv). There are a variety of ways of constructing $(\Omega, \mathscr{F}, P)$ and $X_{[0,\infty)}$ satisfying (i)–(iv). We refer the interested reader to Doob's text [1953, p. 393] for such a construction based on the concept of a separable random function.

Moving to Strassen's work, let $\{T(r), r \geq 0\}$ be Brownian motion with $ET^2(1) = 1$. Let

$$T_n(r) = T(nr)/(2n \log_2 n)^{1/2}$$

for $0 \leq r \leq 1$ and $n \geq 3$ define a sequence of random functions $\{T_n, n \geq 1\}$. Note that the range of each $T_n$ is $C[0, 1]$, the continuous functions of $R^{[0,1]}$. Endow $C[0,1]$ with the sup norm; that is,

$$\| x \| = \sup_{0 \leq r \leq 1} | x(r) |$$

for all $x \in C[0, 1]$.

Let $K$ be the set of absolutely continuous $x \in C[0, 1]$ such that $x(0) = 0$ and

$$\int_0^1 (dx/dr)^2 \, dr \leq 1.$$

$dx/dr$ exists almost everywhere with respect to Lebesgue measure since $x$ is absolutely continuous. Considering $x \in C[0, 1]$ as the motion of a particle of mass 2 from time 0 to time 1,

$$\int_0^1 (dx/dr)^2 \, dr \leq 1$$

says that $K$ contains only those particle motions with kinetic energy less than or equal to one.

**Lemma 5.3.2.**   $K$ is compact.

**Proof.**   We show that the members of $K$ are equicontinuous and uniformly bounded and that $K$ is closed, thus establishing compactness by Ascoli's theorem. Fix $0 \leq a < b \leq 1$ and $x \in K$.

$$| x(b) - x(a) | = \left| \int_a^b \frac{dx}{dr} \, dr \right|$$
$$\leq \left( \int_a^b ds \int_a^b \left( \frac{dx}{dr} \right)^2 dr \right)^{1/2} \leq (b - a)^{1/2}.$$

Thus

$$| x(b) - x(a) | \le (b - a)^{1/2}. \tag{5.3.1}$$

This implies both equicontinuity and uniform boundedness.

Now choose $x_n \in K$, $x \in C[0, 1]$ such that $\| x_n - x \| \to 0$. Choose $0 = r_0 < r_1 < r_2 < \cdots < r_m = 1$ for some $m \ge 1$. Fix $n \ge 1$.

$$\sum_{i=1}^{m} [x_n(r_i) - x_n(r_{i-1})]^2/(r_i - r_{i-1})$$
$$\le \sum_{i=1}^{m} \left( \int_{r_{i-1}}^{r_i} \left| \frac{dx_n}{dr} \right| dr \right)^2 \Big/ (r_i - r_{i-1})$$
$$\le \sum_{i=1}^{m} \left[ \int_{r_{i-1}}^{r_i} \left( \frac{dx_n}{dr} \right)^2 dr \right] = \int_{0}^{1} \left( \frac{dx_n}{dr} \right)^2 dr \le 1.$$

Letting $n \to \infty$,

$$\sum_{i=1}^{m} [x(r_i) - x(r_{i-1})]^2/(r_i - r_{i-1}) \le 1. \tag{5.3.2}$$

Choose $0 \le a_1 < b_1 \le a_2 < b_2 < \cdots \le a_n < b_n \le 1$. By the Holder inequality for sums and the above inequality,

$$\sum_{i=1}^{m} | x(b_i) - x(a_i) | = \sum_{i=1}^{m} \frac{| x(b_i) - x(a_i) |}{(b_i - a_i)^{1/2}} (b_i - a_i)^{1/2}$$
$$\le \left\{ \sum_{i=1}^{m} [x(b_i) - x(a_i)]^2/(b_i - a_i) \right\}^{1/2} \left\{ \sum_{i=1}^{m} (b_i - a_i) \right\}^{1/2}$$
$$\le \left\{ \sum_{i=1}^{m} (b_i - a_i) \right\}^{1/2}.$$

Thus $x$ is absolutely continuous. Thus

$$x(r) = \int_{0}^{r} (dx/dr) \, dr$$

for all $0 < r \le 1$.

Fix $n \ge 1$. Divide $[0, 1]$ into segments $[0, 1/2^n)$, $[1/2^n, 2/2^n)$, $\cdots$, $[1 - 1/2^n, 1]$. Let

$$y_n = \frac{x(i/2^n) - x[(i - 1)/2^n]}{2^{-n}}$$

on the $i$th segment for $0 < i \le 2^n$ define $y_n$ for $n \ge 1$. The step functions $y_n \to dx/dr$. Using Eq. (5.3.2),

$$\int_{0}^{1} y_n^2(r) \, dr = \sum_{i=1}^{2^n} \{x(i/2^n) - x[(i - 1)/2^n]\}^2/2^{-n} \le 1$$

taking $r_i = i/2^n$ for $0 \le i \le 2^n$. By Fatou's lemma,

$$\int_0^1 (dx/dr)^2\, dr = \int_0^1 \liminf y_n{}^2(r)\, dr$$
$$\le \liminf \int_0^1 y_n{}^2(r)\, dr \le 1.$$

Hence $x \in K$. ∎

**Remark.** The closure argument above is essentially a lemma of Riesz [Riesz and Sz-Nagy, 1955, pp. 75–76].

**Lemma 5.3.3.** Let $C$ be an arbitrary Borel set of the metric space $C[0, 1]$ and $Y$ be a random function with continuous paths. Then $[Y \in C] \in \mathscr{F}$, the $\sigma$ field of the probability space on which the random function is defined.

**Proof.** Fix $\varepsilon > 0$. Consider the event $[Y \in (x \mid \| x \| \le \varepsilon)]$.

$$[Y \in (x \mid \| x \| \le \varepsilon)] = \bigcap_{m=1}^{\infty} \bigcap_{i=0}^{2^m} (\mid Y(i/2^m) \mid \le \varepsilon)$$

by the continuity of $Y$. $[Y(i/2^m) \le \varepsilon] \in \mathscr{F}$ for all $i, m$. Thus

$$[Y \in (x \mid \| x \| \le \varepsilon)] \in \mathscr{F}.$$

Thus $\mathscr{F}$ contains $[Y \in C]$ for each Borel set $C$ of $C[0, 1]$ noting $C[0, 1]$ is separable. ∎

Strassen gives a remarkably precise statement of the asymptotic behavior of the random functions $\{T_n, n \ge 1\}$, the functional law of the iterated logarithm for Brownian motion.

**Theorem 5.3.1.** With probability one, the set $\{T_n, n \ge 3\}$ of functions of $C[0, 1]$ is relatively compact (with respect to the norm $\| \cdot \|$) and the set of its limit points coincides with $K$.

**Remark.** It is important to digest what Theorem 5.3.1 says: Given an arbitrary $\varepsilon > 0$, let $K_\varepsilon$ denote the functions having distance $<\varepsilon$ from $K$. Then $P[T_n \in K_\varepsilon \text{ eventually}] = 1$. Moreover, $P[\| T_n - x \| \le \varepsilon \text{ i.o. for each } x \in K] = 1$.

**Proof.** First we show that at most the functions of $K$ can be limit points of $\{T_n, n \ge 1\}$. Fix $\varepsilon > 0$. Let $K_\varepsilon$ be the set of all functions of $C[0, 1]$ having distance $<\varepsilon$ from $K$. Let $m$ be a positive integer and $s > 1$, each to be determined later. $[T_n \notin K_\varepsilon] \in \mathscr{F}$ by Lemma 5.3.3 since $K_\varepsilon$ is a

Borel set of $C[0, 1]$. (Similar applications of Lemma 5.3.3 throughout the remainder of the proof are made without mention.) We find an upper bound for $P[T_n \notin K_\varepsilon]$ and then use the Borel–Cantelli lemma to conclude $P[T_n \notin K \text{ i.o.}] = 0$.

$$P[T_n \notin K_\varepsilon] \leq P\left\{2m \sum_{i=1}^{2m} [T_n(i/(2m)) - T_n((i-1)/(2m))]^2 > s^2\right\}$$

$$+ P\left\{2m \sum_{i=1}^{2m} [T_n(i/(2m)) - T_n((i-1)/(2m))]^2 \leq s^2, T_n \notin K_\varepsilon\right\}$$

$$\equiv \text{I} + \text{II}.$$

We estimate I and II separately.

If $Z_1, Z_2, \ldots, Z_{2m}$ are independent normal random variables with mean zero and variance one, $\sum_{i=1}^{2m} Z_i^2$ has density $2^{-m}\Gamma^{-1}(m)t^{m-1}e^{-t/2}I_{(0,\infty)}(t)$, the chi-square density with $2m$ degrees of freedom. Let $\chi_{2m}^2$ denote a random variable with the above density. Computation shows that

$$\text{I} = P[\chi_{2m}^2 > 2s^2 \log_2 n] \sim (s^2 \log_2 n)^{m-1} \exp(-s^2 \log_2 n)/\Gamma(m) \quad (5.3.3)$$

as $n \to \infty$ for each $s > 1$. Fix $n \geq 1$. Let $Y_n$ be the random function with values in $C[0, 1]$ obtained by linearly interpolating $T_n(i/(2m))$ at $i/(2m)$ for $i = 0, 1, \ldots, 2m$. Then

$$2m \sum_{i=1}^{2m} [T_n(i/(2m)) - T_n((i-1)/(2m))]^2 \leq s^2$$

means just that $Y_n/s \in K$. Hence

$$\text{II} = P[Y_n/s \in K, T_n \notin K_\varepsilon] \leq P[Y_n/s \in K, \| Y_n/s - T_n \| \geq \varepsilon]. \quad (5.3.4)$$

Let $\tau$ be defined by

$$\tau = \min\{u \mid 0 \leq u \leq 1, | Y_n(u)/s - T_n(u) | \geq \varepsilon\}$$

if such a $u$ exists, otherwise let $\tau = 2$.

It is not immediately clear that $\tau$ is a random variable. For each $j \geq 1$, $k \geq 1$, let

$$\tau_{j,k} = \min\{u \mid u = 0, 1/j, 2/j, \ldots, 1, | Y_n(u)/s - T_n(u) | > \varepsilon - 1/k\}$$

if such a $u$ exists, otherwise let $\tau_{j,k} = 2$. Since $Y_n(\cdot)$ and $T_n(\cdot)$ are continuous functions of $u$, a.s.,

$$\tau = \lim_{k \to \infty} \lim_{j \to \infty} \tau_{j,k} \quad \text{a.s.}$$

for all $\omega \in \Omega$.

For each $j$ and $k$, $\tau_{j,k}$ is clearly a random variable. Thus completing $\mathscr{F}$ if necessary, $\tau$ is a random variable. Let $F$ be the distribution of $\tau$.

$$P[Y_n/s \in K, \| Y_n/s - T_n \| \geq \varepsilon] = \int_0^1 P[Y_n/s \in K \mid \tau = u] \, dF(u). \qquad (5.3.5)$$

Fix $u \in [0, 1]$. Letting $i(u)$ be the smallest integer $i$ with $i/(2m) \geq u$, $Y_n/s \in K$ then implies that

$$| Y_n(i(u)/(2m)) - Y_n(u) | \leq s \int_u^{i(u)/(2m)} \left| \frac{1}{s} \frac{dY_n(v)}{dv} \right| \, dv$$

$$\leq s \left( \int_u^{i(u)/(2m)} dv \right)^{1/2} \left[ \int_u^{i(u)/(2m)} \left( \frac{1}{s} \frac{dY_n(v)}{dv} \right)^2 \, dv \right]^{1/2}$$

$$\leq s/(2m)^{1/2}.$$

Suppose $| Y_n(u)/s - T_n(u) | = \varepsilon$ and $Y_n/s \in K$. Then, noting that

$$T_n(i(u)/(2m)) = Y_n(i(u)/(2m)),$$

$| T_n(i(u)/(2m)) - T_n(u) |$
$$\geq | Y_n(u) - T_n(u) | - | Y_n(i(u)/(2m)) - Y_n(u) |$$
$$\geq | Y_n(u) - sT_n(u) | - | sT_n(u) - T_n(u) | - | Y_n(i(u)/(2m)) - Y_n(u) |$$
$$\geq s\varepsilon - (s - 1) | T_n(u) | - s/(2m)^{1/2}$$
$$= s\varepsilon - (s - 1) | Y_n(u)/s \pm \varepsilon | - s/(2m)^{1/2}$$

since $T_n(u) = Y_n(u)/s \pm \varepsilon$.

$Y_n/s \in K$ implies $| Y_n(u) |/s \leq 1$ for $0 \leq u \leq 1$. Thus

$$s\varepsilon - (s - 1) | Y_n(u)/s \pm \varepsilon | - s/(2m)^{1/2}$$
$$\geq s\varepsilon - (s - 1)(1 + \varepsilon) - s/(2m)^{1/2} \geq \varepsilon/2$$

by choosing $s$ sufficiently close to one and $m$ sufficiently large. Thus

$$[Y_n/s \in K, \tau = u] = [Y_n/s \in K, | Y_n(u)/s - T_n(u) | = \varepsilon]$$
$$\subset [| T_n(i(u)/(2m)) - T_n(u) | \geq \varepsilon/2].$$

Thus, using Eq. (5.3.4) and Eq. (5.3.5),

$$\text{II} \leq \int_0^1 P[| T_n(i(u)/(2m)) - T_n(u) | \geq \varepsilon/2 \mid \tau = u] \, dF(u).$$

$$[\tau = u] \qquad \text{and} \qquad [| T_n(i(u)/(2m)) - T_n(u) | \geq \varepsilon/2]$$

are independent because of Assumption (iii) of Definition 5.3.1, the definition of Brownian motion. Thus

$$\text{II} \leq P[|\, T_n(1/(2m))\,| \geq \varepsilon/2] \int_0^1 dF(u)$$
$$\leq P[|\, T_n(1/(2m))\,| \geq \varepsilon/2],$$

using the fact that $T_n(i(u)/(2m)) - T_n(u)$ has the same distribution as $T_n[(i(u) - u)/(2m)]$, $i(u) - u \leq 1$ and $P[|\, T_n(x)\,| \geq \varepsilon/2]$ is an increasing function of $x$ for $x > 0$.

$$P[|\, T_n(1/(2m))\,| \geq \varepsilon/2] = P[|\, T(n/(2m))\,|$$
$$\geq (\varepsilon/2)(2n \log_2 n)^{1/2}]$$
$$= P[\chi_1^2 \geq \varepsilon^2 m \log_2 n],$$

recalling that $\chi_1^2$ denotes a random variable having a chi square density with one degree of freedom. Thus using Eq. (5.3.3),

$$P[\chi_1^2 \geq \varepsilon^2 m \log_2 n]$$
$$\sim \Gamma^{-1}(\tfrac{1}{2})[(\varepsilon^2 m \log_2 n)/2]^{-1/2} \exp[-(\varepsilon^2 m \log_2 n)/2],$$

thus yielding an estimate for II. Recall the estimate Eq. (5.3.3) for I. Choosing $m$ sufficiently large yields

$$P[T_n \notin K_\varepsilon] \leq \exp(-r^2 \log_2 n)$$

for all $n$ sufficiently large and $r$ such that $1 < r < s$.

Fix $c > 1$ and let $n_j$ be the smallest $n$ such that $n \geq c^j$ for each $j \geq 1$. This yields

$$\sum_{j=1}^\infty P[T_{n_j} \notin K_\varepsilon] \leq (\log c)^{-r^2} \sum_{j=1}^\infty j^{-r^2} < \infty.$$

Thus eventually (in $j$) $T_{n_j} \in K_\varepsilon$ with probability one by the Borel–Cantelli lemma. Fix $\omega$ for which there exists $J_0$ such that $j \geq J_0$ implies $T_{n_j} \in K_\varepsilon$. Choose $j \geq J_0$ and $n$ such that $n_j \leq n < n_{j+1}$. There exists $x_{j+1} \in K$ such that

$$\max_{0 \leq u \leq 1} \left| \frac{T(u n_{j+1})}{(2n_{j+1} \log_2 n_{j+1})^{1/2}} - x_{j+1}(u) \right| < \varepsilon. \qquad (5.3.6)$$

Thus

$$\max_{0 \le u \le 1} \left| \frac{T(un)}{(2n \log_2 n)^{1/2}} - x_{j+1}(un/n_{j+1}) \right|$$

$$= \max_{0 \le u \le 1} \left| \frac{T(un)}{(2n_{j+1} \log_2 n_{j+1})^{1/2}} \right.$$

$$\left. - x_{j+1}(un/n_{j+1}) \left( \frac{n \log_2 n}{n_{j+1} \log_2 n_{j+1}} \right)^{1/2} \right| \left( \frac{n_{j+1} \log_2 n_{j+1}}{n \log_2 n} \right)^{1/2}$$

$$\le \left( \frac{n_{j+1} \log_2 n_{j+1}}{n_j \log_2 n_j} \right)^{1/2} \left\{ \max_{0 \le u \le 1} \left| \frac{T(un)}{(2n_{j+1} \log_2 n_{j+1})^{1/2}} - x_{j+1}(un/n_{j+1}) \right| \right.$$

$$+ \left[ 1 - \left( \frac{n_j \log_2 n_j}{n_{j+1} \log_2 n_{j+1}} \right)^{1/2} \right] \max_{0 \le u \le 1} | x_{j+1}(u) | \right\}$$

$$\le \left( \frac{n_{j+1} \log_2 n_{j+1}}{n_j \log_2 n_j} \right)^{1/2} \varepsilon + \left[ 1 - \left( \frac{n_j \log_2 n_j}{n_{j+1} \log_2 n_{j+1}} \right)^{1/2} \right] \cdot 1,$$

using Eq. (5.3.6). Let $y_{j+1}(u) = x_{j+1}(un/n_{j+1})$ define $y_{j+1} \in C[0, 1]$. Choosing $c$ sufficiently close to 1, it thus follows that

$$\| T_n - y_{j+1} \| < 3\varepsilon/2.$$

$$\| y_{j+1} - x_{j+1} \| = \max_{0 \le u \le 1} | x_{j+1}(un/n_{j+1}) - x_{j+1}(u) | \le \left( 1 - \frac{n_j}{n_{j+1}} \right)^{1/2}$$

by Eq. (5.3.1) since $x_{j+1} \in K$. Thus

$$\| y_{j+1} - x_{j+1} \| < \varepsilon/2$$

for $c$ chosen sufficiently close to 1. Hence

$$\| T_n - x_{j+1} \| < 2\varepsilon;$$

that is, $T_n \in K_{2\varepsilon}$ for $n$ sufficiently large for $\omega$ as fixed above. Since $T_{n_j} \in K_\varepsilon$ eventually with probability one, it follows that $T_n \in K_{2\varepsilon}$ eventually with probability one. Hence, with probability one, at most the points of $K$ can be limit points of $\{T_n, n \ge 3\}$. Since $K$ is totally bounded, $\{T_n, n \ge 3\}$ is totally bounded and hence relatively compact with probability one.

Fix $\varepsilon > 0$. To finish the proof, it suffices to show for any $x \in K$ and $\varepsilon > 0$ that the probability that $T_n$ is infinitely often in the open $\varepsilon$ sphere about $x$ equals one; that is,

$$P[\| T_n - x \| < \varepsilon \text{ i.o.}] = 1.$$

Note that we have used the compactness of $K$ in making this statement since each $x$ can have an exceptional set of probability zero associated with it and there are uncountably many $x \in K$. Thus it could be that

$$P[\| T_n - x \| < \varepsilon \text{ i.o. for every } x \in K] = 0.$$

However, compactness allows us to work with finitely many spheres. Then letting $\varepsilon \to 0$ through a rational sequence establishes the result. Fix $n \geq 1$. Let $m \geq 1$ be an integer, $0 < \delta < 1$, $x \in K$, and

$$A_n = \{| [T_n(i/m) - T_n((i-1)/m)] - [x(i/m) - x((i-1)/m)] |$$
$$< \delta \text{ for } 2 \leq i \leq m\}.$$

$$P(A_n) \geq \prod_{i=2}^{m} (2\pi)^{-1/2} \int_{(2m\log_2 n)^{1/2}|x(i/m)-x((i-1)/m)|}^{(2m\log_2 n)^{1/2}|x(i/m)-x((i-1)/m)|+\delta} \exp(-u^2/2)\, du.$$

For $0 \leq a < b$,

$$\int_a^b (2\pi)^{-1/2} \exp(-u^2/2)\, du \geq (2\pi b^2)^{-1/2} \exp(-a^2/2)\{1 - \exp[(b^2 - a^2)/2]\}.$$

Applying this,

$$P(A_n) \geq C \prod_{i=2}^{m} \exp\{-m[x(i/m) - x((i-1)/m)]^2 \log_2 n\}/(m \log_2 n)^{1/2}$$
$$\geq C/[\log n(m \log_2 n)^{1/2}]$$

for $n$ sufficiently large and some $C > 0$ dependent only on $\delta$. Let $n_j = m^j$ for $j \geq 1$. $\sum_{j=1}^{\infty} P(A_{n_j}) = \infty$ since $\sum_{j=1}^{\infty} 1/[j(\log j)^{1/2}] = \infty$. The $A_{n_j}$ are independent by the choice of the $n_j$ and thus by the converse to the Borel–Cantelli lemma, $P[A_n \text{ i.o.}] = 1$. By the first part of the proof, eventually $T_n$ stays close to $K$, say within length $\delta/2$, with probability one. Thus

$$| T_n(u) - T_n(v) | \leq (u - v)^{1/2} + \delta \qquad (5.3.7)$$

for all $0 \leq u < v < 1$, using Eq. (5.3.1). Given

$$y \in C[0, 1], \qquad \text{if } | y(u) - y(v) | \leq (u - v)^{1/2} + \delta$$

and

$$| y(i/m) - y((i-1)/m) - (x(i/m) - x((i-1)/m)) | < \delta$$

for $2 \leq i \leq m$, all $m \geq 1$ and all $\delta > 0$, and $y(0) = 0$, then $\| y - x \| < \varepsilon$ for $m$ sufficiently large and $\delta$ sufficiently small (the choice of $\delta$ depends also on $m$). Thus $P(A_n \text{ i.o.}) = 1$ and Eq. (5.3.7) imply

$$P[\| T_n - x \| < \varepsilon \text{ i.o.}] = 1,$$

thus completing the proof. ∎

**Corollary 5.3.1.** With probability one, $\{T_r, r > e\}$ is relatively compact and the set of its limit points coincides with $K$.

**Proof.** The discreteness of $n$ was inessential in the proof of Theorem 5.3.1. ∎

**Corollary 5.3.2.** (Law of the iterated logarithm for Brownian motion)

$$\limsup_{r \to \infty} T(r)/(2r \log_2 r)^{1/2} = 1 \qquad \text{a.s.}$$

**Proof.** Define $\phi$ on $C[0, 1]$ by $\phi(x) = x(1)$ for $x \in C[0, 1]$. $\phi$ is clearly continuous since $x_n \in C[0, 1]$ such that $x_n \to x \in C[0, 1]$ implies $x_n(1) \to x(1)$. By Corollary 5.3.1, the limit points of $\{T_r, r > e\}$ coincide with $K$ on a set $A$ such that $P(A) = 1$. Since $\phi$ is continuous on $C[0, 1]$ and $K$ is compact the limit points of $\{\phi(T_r), r > e\}$ coincide with $\phi(K)$ on some $B \supset A$. Completing the probability space if necessary so that $B$ is measurable, it follows that $P(B) = 1$. Hence $\limsup \phi(T_r) = \sup_{x \in K} \phi(x)$ a.s. But $\limsup \phi(T_r) = \limsup T_r(1) = \limsup T(r)/(2r \log_2 r)^{1/2}$. Since $x(s) - x(r) \leq (s - r)^{1/2}$ for all $1 \geq s > r \geq 0$, $x \in K$ and since $x(0) = 0$, $\sup_{x \in K} \phi(x) = \sup_{x \in K} x(1) \leq 1$. But $x$ defined by $x(r) = r$ is in $K$ and satisfies $x(1) = 1$. Thus $\sup_{x \in K} \phi(x) = 1$. ∎

The argument used to prove Corollary 5.3.2 can be modified to prove a result stronger than the law of the iterated logarithm.

**Exercise 5.3.1.** Prove that with probability one the limit points of

$$T(r)/(2r \log_2 r)^{1/2}$$

coincide with $[-1, 1]$. ∎

Other results for Brownian motion could be stated by choosing other continuous functionals on $C[0, 1]$ and giving an argument like that given in the proof of Corollary 5.3.2. However, our primary purpose is to study the almost sure behavior of partial sums of independent random variables, not the properties of Brownian motion. Suppose throughout the remainder of Section 5.3 that $EX_1^2 = 1$. Let $S_0 = 0$. We define the random function $\{S(r), r \geq 0\}$ by linearly interpolating the random variables $\{S_n, n \geq 0\}$

at $n$ for each $n \geq 0$, that is,

$$S(r) = S_{[r]} + (S_{[r]+1} - S_{[r]})(r - [r]).$$

This random function behaves very similarly to Brownian motion as the next theorem shows. Moreover, this random function carries information about the behavior of $\{S_n, n \geq 1\}$, hence the rather apt name "bookkeeping functions" applied by Chover [1967] to the $S(r)$.

**Theorem 5.3.2.** $\{S(r), r \geq 0\}$ and $\{T(r), r \geq 0\}$ can be defined on a common probability space such that

$$P[\lim_{r \to \infty} \sup_{\varrho \leq r} | S(\varrho) - T(\varrho) |/(2r \log_2 r)^{1/2} = 0] = 1. \qquad (5.3.8)$$

**Remarks.** This is an invariance principle in the sense that the closeness of $S$ and $T$ as expressed by Eq. (5.3.8) is invariant with respect to the distribution of $\{X_i, i \geq 1\}$. It seems rather clear that Corollary 5.3.2 together with Theorem 5.3.2 will at last furnish us with a proof of the Hartman–Wintner law of the iterated logarithm.

The proof of Theorem 5.3.2 depends on being able to "represent" $S_1, S_2, \ldots$ in terms of $\{T(r), r > 0\}$. This by now rather famous and very useful result is due to Skorokhod. Rather than develop the machinery to prove this result, we merely state it:

**Theorem 5.3.3.** Let $\{Y_i, i \geq 1\}$ be a sequence of independent identically distributed random variables with $EY_1 = 0$. Then there exists a probability space with Brownian motion $\{T(r), r \geq 0\}$ defined on it and a sequence $\{\tau_i, i \geq 1\}$ of non-negative independent identically distributed random variables such that $\{T(\sum_{j=1}^n \tau_j), n \geq 1\}$ and $\{S_n, n \geq 1\}$ have the same distribution and $E\tau_1 = EY_1^2$ (possibly infinite). Here $S_n = \sum_{i=1}^n Y_i$.

**Proof of Theorem 5.3.3.** (See Skorokhod [1957, p. 163] or Breiman [1968, pp. 276–278]).

**Proof of Theorem 5.3.2.** Let $\{\tau_i, i \geq 1\}$ be as in Skorokhod's representation theorem. Let

$$X_i' = T\left(\sum_{j=1}^i \tau_j\right) - T\left(\sum_{j=1}^{i-1} \tau_j\right)$$

define $\{X_i', i \geq 1\}$. Let $S_0' = 0$ a.s. Define the random function $\{S'(r), r \geq 0\}$ by linearly interpolating $S_n' = \sum_{i=1}^n X_i'$ at $n$ for each $n \geq 0$. Thus $\{S'(r), r \geq 0\}$ and $\{S(r), r \geq 0\}$ each have the same distribution. Fix

$\varepsilon > 0$ and $c > 1$. Let $n_j$ be the smallest $n \geq 1$ such that $n \geq c^j$ for $j \geq 1$. By the strong law of large numbers (recalling that part of the conclusion of Skorokhod's representation theorem was that $E\tau_1 = EX_1^2 = 1$) for sufficiently large $N$,

$$P\left[-\varepsilon n < \sum_{i=1}^{n} \tau_i - n < \varepsilon n \text{ for some } n > N\right] > 1 - \varepsilon.$$

Now, $|S'(r) - T(r)| \leq \max\{|T(\sum_{i=1}^{[r]} \tau_i) - T(r)|, |T(\sum_{i=1}^{[r]+1} \tau_i - T(r)|\}$. Thus with probability $1 - \varepsilon$, for $j$ sufficiently large,

$$
\begin{aligned}
\sup\{|\,S'(r) - T(r)\,| &: n_j \leq r \leq n_{j+1}\} \\
&\leq \sup\{|\,T(s) - T(r)\,| : n_j(1 - \varepsilon) \leq r, s \leq n_{j+1}(1 + \varepsilon)\} \\
&\leq 2 \sup\{|\,T(s) - T(n_j(1 - \varepsilon))\,| : n_j(1 - \varepsilon) \leq s \leq n_{j+1}(1 + \varepsilon)\} \\
&\equiv M_j.
\end{aligned}
\tag{5.3.9}
$$

Since Brownian motion has independent increments, it is easy to show using the second Levy maximal inequality (Theorem 2.13.1 (ii)) that

$$P[\sup_{r \leq r_0} |\,T(r)\,| > x] \leq 2P[|\,T(r_0)\,| > x] \tag{5.3.10}$$

for every $x > 0$ and $r_0 > 0$. Here we have also used the fact that

$$\max_{1 \leq j \leq r_0 2^n} \left|\sum_{i=1}^{j} T(i2^{-n}) - T((i-1)2^{-n})\right| \to \sup_{r \leq r_0} |\,T(r)\,| \qquad \text{a.s. as} \quad n \to \infty.$$

Fix $\delta > 0$. Applying Eq. (5.3.10),

$$
\begin{aligned}
P[M_j &> \delta(c^j \log_2 c^j)^{1/2}] \\
&\leq 2P[T(c^{j+2}(1 + \varepsilon) - c^j(1 - \varepsilon)) > \delta(c^j \log_2 c^j)^{1/2}/2] \\
&= 2P\left[T(1) > \frac{\delta(\log_2 c^j)^{1/2}}{2(c^2 + \varepsilon c^2 - 1 + \varepsilon)}\right].
\end{aligned}
\tag{5.3.11}
$$

By Lemma 5.1.1, choosing $\varepsilon$ sufficiently small and $c$ sufficiently close 1 so that

$$\delta/[2(c^2 + \varepsilon c^2 - 1 + \varepsilon)] = (2 + \delta')^{1/2} \text{ for some } \delta' > 0,$$

$$P\left[T(1) > \frac{\delta(\log_2 c^j)^{1/2}}{2(c^2 + \varepsilon c^2 - 1 + \varepsilon)}\right] \leq Cj^{-(2+\delta')/2}$$

for some $C > 0$ and all $j$ sufficiently large.

Thus, applying the Borel–Cantelli lemma, Eq. (5.3.9), and Eq. (5.3.11),

$$
\begin{aligned}
P[\sup\{|\,S'(r) - T(r)\,| &: n_j \leq r \leq n_{j+1}\} \\
&> \delta(c^j \log_2 c^j)^{1/2} \text{ i.o. in } j] \leq \varepsilon.
\end{aligned}
$$

But this implies

$$\frac{S'(r) - T(r)}{(r \log_2 r)^{1/2}} \to 0 \qquad \text{a.s.}$$

since $\varepsilon > 0$ can be chosen arbitrarily small. Let $b_r \uparrow \infty$. $a_r/b_r \to 0$ implies $\max_{\varrho \leq r} a_\varrho/b_r \to 0$. Applying this to the preceding statement establishes the theorem. ∎

**Remark.**  Strassen used Theorem 5.3.1 in proving Theorem 5.3.2. We have substituted a proof which does not depend on such a deep property of Brownian motion as that given by Theorem 5.3.1.

Let

$$U_n(r) = (2n \log_2 n)^{-1/2} S(nr)$$

for $0 \leq r \leq 1$ define the sequence of random functions $\{U_n, n \geq 3\}$, another collection of "bookkeeping functions."

Let $\{Z_i, i \geq 1\}$ be independent normal random variables with mean 0 and variance 1. One way to express the conclusion of Theorem 5.3.2 is to say that

$$\left(S_n - \sum_{i=1}^{n} Z_i\right)\Big/ a_n \to 0 \qquad \text{a.s.},$$

where $a_n = (2n \log_2 n)^{1/2}$. That is, $\{S_n, n \geq 1\}$ "imitates" normal behavior to within $\{a_n, n \geq 1\}$. In an interesting paper, Breiman [1967] considers other choices of $\{a_n, n \geq 1\}$, under the additional assumption that certain moments higher than second are finite. We will not look at these results.

**Theorem 5.3.4.**  With probability one, $\{U_n, n \geq 3\}$ is relatively compact (with respect to the norm $\| \cdot \|$) and the set of its limit points coincide with $K$.

**Proof.**  This is immediate from Theorem 5.3.2 and Theorem 5.3.1. ∎

Under the more restrictive assumption that $E \mid X_1 \mid^{2+\delta} < \infty$ for some $\delta > 0$, Chover [1967] establishes the conclusion of Theorem 5.3.4 using only classical results. A central limit theorem with error estimate (due to Esseen) plays the role of Skorokhod's representation theorem.

**Corollary 5.3.3.**  (Hartman–Wintner law of the iterated logarithm)

$$\limsup S_n/(2n \log_2 n)^{1/2} = 1 \qquad \text{a.s.}$$

***Proof.***  Theorem 5.3.4 is applied exactly as Theorem 5.3.1 was applied to prove the law of the iterated logarithm for Brownian motion.  ∎

**Corollary 5.3.4.**  (A sharpening of the Hartman–Wintner law of the iterated logarithm).  With probability one, the limit points of $S_n/(2n \log_2 n)^{1/2}$ coincide with $[-1, 1]$.

***Remarks.***  Corollary 5.3.3 only says that the set of limit points of $S_n/(2n \log_2 n)^{1/2}$ includes $-1$ and $1$ and is contained in $[-1, 1]$. From a pedagogical viewpoint, it is important to note that the Hartman–Wintner law of the iterated logarithm can be proved from Theorem 5.3.2 and the law of the iterated logarithm for Brownian motion. That is, the deep Theorem 5.3.1 is *not* needed.

***Proof of Corollary 5.3.4.***  Omitted. See Exercise 5.3.1.  ∎

If the only use of Strassen's work were proving Corollary 5.3.4, the more elementary method (also computationally involved) of Hartman and Wintner would have sufficed. Hartman and Wintner truncate the $X_i$ in a very special way and then apply Kolmogorov's law of the iterated logarithm. However, Strassen's work has many other applications as well. The following continuity principle indicates one way in which it is applied.

**Corollary 5.3.5.**  Let $\phi : C[0, 1] \to R$ be continuous. Then

$$P[\limsup \phi(U_n) = \sup \phi(K)] = 1.$$

***Proof.***  Since with probability one the limit points of $\{U_n, n \geq 3\}$ coincide with $K$ and since $\phi$ is continuous and $K$ is compact the result follows provided the event in question is measurable. Its complement is clearly a subset of a set of measure zero; hence, by completing the underlying probability space if necessary, measurability follows.  ∎

Different choices of continuous $\phi$ thus produce results of the law of the iterated logarithm type, provided one can compute $\sup \phi(K)$ and $\limsup \phi(U_n)$. Strassen presents a number of such results. One would labor long (and possibly in vain) to produce those results by classical methods.

***Example 5.3.1.***  Let $f$ be any continuous real function defined on $[0, 1]$. Let $F(r) = \int_r^1 f(x) \, dx$ for $0 \leq r \leq 1$. We show that

$$P\left[\limsup (2n^3 \log_2 n)^{-1/2} \sum_{i=1}^n f(i/n)S_i = \left(\int_0^1 F^2(t) \, dt\right)^{1/2}\right] = 1.$$

In particular $f \equiv 1$ yields

$$P\left[\lim \sup(2n^3 \log_2 n)^{-1/2} \sum_{i=1}^{n} S_i = 3^{-1/2}\right] = 1.$$

Let $\phi : C[0, 1] \to R$ be defined by

$$\phi(x) = \int_0^1 x(r)f(r) \, dr.$$

$\phi$ is continuous. Thus, by Corollary 5.3.5

$$P\left[\lim \sup \int_0^1 U_n(r)f(r) \, dr = \sup \phi(K)\right] = 1.$$

$$\sup \phi(K) = \sup_{x \in K} \int_0^1 x(r)f(r) \, dr.$$

Integrating by parts,

$$\sup_{x \in K} \int_0^1 \frac{dx(r)}{dr} F(r) \, dr = \sup_{y \in L_2, \|y\|_2 \le 1} \int_0^1 y(r)F(r) \, dr.$$

Now $L_F(y) = \int_0^1 F(r)y(r) \, dr$ is a linear functional defined on $L_2$. Let $L_2^*$ denote the continuous linear functionals defined on $L_2$. By the isometric isomorphism between $L_2$ and $L_2^*$,

$$\sup_{y \in L_2, \|y\|_2 \le 1} \int_0^1 y(r)F(r) \, dr = \| L_F \| = \| F \|_2 = \left(\int_0^1 F^2(r) \, dr\right)^{1/2}.$$

Hence

$$\sup \phi(K) = \left(\int_0^1 F^2(r) \, dr\right)^{1/2}.$$

Fix $n \ge 1$. Let $S^*(r) = S_{[r]+1}$ for all $0 \le r < n$ define $S^*$ and $f_n(r) = f[([nr] + 1)/n]$ for all $0 \le r < 1$ define $f_n$.

$$\int_0^1 S^*(nr)f_n(r)(2n \log_2 n)^{-1/2} \, dr = \frac{1}{n} \sum_{i=1}^{n} f(i/n)(2n \log_2 n)^{-1/2} S_i.$$

$$\int_0^1 U_n(r)f(r) \, dr - \frac{1}{n} \sum_{i=1}^{n} f(i/n)(2n \log_2 n)^{-1/2} S_i$$

$$= \int_0^1 f(r)[S(nr) - S^*(nr)](2n \log_2 n)^{-1/2} \, dr$$

$$+ \int_0^1 S^*(nr)[f(r) - f_n(r)](2n \log_2 n)^{-1/2} \, dr.$$

$$| S(nr) - S^*(nr) | \leq | X_{[nr]+1} |$$

for $0 \leq r < 1$. $EX_1^2 < \infty$ implies

$$\sum_{i=1}^{\infty} P[X_i^2 > \varepsilon(2i \log_2 i)] < \infty$$

for all $\varepsilon > 0$. Hence, by the Borel–Cantelli lemma,

$$\sup_{0 \leq r < 1} | X_{[nr]+1} | (2n \log_2 n)^{-1/2} \to 0 \qquad \text{a.s.}$$

as $n \to \infty$. Thus by the Lebesgue dominated convergence theorem,

$$\int_0^1 f(r)[S(nr) - S^*(nr)](2n \log_2 n)^{-1/2} \, dr \to 0 \qquad \text{a.s.}$$

as $n \to \infty$. By the law of the iterated logarithm,

$$\sup_{n \geq 1, 0 \leq r \leq 1} S^*(nr)(2n \log_2 n)^{-1/2} \leq K < \infty$$

for all $n \geq 1$. (Here $K$ is a random variable.) $f_n(r) - f(r) \to 0$ as $n \to \infty$ uniformly in $r$. Thus

$$\int_0^1 S^*(nr)[f(r) - f_n(r)](2n \log_2 n)^{-1/2} \, dr \to 0 \qquad \text{a.s.}$$

as $n \to \infty$. Hence

$$P\left[ \limsup \int_0^1 U_n(r)f(r) \, dr = \limsup \sum_{i=1}^{n} f(i/n)S_i(2n^3 \log_2 n)^{-1/2} \right] = 1,$$

completing the example. ∎

For $a \geq 1$, Strassen also obtains

$$P\left[ \limsup n^{-1-a/2}(2 \log_2 n)^{-a/2} \sum_{i=1}^{n} | S_i |^a \right.$$
$$\left. = 2(a + 2)^{a/2-1} a^{-a/2}\left( \int_0^1 (1 - r^a)^{-1/2} \right)^{-a} dr \right] = 1,$$

yielding in particular

$$P\left[ \limsup n^{-3/2}(2 \log_2 n)^{-1/2} \sum_{i=1}^{n} | S_i | = 3^{-1/2} \right] = 1$$

and

$$P\left[ \limsup n^{-2}(2 \log_2 n)^{-1} \sum_{i=1}^{n} S_i^2 = 4/\pi^2 \right] = 1.$$

Just as $\{Y_i, i \geq 1\}$ independent identically distributed with $\sum_{i=1}^{n} Y_i/n$ $\rightarrow 0$ a.s. implies $E \mid Y_i \mid < \infty$ and $EY_1 = 0$ (the converse of the strong law) the law of the iterated logarithm for independent identically distributed random variables has a converse too.

**Theorem 5.3.5.** [Strassen, 1966]. Let $\{Y_i, i \geq 1\}$ be a sequence of independent identically distributed random variables for which

$$P\left[\limsup \left|\sum_{i=1}^{n} Y_i\right| \middle/ (2n \log_2 n)^{1/2} < \infty\right] > 0.$$

Then $EY_1 = 0$ and $EY_1^2 < \infty$.

**Proof.** (Due to Feller [1968b]).  By the Kolmogorov 0–1 law,

$$P\left[\limsup \left|\sum_{i=1}^{n} Y_i\right| \middle/ (2n \log_2 n)^{1/2} < \infty\right] = 1.$$

Hence $\sum_{i=1}^{n} Y_i/n \rightarrow 0$ a.s. Thus $EY_1 = 0$ by the converse to the strong law. By symmetrization, without loss of generality suppose the $Y_i$ are symmetric.

To finish the proof, we assume $EY_1^2 = \infty$ and show that $\limsup$ $\sum_{i=1}^{n} Y_i/(2n \log_2 n)^{1/2} = \infty$ a.s. Fix $M > 0$. Let $Y_i' = Y_i I(\mid Y_i \mid \leq c_M)$ where $c_M$ is chosen such that $E(Y_i')^2 \geq M$. Let $T_n = \sum_{i=1}^{n} Y_i$ and $T_n' = \sum_{i=1}^{n} Y_i'$ for $n \geq 1$. By symmetry, $Y_i$ and $Y_i I(\mid Y_i \mid \leq c_M) - Y_i I(\mid Y_i \mid > c_M)$ have the same distribution for each $i \geq 1$. Thus the two events

$$[T_n' > (Mn \log_2 n)^{1/2}, \qquad T_n - T_n' \geq 0 \text{ i.o.}]$$

and

$$[T_n' > (Mn \log_2 n)^{1/2}, \qquad T_n - T_n' \leq 0 \text{ i.o.}]$$

have the same probability. By the Hewitt–Savage 0–1 law (Theorem 2.12.4) both events either have probability 0 or probability 1. By the Kolmogorov law of the iterated logarithm

$$P[T_n' > (Mn \log_2 n)^{1/2} \text{ i.o.}] = 1.$$

But

$$[T_n' > (Mn \log_2 n)^{1/2} \text{ i.o.}]$$

is the union of the two events given above and hence

$$P[T_n' > (Mn \log_2 n)^{1/2}, \qquad T_n - T_n' \geq 0 \text{ i.o.}] = 1.$$

This implies

$$P[T_n > (Mn \log_2 n)^{1/2} \text{ i.o.}] = 1$$

for all $M > 0$, that is

$$\limsup T_n/(2n \log_2 n)^{1/2} = \infty \qquad \text{a.s.,}$$

as desired. ∎

Feller's proof is impressively simple when one compares it with two other proofs in the literature. Strassen [1966] established the result using the Skorokhod representation theorem. Heyde [1968a,b] gave the first proof using only results from classical probability, his proof based on a mixing technique of Charles Stone which we consider in Chapter 6 (pp. 358–362). More recently, Steiger and Zaremba [1972] have given another sample proof.

## 5.4.  A Survey of Further Results Related to the Law of the Iterated Logarithm

Many laws of the iterated logarithm have been proved for noninde-pendent $X_i$. We survey some of these, proving some results and stating others without proof.

First, we look at the martingale case. Lévy [1954, pp. 258–268] gave the first law of the iterated logarithm for martingales, using a central limit theorem with error estimate. Recently Stout [1970a] has extended the Kolmogorov law of the iterated logarithm to the martingale case. This result is proved by extending the classical Kolmogorov exponential in-equalities to randomly stopped sums of martingale differences. Here we prove only the upper class part of this law of the iterated logarithm. We extend a method used by Meyer [1972, pp. 70–73] for supermartingales with uniformly bounded differences. This produces a shorter and more easily understood proof than that given by Stout [1970a].

**Definition 5.4.1.**  $\{U_n, \mathscr{F}_n, n \geq 1\}$ is called a supermartingale if $\{-U_n, \mathscr{F}_n, n \geq 1\}$ is a submartingale; that is, if $E[U_n \mid \mathscr{F}_{n-1}] \leq U_{n-1}$ a.s. for all $n \geq 2$. ∎

The key to the extension of the exponential inequality Theorem 5.2.2 (i) to the martingale case lies in the following simple observation.

**Lemma 5.4.1.**   Let $\{U_n, \mathscr{F}_n, n \geq 1\}$ be a supermartingale with $EU_1 = 0$. Let $U_0 = 0$ and $Y_i = U_i - U_{i-1}$ for $i \geq 1$. Suppose $Y_i \leq c$ a.s. for some $0 \leq c < \infty$ and all $i \geq 1$. Fix $\lambda > 0$ such that $\lambda c \leq 1$. Let

$$T_n = \exp(\lambda U_n)\exp\left[-(\lambda^2/2)(1 + \lambda c/2)\sum_{i=1}^{n} E(Y_i^2 \mid \mathscr{F}_{i-1})\right]$$

for $n \geq 1$ and $T_0 = 1$ a.s. Then $\{T_n, \mathscr{F}_n, n \geq 0\}$ is a nonnegative supermartingale.

**Proof.**   Fix $j \geq 1$. Series expansion yields

$$\exp(\lambda Y_j) \leq 1 + \lambda Y_j + Y_j^2(\lambda^2/2)(1 + \lambda c/2),$$

using $\lambda c \leq 1$. Thus

$$E[\exp(\lambda Y_j) \mid \mathscr{F}_{j-1}] \leq 1 + (\lambda^2/2)(1 + \lambda c/2)E(Y_j^2 \mid \mathscr{F}_{j-1})$$
$$\leq \exp[(\lambda^2/2)(1 + \lambda c/2)E(Y_j^2 \mid \mathscr{F}_{j-1})] \qquad \text{a.s.} \qquad (5.4.1)$$

Here we have used the simple inequality $1 + x \leq \exp x$ for all $x \in R_1$. Fix $n \geq 2$.

$$E[T_n \mid \mathscr{F}_{n-1}] = \exp\left(\lambda \sum_{i=1}^{n-1} Y_i\right)\exp\left[-(\lambda^2/2)(1 + \lambda c/2)\sum_{i=1}^{n} E(Y_i^2 \mid \mathscr{F}_{i-1})\right]$$
$$\times E[\exp(\lambda Y_n) \mid \mathscr{F}_{n-1}] \leq T_{n-1} \qquad \text{a.s.},$$

using (5.4.1). ∎

**Corollary 5.4.1.**   Under the assumptions of Lemma 5.4.1, $P[\sup_{n \geq 0} T_n > \alpha] \leq \alpha^{-1}$ for each $\alpha > 0$.

**Proof.**   Fix $\alpha > 0$. Let $t$ be the smallest integer $n \geq 0$ such that $T_n > \alpha$ if such an $n$ exists, otherwise let $t = \infty$. $(T_{\min(t,n)}, n \geq 0)$ is also a non-negative supermartingale with first element equal to one. Hence, for each $n \geq 1$,

$$1 \geq ET_{\min(t,n)} \geq \alpha P[t \leq n].$$

Letting $n \to \infty$, we obtain

$$1 \geq \alpha P[t < \infty] = \alpha P[\sup_{n \geq 0} T_n > \alpha],$$

completing the proof. ∎

**Theorem 5.4.1.**   Let $\{U_n, \mathscr{F}_n, n \geq 1\}$ be a supermartingale. Let $U_0 = 0$ a.s., $EU_1 = 0$ and $Y_i = U_i - U_{i-1}$ for $i \geq 1$. Let $s_n^2 = \sum_{i=1}^{n}$

$E[Y_i^2 \mid \mathscr{F}_{i-1}]$ and $u_n = (2 \log_2 s_n^2)^{1/2}$ for $n \geq 1$. Suppose $s_n^2 \to \infty$ a.s. Let $K_i$ be $\mathscr{F}_{i-1}$ measurable for $i \geq 1$. Suppose for some constant $0 < K \leq \frac{1}{2}$ that $\lim \sup K_i < K$ and

$$Y_i \leq K_i s_i / u_i \qquad \text{a.s.} \tag{5.4.2}$$

for each $i \geq 1$. Then there exists a function $\varepsilon(\cdot)$ such that $\varepsilon(K) < 1$ and $\varepsilon(x) \downarrow 0$ as $x \downarrow 0$ for which

$$\lim \sup U_n/(s_n u_n) \leq 1 + \varepsilon(K) \qquad \text{a.s.}$$

**Proof.** Let $Y_i' = Y_i I(K_i \leq K)$ for $i \geq 1$ and $U_n' = \sum_{i=1}^n Y_i'$ for $n \geq 1$. Since $P[Y_i' \neq Y_i \text{ i.o.}] = 0$, it suffices to show $\lim \sup U_n'/(s_n u_n) \leq 1 + \varepsilon(K)$ a.s. in order to prove the theorem. Fix a number $p > 1$ to be specified later and an integer $k \geq 1$. Let $t_k$ be the smallest integer $n \geq 0$ such that $s_{n+1}^2 \geq p^{2k}$. Since $s_{n+1}^2$ is $\mathscr{F}_n$ measurable for each $n \geq 0$, it follows that $t_k$ is a stopping rule. Hence

$$U_n^{(k)} = \begin{cases} U_n' & \text{if } n \leq t_k \\ U_{t_k}' & \text{if } n > t_k \end{cases}$$

defines a supermartingale $\{U_n^{(k)}, n \geq 0\}$.

Fix $\delta > 0$. Let $a \vee b$ denote $\max(a, b)$.

$$P[U_n' > (1 + \delta)s_n u_n \text{ i.o.}]$$

$$\leq P[\sup_{t_k \geq n \geq 0} U_n' > (1 + \delta)s_{t_{k-1}+1} u_{t_{k-1}+1} \text{ i.o. in } k]$$

$$= P[\sup_{n \geq 0} U_n^{(k)} > (1 + \delta)s_{t_{k-1}+1} u_{t_{k-1}+1} \text{ i.o. in } k].$$

$$s_{t_{k-1}+1}^2 u_{t_{k-1}+1}^2 / [2p^{2k} \log_2 (e^2 \vee p^{2k})]$$

$$\geq p^{-2} \log_2 (e^2 \vee p^{2(k-1)})/\log_2 (e^2 \vee p^{2k}) \sim p^{-2}.$$

Choose $\delta' > 0$ and $p > 1$ such that $(1 + \delta) > p(1 + \delta')$.

$$P[\sup_{n \geq 0} U_n^{(k)} > (1 + \delta)s_{t_{k-1}+1} u_{t_{k-1}+1} \text{ i.o. in } k]$$

$$\leq P[\sup_{n \geq 0} U_n^{(k)} > (1 + \delta')[2p^{2k} \log_2(e^2 \vee p^{2k})]^{1/2} \text{ i.o. in } k].$$

We will produce $1 > \varepsilon(x) \downarrow 0$ as $x \downarrow 0$ such that

$$\sum_{k=1}^{\infty} \{P \sup_{n \geq 0} U_n^{(k)} > (1 + \delta')[2p^{2k} \log_2 (e^2 \vee p^{2k})]^{1/2}\} < \infty \tag{5.4.3}$$

for $\delta' > \varepsilon(K)$. It will then follow by the Borel–Cantelli lemma and the preceding remarks that

$$P[U_n' > (1 + \delta)s_n u_n \text{ i.o.}] = 0$$

for all $\delta > [\varepsilon(K) + 1]p - 1$. Since $p > 1$ is arbitrary, the result will then follow for all $\delta > \varepsilon(K)$. Thus

$$\limsup U_n'/(s_n u_n) \leq 1 + \varepsilon(K) \qquad \text{a.s.},$$

the desired result, will follow from Eq. (5.4.3).

For each $n \geq 0$,

$$U_{n+1}^{(k)} - U_n^{(k)} \leq Ks_{t_k}/u_{t_k} \leq Kp^k/[2 \log_2 (e^2 \vee p^{2k})]^{1/2} \qquad \text{a.s.}$$

by Eq. (5.4.2), the definition of $t_k$, and the fact that $s_n/u_n$ is nondecreasing in $n$ (without loss of generality taking $EU_1^2$ large). Let

$$(s_n^{(k)})^2 = \sum_{i=1}^n E[(U_i^{(k)} - U_{i-1}^{(k)})^2 \mid \mathscr{F}_{i-1}]$$

for $n \geq 1$. We apply Corollary 5.4.1 to $\{U_n^{(k)}, \mathscr{F}_n, n \geq 0\}$, taking

$$c = Kp^k/[2 \log_2 (e^2 \vee p^{2k})]^{1/2} \qquad \text{and} \qquad \lambda = (1 + \delta')[2 \log_2 (e^2 \vee p^{2k})]^{1/2}/p^k$$

with

$$T_n = \exp(\lambda U_n^{(k)}) \exp[-(\lambda^2/2)(1 + \lambda c/2)(s_n^{(k)})^2]$$

for $n \geq 1$ and $T_0 = 0$ a.s. $\lambda c = (1 + \delta')K$. Choose $\delta' \leq 1$, thus implying $\lambda c \leq 1$ as required in order to apply Corollary 5.4.1. Note that $\sup_{n \geq 1} (s_n^{(k)})^2 \leq s_{t_k}^2$ a.s.

$$P\{\sup_{n \geq 0} U_n^{(k)} > (1 + \delta')[2p^{2k} \log_2 (e^2 \vee p^{2k})]^{1/2}\}$$

$$= P[\sup_{n \geq 0} U_n^{(k)} > \lambda p^{2k}]$$

$$= P[\sup_{n \geq 0} \exp(\lambda U_n^{(k)}) > \exp(\lambda^2 p^{2k})]$$

$$\leq P\{\sup_{n \geq 0} T_n > \exp[\lambda^2 p^{2k} - (\lambda^2/2)(1 + \lambda c/2)s_{t_k}^2]\}$$

$$\leq P\{\sup_{n \geq 0} T_n > \exp[\lambda^2 p^{2k} - (\lambda^2/2)(1 + \lambda c/2)p^{2k}]\}$$

$$\leq \exp[-\lambda^2 p^{2k} + (\lambda^2/2)(1 + \lambda c/2)p^{2k}]$$

by Corollary 5.4.1. Substituting for $\lambda$ and $c$,

$$-\lambda^2 p^{2k} + (\lambda^2/2)(1 + \lambda c/2)p^{2k}$$
$$= -(1 + \delta')^2[1 - K(1 + \delta')/2] \log_2 (e^2 \vee p^{2k}).$$

Let

$$g_K(x) = (1 + x)^2[1 - K(1 + x)/2] - 1$$

for $0 < x \leq 1$, $0 < K \leq \tfrac{1}{2}$. $g_K(\cdot)$ is increasing. Moreover $g_K(0) < 0$, $g_K(1) > 0$ for each $0 < K \leq \tfrac{1}{2}$. Let $\varepsilon(K)$ be the zero of $g_K(\cdot)$ for each $0 < K \leq \tfrac{1}{2}$. $g_K(x)$ increases to a strictly positive number as $K \downarrow 0$ for each fixed $x$. Thus $1 > \varepsilon(K)$ for each $0 < K \leq \tfrac{1}{2}$ and $\varepsilon(K) \downarrow 0$ as $K \downarrow 0$. Thus

$$(1 + \delta')^2[1 - K(1 + \delta')/2] - 1 > 0$$

for all $1 \geq \delta' > \varepsilon(K)$. Choose such a $\delta'$. Then, there exists $\beta > 1$ such that

$$P\{\sup_{n \geq 0} U_n^{(k)} > (1+\delta')[2p^{2k} \log_2 (e^2 \vee p^{2k})]^{1/2}\} \leq \exp[-\beta \log_2 (e^2 \vee p^{2k})]$$
$$= (2k \log p)^{-\beta}$$

for $k$ sufficiently large.

$$\sum_{k=1}^{\infty} (2k \log p)^{-\beta} < \infty,$$

establishing Eq. (5.4.3) and thus completing the proof. ∎

**Corollary 5.4.2.** [Stout, 1970a]. Let $\{U_n, \mathscr{F}_n, n \geq 1\}$ be a super-martingale. Let $U_0 = 0$, $EU_1 = 0$, and $Y_i = U_i - U_{i-1}$ for $i \geq 1$. Let $s_n^2 = \sum_{i=1}^{n} E[Y_i^2 \mid \mathscr{F}_{i-1}]$ and $u_n = (2 \log_2 s_n^2)^{1/2}$ for $n \geq 1$. Suppose $s_n^2 \to \infty$ a.s. Let $K_n$ be $\mathscr{F}_{n-1}$ measurable for all $n \geq 1$ with $K_n \to 0$ a.s. Suppose

$$Y_i \leq K_i s_i/u_i \qquad \text{a.s.} \tag{5.4.4}$$

for $i \geq 1$. Then

$$\limsup U_n/(s_n u_n) \leq 1 \qquad \text{a.s.}$$

**Proof.** Lim sup $K_i = 0$ a.s. Thus Corollary 5.4.2 follows immediately from Theorem 5.4.1. ∎

When specialized to the independent case, Theorem 5.4.1 applies to random variables which lie outside the scope of Kolmogorov's law of the iterated logarithm; that is, independent $X_i$ which are bounded above by $Ks_i/u_i$ with $K < \tfrac{1}{2}$ rather than bounded above by $K_i s_i/u_i$ for constants $K_i \to 0$. Using the Kolmogorov 0–1 law we obtain

$$\limsup S_n/(s_n u_n) = C \qquad \text{a.s.}$$

for some $0 \leq C \leq 1 + \varepsilon(K)$. Of course, this conclusion is weaker than the conclusion that $C = 1$ in the Kolmogorov law of the iterated logarithm. One might be tempted to guess that the stronger conclusion holds under the weaker hypothesis. However Feller's remarks [1943] show that this is indeed not so. In Stout [1970b], the hypotheses of Corollary 5.4.2 are shown to imply the full conclusion of the law of the iterated logarithm; that is,

$$\lim \sup U_n/(s_n u_n) = 1 \qquad \text{a.s.}$$

The proof of the lower class result; that is,

$$\lim \sup U_n/(s_n u_n) \geq 1 \qquad \text{a.s.}$$

is technically involved and for this reason we omit it. This author [1970a] applies his martingale law of the iterated logarithm to show that $\{X_i, i \geq 1\}$, a stationary ergodic martingale difference sequence with $EX_1 = 0$ and $EX_1^2 < \infty$ implies

$$\lim \sup S_n/[2n \, EX_1^2 \log_2 (nEX_1^2)]^{1/2} = 1 \qquad \text{a.s.}$$

thus extending the Hartman–Wintner law of the iterated logarithm to the martingale case. Steiger [1972] shows that the converse to the law of the iterated logarithm (Theorem 5.3.5) extends to the martingale case. Basu [1973] extends Strassen's Theorem 5.3.4 to the stationary ergodic martingale difference case.

Next, using one of Serfling's maximal inequalities we derive a law of the iterated logarithm for generalized Gaussian random variables. (Recall that a random variable $X$ is generalized Gaussian with constant $\alpha > 0$ if

$$E \exp(uX) \leq \exp(u^2 \alpha^2/2)$$

for all $u \in R_1$.) We then apply this result to obtain laws of the iterated logarithm for multiplicative random variables. Let $S_{m,n} = \sum_{i=m+1}^{m+n} X_i$ for $m \geq 0$ and $n \geq 1$.

**Lemma 5.4.2.** Let $S_{m,n}$ be generalized Gaussian with parameter $\alpha = An$ for some positive number $A$ and all $m \geq 0$ and $n \geq 1$. Then for each $\nu > 2$, there exists a positive constant $K_\nu$ such that

$$E \max_{i \leq n} | S_{m,i} |^\nu \leq K_\nu (An)^{\nu/2} \qquad (5.4.5)$$

for all $m \geq 0$ and $n \geq 1$.

**Proof.** Fix $m \geq 0$ and $n \geq 1$.

$$E \mid S_{m,n} \mid^\nu = \nu \int_0^\infty x^{\nu-1} P[\mid S_{m,n} \mid > x] \, dx. \qquad (5.4.6)$$

$$P[\mid S_{m,n} \mid > x] \leq [E \exp(u S_{m,n}) + E \exp(-u S_{m,n})] \exp(-ux)$$
$$\leq 2 \exp(u^2 An/2) \exp(-ux)$$

for all real $u$ and all positive $x$ by hypothesis. Taking $u = x/(An)$, it follows that

$$P[\mid S_{m,n} \mid > x] \leq 2 \exp[-x^2/(2An)] \qquad (5.4.7)$$

for all positive $x$. Combining this inequality with Eq. (5.4.6), it follows that

$$E \mid S_{m,n} \mid^\nu \leq 2\nu \int_0^\infty x^{\nu-1} \exp[-x^2/(2An)] \, dx = J_\nu (An)^{\nu/2},$$

where $J_\nu = \nu 2^{\nu/2} \Gamma(\nu/2)$.

Fix $\nu > 2$. According to Serfling's Theorem 3.7.5 with

$$g(n) = J_\nu^{2/\nu} An,$$

there exists $K_\nu > 0$ such that

$$E \max_{i \leq n} \mid S_{m,i} \mid^\nu \leq K_\nu (An)^{\nu/2}$$

for each $m \geq 0$ and $n \geq 1$ as desired. ∎

It is now an easy matter to state a law of the iterated logarithm for generalized Gaussian random variables.

**Theorem 5.4.2.** Let $S_{m,n}$ be generalized Gaussian with $\alpha = An$ for some positive number $A$ and all $m \geq 0$ and $n \geq 1$. Then

$$\limsup \mid S_n \mid/(2nA \log_2 n)^{1/2} \leq 1 \qquad \text{a.s.}$$

**Proof.** Fix $n \geq 1$ and $\delta > 0$. By Eq. (5.4.7) with $m = 0$

$$P[\mid S_n \mid > (1 + \delta)(2nA \log_2 n)^{1/2}] \leq 2(\log n)^{-(1+\delta)^2}.$$

Fix $a < 1$ such that $a(1 + \delta)^2 > 1$. Let $n_k = [\exp k^a]$ for $k \geq 1$. $\sum_{k=1}^\infty (\log n_k)^{-(1+\delta)^2} < \infty$ and hence

$$\sum_{k=1}^\infty P[\mid S_{n_k} \mid > (1 + \delta)(2n_k A \log_2 n_k)^{1/2}] < \infty.$$

By the Borel–Cantelli lemma, with probability one

$$| S_{n_k} |/(2n_k A \log_2 n_k)^{1/2} \leq 1 + \delta$$

for $k$ sufficiently large. Let

$$M_k = \max_{n_k \leq n < n_{k+1}} | S_n - S_{n_k} |/(2n_k A \log_2 n_k)^{1/2}$$

for $k \geq 1$. For each $k \geq 1$,

$$| S_n |/(2nA \log_2 n)^{1/2} \leq | S_{n_k} |/(2n_k A \log_2 n_k)^{1/2} + M_k$$

for $n_k \leq n < n_{k+1}$. Thus it suffices to show that $M_k \to 0$ a.s. to complete the proof. Fix $\nu > 2$ such that $(1 - a)\nu/2 > 1$. By Lemma 5.4.2,

$$E | M_k |^\nu \leq \frac{K_\nu(n_{k+1} - n_k)^{\nu/2}}{(2n_k \log_2 n_k)^{\nu/2}}$$

for each $k \geq 1$. Computation shows that the right hand side of the inequality is bounded above by a constant times $k^{-(1-a)\nu/2}(\log k)^{-\nu/2}$. Thus

$$\sum_{k=1}^{\infty} E | M_k |^\nu < \infty$$

since $(1 - a)\nu/2 > 1$ and hence $M_k \to 0$ a.s. as desired. ∎

In the spirit of Serfling's work presented in Chapters 2 and 3, the essential characteristic of Theorem 5.4.2 is that the $X_i$ are not assumed to have a specific dependence structure. That $S_{m,n}$ is generalized Gaussian with parameter $An$ for $m > 0$ and $n \geq 1$ is the *only* assumption. A variety of dependence structures satisfy this assumption. Indeed, it follows immediately from Lemma 4.2.3 that the assumption holds for $\{X_i, i \geq 1\}$ multiplicative with $| X_i | \leq A^{1/2}$ a.s. for each $i \geq 1$. (This, of course, includes the special cases of $\{X_i, i \geq 1\}$ a martingale difference sequence and $\{X_i, i \geq 1\}$ a sequence of independent mean zero random variables.) Thus:

**Corollary 5.4.3.** [Takahashi, 1972]. Let $\{X_i, i \geq 1\}$ be multiplicative with $| X_i | \leq A^{1/2}$ a.s. for some constant and all $i \geq 1$. Then

$$\limsup | S_n |/(2An \log_2 n)^{1/2} \leq 1 \qquad \text{a.s.}$$

*Proof.* Immediate from Theorem 5.4.2 and the remark preceding Corollary 5.4.3. ∎

**Definition 5.4.2.** A multiplicative sequence $\{X_i, i \geq 1\}$ is said to be equinormed strongly multiplicative (ESMS) if $EX_i = 0$ and $EX_i^2 = 1$ for all $i \geq 1$ and

$$E(X_{i_1}^{r_1} X_{i_2}^{r_2} \cdots X_{i_n}^{r_n}) = \prod_{j=1}^{n} EX_{i_j}^{r_j}$$

for all $1 \leq i_1 < i_2 < \cdots < i_n$, all $r_j$ such that $r_j = 1$ or $2$ for $j = 1, 2, \ldots, n$, and all $n \geq 1$. ∎

**Lemma 5.4.3.** [Serfling, 1969]. Let $\{X_i, i \geq 1\}$ be a uniformly bounded ESMS sequence. Then, given $\delta > 0$,

$$E \exp(uS_n) \leq \exp[(1 + \delta)u^2 n/2]$$

for all $n \geq 1$ and $|u|$ sufficiently small.

*Proof.* Series expansion shows that

$$e^x \leq 1 + x + (1 + \delta)x^2/2$$

for $|x|$ sufficiently small. Fix $n \geq 1$. Thus

$$E \exp(uS_n) = E \prod_{i=1}^{n} \exp(uX_i) \leq E \prod_{i=1}^{n} [1 + uX_i + (1 + \delta)u^2 X_i^2/2]$$

for $|u|$ sufficiently small, using the uniform boundedness of the $X_i$. Using the ESMS assumption,

$$E \exp(uS_n) \leq [1 + (1 + \delta)u^2/2]^n \leq \exp[(1 + \delta)u^2 n/2]. \quad ∎$$

**Theorem 5.4.3.** [Takahashi, 1972]. Let $\{X_i, i \geq 1\}$ be a uniformly bounded ESMS sequence. Then

$$\limsup |S_n|/(2n \log_2 n)^{1/2} \leq 1 \qquad \text{a.s.}$$

*Exercise 5.4.1.* Prove Theorem 5.4.3. Hint: Mimic the proof of Theorem 5.4.2, using Lemma 5.4.3 and taking $n_k = [\exp k^a]$ for $(1 + \delta)^{-1} < a < 1$ and $A = 1$. ∎

Revesz [1972] recently proved the corresponding lower class result to Theorem 5.4.2, namely that

$$\limsup S_n/(2n \log_2 n)^{1/2} \geq 1 \qquad \text{a.s.}$$

To prove his result Revesz derives a central limit theorem with error estimate, using Fourier–Stieltjes transforms.

As remarked in Section 5.2, one useful approach to proving a law of the iterated logarithm is the establishment of a central limit theorem with a good estimate of error. This approach is well suited to the analysis of mixing sequences and has been exploited by several authors. We look in particular at a law of the iterated logarithm for mixing random variables due to Reznik [1968]. Suppose throughout that $\{X_i, i \geq 1\}$ is a stationary sequence with $EX_1 = 0$ and $E|X_1|^3 < \infty$. Let

$$\mathscr{B}_{ij} = \mathscr{B}(X_k, i \leq k \leq j)$$

for all $1 \leq i \leq j \leq \infty$. Suppose further that $\{X_i, i \geq 1\}$ is $\phi$-mixing (see Definition 3.7.3), that is there exists a function $\phi$ for which $\phi(m) \to 0$ as $m \to \infty$ and $A \in \mathscr{B}_{1n}$, $B \in \mathscr{B}_{n+m,\infty}$ implies

$$|P(A \cap B) - P(A)P(B)| \leq \phi(m)P(A) \qquad (5.4.8)$$

for all $m \geq 1$ and all $1 \leq n < \infty$. Let $\sigma_n^2 = E(\sum_{i=1}^n X_i)^2$. Let $\sigma^2 = EX_1^2 + 2\sum_{i=1}^\infty EX_1 X_{1+i}$. Suppose throughout that $\sigma^2 \neq 0$.

**Theorem 5.4.4.**   (Reznik's law of the iterated logarithm).   In addition to the above assumptions, suppose $\phi$ is decreasing with $\sum_{n=1}^\infty \phi^{1/2}(n) < \infty$. Then

$$\limsup S_n/(2n\sigma^2 \log_2 n\sigma^2)^{1/2} = 1 \qquad \text{a.s.}$$

**Remarks.**   Reznik obtains his result under the weaker assumption $E|X_1|^{2+\delta} < \infty$ for some $\delta > 0$. We content ourselves with proving the result under the assumption $E|X_1|^3 < \infty$.

We will presently see that $\sum_{n=1}^\infty \phi^{1/2}(n) < \infty$ implies that $\infty > \sigma^2 \geq 0$. Thus the hypothesis of the theorem concerning $\sigma^2$ is really that $\sigma^2 > 0$.

In order to prove Theorem 5.4.4, we need to introduce some notation and prove several lemmas. Essential to the proof is a clever splitting procedure standard to the analysis of mixing sequences:

Fix $n \geq 1$ and $0 < \varepsilon < 1/8$. Let

$$p = [n^{1/2+\varepsilon}], \qquad q = [n^{1/2-\varepsilon}]$$

and $k \geq 1$ satisfy

$$kp + (k-1)q \leq n < (k+1)p + kq.$$

Note that $k = [(n+q)/(p+q)]$.

Let

$$\zeta_1 = X_1 + \cdots + X_p \qquad\qquad \eta_1 = X_{p+1} + \cdots + X_{p+q}$$

$$\zeta_2 = X_{p+q+1} + \cdots + X_{2p+q} \qquad \eta_2 = X_{2p+q+1} + \cdots + X_{2p+2q}$$

$$\vdots \qquad\qquad\qquad\qquad \vdots$$

$$\zeta_k = X_{(k-1)(p+q)+1} + \cdots + X_{kp+(k-1)q} \quad \eta_k = \begin{cases} X_{kp+(k-1)q+1} + \cdots + X_n & \text{if} \\ \quad kp + (k-1)q + 1 \le n \\ 0 & \text{otherwise.} \end{cases}$$

This is the "splitting procedure" referred to above. Note that

$$S_n = (\zeta_1 + \zeta_2 + \cdots + \zeta_k) + (\eta_1 + \eta_2 + \cdots + \eta_k)$$

follows. The idea is that for large $n$ the $\zeta_i$ by the mixing and stationarity assumptions, are approximately independent identically distributed while $\sum_{i=1}^k \eta_i$ has very little influence on the magnitude of $S_n$. Thus $S_n$ is approximately the partial sum of independent identically distributed random variables. Assume $\sum_{n=1}^\infty \phi^{1/2}(n) < \infty$ throughout.

**Lemma 5.4.4.** (Ibragimov [1962], see Billingsley [1968, p. 170]). Let $U$ be $\mathscr{B}_{1n}$ measurable and $V$ be $\mathscr{B}_{n+m,\infty}$ measurable for integers $n$ and $m$, $\phi$ as in Eq. (5.4.8). Let $r^{-1} + s^{-1} = 1$ with $r > 1$. Suppose $E\,|\,U\,|^r < \infty$ and $E\,|\,V\,|^s < \infty$. Then

$$|\,EUV - EUEV\,| \le 2\phi^{1/r}(m)(E\,|\,U\,|^r)^{1/r}(E\,|\,V\,|^s)^{1/s}.$$

**Proof.** It is possible to choose $\{U_j, j > 1\}$ and $\{V_j, j \ge 1\}$ simple such that $|\,U_j\,| \uparrow$, $|\,V_j\,| \uparrow$, $U_j \to U$, and $V_j \to V$ a.s. as $j \to \infty$. Hence it suffices to prove the result for $U = \sum_i a_i I(A_i)$ and $V = \sum_j b_j I(B_j)$ where $A_i \in \mathscr{B}_{1n}$ for all $i$, $B_j \in \mathscr{B}_{n+m,\infty}$ for all $j$, $\bigcup_i A_i = \bigcup_j B_j = \Omega$,

$$A_{i_1} \cap A_{i_2} = \phi \quad \text{for all} \quad i_1 \ne i_2$$
$$B_{j_1} \cap B_{j_2} = \phi \quad \text{for all} \quad j_1 \ne j_2$$

and the sums are finite sums.

$$|\,EUV - EUEV\,| = \left|\sum_{i,j} a_i b_j P(A_i B_j) - \sum_{i,j} a_i b_j P(A_i)P(B_j)\right|.$$

Applying Holder's inequality twice,

$$
\begin{aligned}
& | EUV - EUEV | \\
&= \left| \sum_{i,j} a_i b_j P(A_i)[P(B_j \mid A_i) - P(B_j)] \right| \\
&= \left| \sum_i \{ a_i P^{1/r}(A_i) \} \Big\{ P^{1/s}(A_i) \sum_j [b_j P(B_j \mid A_i) - P(B_j)] \Big\} \right| \\
&\le \left\{ \sum_i | a_i |^r P(A_i) \right\}^{1/r} \left\{ \sum_i P(A_i) \Big| \sum_j b_j [P(B_j \mid A_i) - P(B_j)] \Big|^s \right\}^{1/s} \\
&\le (E \mid U \mid^r)^{1/r} \left[ \sum_i P(A_i) \Big| \sum_j \{ | b_j | \mid P(B_j \mid A_i) - P(B_j) \mid^{1/s} \} \right. \\
&\qquad \left. \times \{ | P(B_j \mid A_i) - P(B_j) |^{1/r} \} \Big|^s \right]^{1/s} \\
&\le (E \mid U \mid^r)^{1/r} \left[ \sum_i P(A_i) \sum_j | b_j |^s [P(B_j \mid A_i) + P(B_j)] \right. \\
&\qquad \left. \times \Big( \sum_j | P(B_j \mid A_i) - P(B_j) | \Big)^{s/r} \right]^{1/s}.
\end{aligned}
$$

Let

$$
J_+ = \{ j \mid P(B_j \mid A_i) \ge P(B_j) \} \quad \text{and} \quad J_- = \{ j \mid P(B_j \mid A_i) < P(B_j) \}.
$$

$$
\begin{aligned}
\sum_j & | P(B_j \mid A_i) - P(B_j) | \\
&= \left[ P\Big( \bigcup_{J_+} B_j \mid A_i \Big) - P\Big( \bigcup_{J_+} B_j \Big) \right] + \left[ P\Big( \bigcup_{J_-} B_j \Big) - P\Big( \bigcup_{J_-} B_j \mid A_i \Big) \right] \\
&\le 2\phi(m)
\end{aligned}
$$

by Eq. (5.4.8). Thus

$$
\begin{aligned}
\left[ \sum_i P(A_i) \sum_j | b_j |^s [P(B_j \mid A_i) + P(B_j)] \Big( \sum_j | P(B_j \mid A_i) - P(B_j) | \Big)^{s/r} \right]^{1/s} \\
\le [2\phi(m)]^{1/r} \Big\{ \sum_i P(A_i) \sum_j | b_j |^s [P(B_j \mid A_i) + P(B_j)] \Big\}^{1/s} \\
= [2\phi(m)]^{1/r} \Big\{ 2 \sum_j | b_j |^s P(B_j) \Big\}^{1/s} \\
\le 2\phi^{1/r}(m)(E \mid V \mid^s)^{1/s}.
\end{aligned}
$$

Thus

$$
| EUV - EUEV | \le 2\phi^{1/r}(m)(E \mid U \mid^r)^{1/r}(E \mid V \mid^s)^{1/s}. \quad \blacksquare
$$

**Lemma 5.4.5.** (Ibragimov [1962], see Billingsley [1968, p. 171]). Let $U$ be $\mathscr{B}_{1n}$ measurable and $V$ be $\mathscr{B}_{n+m,\infty}$ measurable for integers $n$ and $m$,

$\phi$ as in Eq. (5.4.8). Suppose $|U| \leq 1$ a.s. and $|V| \leq 1$ a.s. Then

$$|EUV - EUEV| \leq 4\phi(m).$$

**Proof.** Let

$$A = \{E[V \mid \mathscr{B}_{1n}] > EV\}$$

and $U' = 2I(A) - 1$.

$$
\begin{aligned}
|EUV - EUEV| &= |E\{U(E[V \mid \mathscr{B}_{1n}] - EV)\}| \\
&\leq |E\{U'(E[V \mid \mathscr{B}_{1n}] - EV)\}| \\
&= |EU'V - EU'EV|.
\end{aligned}
$$

Let

$$B = \{E[U' \mid \mathscr{B}_{n+m,\infty}] > EU'\}$$

and

$$V' = 2I(B) - 1.$$

Arguing as above,

$$|EU'V - EU'EV| \leq |EU'V' - EU'EV'|.$$

We thus have

$$
\begin{aligned}
|EUV - EUEV| &\leq |EU'V' - EU'EV'| \\
&= 4|P(A \cap B) - P(A)P(B)| \\
&\leq 4\phi(m)
\end{aligned}
$$

by Eq. (5.4.8). $\blacksquare$

Recall that $x_n = o(y_n)$ means that $x_n/y_n \to 0$.

**Lemma 5.4.6.** $\sigma_n^2 = n\sigma^2(1 + o(1))$ and $\infty > \sigma^2 > 0$.

**Proof.** By Lemma 5.4.4, $\sum_{i=1}^{\infty} |EX_1X_{1+i}| \leq 2\sum_{i=1}^{\infty} \phi^{1/2}(i)EX_1^2 < \infty$. Thus $|\sigma^2| < \infty$.

$$
\begin{aligned}
E(X_1 + \cdots + X_n)^2/n &= EX_1^2 + 2\sum_{i=1}^{n-1}(1 - i/n)EX_1X_{i+1} \\
&= EX_1^2 + 2\sum_{i=1}^{\infty}EX_1X_{i+1} - 2\sum_{i=n}^{\infty}EX_1X_{i+1} - 2n^{-1}\sum_{i=1}^{n-1}iEX_1X_{i+1} \\
&= \sigma^2 + o(1),
\end{aligned}
$$

using $\sum_{i=1}^{\infty} i\phi^{1/2}(i)/i < \infty$ and the fact that $\sum_{i=1}^{\infty} EX_1X_{i+1}$ is a convergent series. $\blacksquare$

Note that $\{Y_i, i \geq 1\}$ independent identically distributed with $EY_1 = 0$, $EY_1^2 = 1$, and $X_i = Y_{i+1} - Y_i$ yields an example where $\sigma^2 = 0$.

**Lemma 5.4.7.**  $\sigma_n^2/(kE\zeta_1^2) - 1 = O(n^{-\varepsilon})$.

**Remark.**  If the $\zeta_i$ were independent, $\sum_{i=1}^{k} \zeta_i$ would of course have variance $kE\zeta_1^2$. This helps in interpreting Lemma 5.4.7.

**Proof of Lemma 5.4.7.**  $E\zeta_1^2 \neq 0$ for sufficiently large $n$ by Lemma 5.4.6.

$$E(\zeta_1 + \zeta_2 + \cdots + \zeta_k)^2 = kE\zeta_1^2 + 2 \sum_{i=1}^{k-1} (k - i)E\zeta_1\zeta_{1+i}.$$

$$E\zeta_1\zeta_{1+i} = O[E\zeta_1^2\phi^{1/2}(iq + (i - 1)p)]$$

for $1 \leq i < k$ by Lemma 5.4.4 and the stationarity of $\{X_i, i \geq 1\}$. Hence

$$E(\zeta_1 + \cdots + \zeta_k)^2/(kE\zeta_1^2) = 1 + O\left[\sum_{i=1}^{k-1} (1 - i/k)\phi^{1/2}(iq + (i - 1)p)\right]$$

$$= 1 + O\left[\sum_{i=1}^{k-1} (1 - i/k)\{iq + (i - 1)p\}^{-1}\right]$$

$$= 1 + O(n^{-1/2+\varepsilon}) + O\left[\sum_{i=1}^{k-1} (1 - i/k)(n^{1/2+\varepsilon})^{-1}\right]$$

$$= 1 + O(n^{-1/4}) + O[k(n^{1/2+\varepsilon})^{-1}]$$

$$= 1 + O(n^{-2\varepsilon}).$$

Here we have used the fact that $\sum_{i=1}^{\infty} \phi^{1/2}(i) < \infty$ and $\phi$ decreasing implies $i\phi^{1/2}(i) \to 0$ as $i \to \infty$ and that $k = O(n^{1/2-\varepsilon})$.

$$E\eta_1\eta_{1+i} = O[E\eta_1^2\phi^{1/2}(ip + (i - 1)q]$$

for $1 \leq i < k - 1$.

$$E\eta_i\eta_k = O[(E\eta_1^2)^{1/2}(E\eta_k^2)^{1/2}\phi^{1/2}((k - i)p + (k - i - 1)q)] \quad \text{for } 1 \leq i < k.$$

Arguing as above,

$$E(\eta_1 + \cdots + \eta_k)^2/\sigma_n^2$$

$$= \frac{(k - 1)E\eta_1^2 + E\eta_k^2 + 2\sum_{i=1}^{k-2} (k - 1 - i)E\eta_1\eta_{1+i} + 2\sum_{i=1}^{k-1} E\eta_i\eta_k}{\sigma_n^2}$$

$$\leq \frac{(k - 1)E\eta_1^2 + E\eta_k^2 + C\sum_{i=1}^{k-2} E\eta_1^2(k - 1 - i)(in^{1/2+\varepsilon})^{-1}}{\sigma_n^2}$$

$$+ \frac{Ck(E\eta_1^2)^{1/2}(E\eta_k^2)^{1/2}(\eta^{1/2+\varepsilon})^{-1}}{\sigma_n^2} \tag{5.4.9}$$

for some $C > 0$. $E\eta_1^2 = O(n^{1/2-\varepsilon})$ and $E\eta_k^2 = O(n^{1/2+\varepsilon})$ follow easily from Lemma 5.4.6. Using Lemma 5.4.6 it thus follows that

$$E(\eta_1 + \cdots + \eta_k)^2/\sigma_n^2 = O(n^{-2\varepsilon}). \qquad (5.4.10)$$

Let $S_k' = \sum_{i=1}^{k} \zeta_i$ and $S_k'' = \sum_{i=1}^{k} \eta_i$.

$$\left| \frac{E(\zeta_1 + \cdots + \zeta_k)^2}{\sigma_n^2} - 1 \right| = \left| \frac{E(S_n - S_k'')^2}{\sigma_n^2} - 1 \right|$$

$$\leq \frac{2E \mid S_n S_k'' \mid}{\sigma_n^2} + \frac{E(S_k'')^2}{\sigma_n^2}$$

$$\leq 2[E(S_k'')^2/\sigma_n^2]^{1/2} + E(S_k'')^2/\sigma_n^2,$$

using the Holder inequality. Thus, by Eq. (5.4.10),

$$\frac{E(\zeta_1 + \cdots + \zeta_k)^2}{\sigma_n^2} - 1 = O(n^{-\varepsilon}). \qquad (5.4.11)$$

Above we saw that

$$E(\zeta_1 + \cdots + \zeta_k)^2/(kE\zeta_1^2) = 1 + O(n^{-2\varepsilon}).$$

It is easy to see that $x_n - 1 = O(n^{-\delta})$ implies $x_n^{-1} - 1 = O(n^{-\delta})$ for $\delta > 0$. Thus

$$kE\zeta_1^2/E(\zeta_1 + \cdots + \zeta_k)^2 = 1 + O(n^{-2\varepsilon}).$$

Combining this with Eq. (5.4.11), it follows that

$$kE\zeta_1^2/\sigma_n^2 = 1 + O(n^{-\varepsilon})$$

and hence that

$$\sigma_n^2/(kE\zeta_1^2) = 1 + O(n^{-\varepsilon}). \quad \blacksquare$$

**Lemma 5.4.8.**  $E \mid S_n \mid^3 = O(ES_n^2)^{3/2}$.

**Proof.**  Let $S_{n+j,n} = \sum_{i=n+j+1}^{2n+j} X_i$ for $j \geq 1$, $n \geq 1$. Fix $\varepsilon_1 > 0$. We first show that there exists $C_1 > 0$ and $j \geq 1$ such that

$$E \mid S_n + S_{n+j,n} \mid^3 \leq (2 + \varepsilon_1)E \mid S_n \mid^3 + C_1(ES_n^2)^{3/2} \qquad (5.4.12)$$

for all $n \geq 1$. Using the assumption of stationarity,

$$E \mid S_n + S_{n+j,n} \mid^3 \leq E(S_n + S_{n+j,n})^2(\mid S_n \mid + \mid S_{n+j,n} \mid)$$

$$\leq 2E \mid S_n \mid^3 + 3E \mid S_n \mid^2 \mid S_{n+j,n} \mid + 3E \mid S_n \mid \mid S_{n+j,n} \mid^2.$$

$$(5.4.13)$$

For $2 \geq u > 0$, $v > 0$ such that $u + v = 3$, it follows by Lemma 5.4.4 and stationarity that

$$E \mid S_n \mid^u \mid S_{n+j,n} \mid^v \leq 2[\phi(j+1)]^{u/3}E \mid S_n \mid^3 + E \mid S_n \mid^u E \mid S_{n+j,n} \mid^v. \qquad (5.4.14)$$

Jensen's inequality yields

$$E \mid S_n \mid \leq (ES_n^2)^{w/2} \qquad (5.4.15)$$

for $0 < w \leq 2$. Combining Eqs. (5.4.13), (5.4.14), and (5.4.15), and noting that $\phi \leq 1$ we obtain

$$E \mid S_n + S_{n+j,n} \mid^3 \leq (2 + 12[\phi(j+1)]^{1/3})E \mid S_n \mid^3 + 6(ES_n^2)^{3/2}.$$

Choosing $j$ sufficiently large establishes Eq. (5.4.12).

Fix $\varepsilon_2 > 0$. We now show that there exists $C_2 > 0$ such that

$$E \mid S_{2n} \mid^3 \leq (2 + \varepsilon_2)E \mid S_n \mid^3 + C_2(ES_n^2)^{3/2}. \qquad (5.4.16)$$

Using Minkowski's inequality and Eq. (5.4.12), it follows that

$$
\begin{aligned}
E \mid S_{2n} \mid^3 &= E \left| S_n + \sum_{i=n+1}^{n+j} X_i + S_{n+j,n} - \sum_{i=2n+1}^{2n+j} X_i \right|^3 \\
&\leq \left[ (E \mid S_n + S_{n+j,n} \mid^3)^{1/3} + \sum_{i=n+1}^{n+j} (E \mid X_i \mid^3)^{1/3} \right. \\
&\qquad \left. + \sum_{i=2n+1}^{2n+j} (E \mid X_i \mid^3)^{1/3} \right]^3 \\
&\leq \{[(2 + \varepsilon_1)E \mid S_n \mid^3 + C_1(ES_n^2)^{3/2}]^{1/3} + 2j(E \mid X_1 \mid^3)^{1/3}\}^3 \\
&= (1 + \varepsilon^{(n)})^3(2 + \varepsilon_1)E \mid S_n \mid^3 + C_1(1 + \varepsilon^{(n)})^3(ES_n^2)^{3/2},
\end{aligned}
$$

where

$$\varepsilon^{(n)} = 2j \left[ \frac{E \mid X_1 \mid^3}{(2 + \varepsilon_1)E \mid S_n \mid^3 + C_1(ES_n^2)^{3/2}} \right]^{1/3}.$$

By Lemma 5.4.6, $ES_n^2 \to \infty$. Thus, choosing $\varepsilon_1$ sufficiently small, for $n$ sufficiently large ($n \geq n_0$ say)

$$E \mid S_{2n} \mid^3 \leq (2 + \varepsilon_2)E \mid S_n \mid^3 + 2C_1(ES_n^2)^{3/2}.$$

Clearly there exists $C_2(\geq 2C_1)$ such that Eq. (5.4.16) holds for all $n \leq n_0$, thus establishing Eq. (5.4.16).

By Lemma 5.4.6, $ES_n^2 = n(\sigma^2 + \varrho_n)$ where $\varrho_n \to 0$. Applying Eq. (5.4.16) inductively, it follows for every integer $r \geq 1$ and $\varepsilon > 0$ that

$$E \mid S_{2^r} \mid^3 \leq (2 + \varepsilon)^r E \mid X_1 \mid^3 + C_2 (ES_{2^{r-1}}^2)^{3/2}$$

$$\times \left[ \sum_{i=0}^{r-1} \left( \frac{2 + \varepsilon}{2^{3/2}} \right)^{r-1-i} \left( \frac{\sigma^2 + \varrho_{2^i}}{\sigma^2 + \varrho_{2^{r-1}}} \right)^{3/2} \right].$$

Thus, choosing $\varepsilon$ sufficiently small, there exists $C_3$ such that

$$E \mid S_{2^r} \mid^3 \leq (2 + \varepsilon)^r E \mid X_1 \mid^3 + C_3 (ES_{2^{r-1}}^2)^{3/2}$$

for all integers $r \geq 1$. It follows that

$$E \mid S_{2^r} \mid^3 \leq (ES_{2^r}^2)^{3/2} \left[ C_3 \left( \frac{\sigma^2 + \varrho_{2^{r-1}}}{\sigma^2 + \varrho_{2^r}} \right)^{3/2} + \frac{E \mid X_1 \mid^3 (2 + \varepsilon)^r}{[2^r(\sigma^2 + \varrho_{2^r})]^{3/2}} \right].$$

Thus, there exists $C_4 > 0$ such that

$$E \mid S_{2^r} \mid^3 \leq C_4 (ES_{2^r}^2)^{3/2} \tag{5.4.17}$$

for all integers $r \geq 1$.

Let $2^r \leq n < 2^{r+1}$. Expand $n = v_0 2^r + v_1 2^{r-1} + \cdots + v_r$ with $v_0 = 1$, $v_j = 0$ or $1$ for $1 \leq j \leq r$. Thus $S_n$ may be split into $r + 1$ sums (note the similarity to the proof of the basic maximal inequality for orthogonal random variables), the $i$th sum containing $v_{i-1} 2^{r-i+1}$ terms. Thus, it follows by stationarity, Eq. (5.4.17), and Minkowski's inequality that

$$E \mid S_n \mid^3 \leq \left[ \sum_{i=0}^{r} E^{1/3} \mid S_{2^i} \mid^3 \right]^3$$

$$\leq C_4 (ES_n^2)^{3/2} \left[ \sum_{i=0}^{r} (ES_{2^i}^2 / ES_n^2)^{1/2} \right]^3$$

$$\leq C_4 (ES_n^2)^{3/2} \left[ \sum_{i=0}^{r} \left( \frac{\sigma^2 + \varrho_{2^i}}{\sigma^2 + \varrho_n} \right)^{1/2} 2^{(i-r)/2} \right]^3$$

$$= O(ES_n^2)^{3/2}. \quad \blacksquare$$

Lemma 5.4.9 below is the basic analytic lemma. It is a version of Esseen's lemma.

**Lemma 5.4.9.** Let $F$ and $G$ be distribution functions with characteristic functions $\phi_F$ and $\phi_G$, respectively.

Suppose that $F$ and $G$ each have mean 0.
Suppose $G$ has a derivative $g$ such that $|g| \leq M$. Then

$$\sup_x |F(x) - G(x)| \leq \frac{1}{\pi} \int_{-T}^{T} \left| \frac{\phi_F(t) - \phi_G(t)}{t} \right| dt + \frac{24M}{\pi T} \qquad (5.4.18)$$

for every $T > 0$.

**Proof.** [Feller, 1966, pp. 511–512]. Fix $T > 0$. Let $V_T$ be the distribution function with density

$$v_T(x) = \frac{1 - \cos Tx}{\pi T x^2}, \qquad -\infty < x < \infty.$$

The characteristic function of $V_T$ is given by

$$\phi_V(t) = \begin{cases} 1 - |t|/T & \text{if } |t| \leq T \\ 0 & \text{elsewhere.} \end{cases}$$

Let

$$H_T(y) = \int_{-\infty}^{\infty} [F(y - x) - G(y - x)] v_T(x) \, dx. \qquad (5.4.19)$$

We first show that

$$\varDelta \equiv \sup |F(x) - G(x)| \leq 2 \sup |H_T(y)| + \frac{24M}{\pi T}. \qquad (5.4.20)$$

Either there exists $\{x_n\}$ converging such that $F(x_n) - G(x_n)$ converges to $\varDelta$ or $-\varDelta$. Thus either there exists $x_0$ such that $F(x_0) - G(x_0) = \varDelta$ or there exists $x_0$ such that $F(x_0^-) - G(x_0) = -\varDelta$. The two cases are similar to analyze; hence we analyze the first case only. Since $F$ does not decrease and $G$ grows at a rate $\leq M$ this implies

$$F(x_0 + s) - G(x_0 + s) \geq \varDelta - Ms \qquad \text{for } s > 0.$$

Putting $h = \varDelta/(2M)$, $y = x_0 + h$, and $x = h - s$, we obtain

$$F(y - x) - G(y - x) \geq \varDelta/2 + Mx \qquad \text{for } |x| \leq h.$$
$$F(y - x) - G(y - x) \geq -\varDelta \qquad \text{for } |x| > h.$$

The density $v_T$ distributes to $|x| > h$ a mass $\leq 4/(\pi T h)$. Thus using Eq.

(5.4.19) we obtain

$$H_T(x_0) \geq \frac{\Delta}{2}\left[1 - \frac{4}{\pi Th}\right] - \Delta\frac{4}{\pi Th} = \frac{\Delta}{2} - \frac{6\Delta}{\pi Th}.$$

Using the definition of $h$, this implies Eq. (5.4.20).

Consider the densities

$$g_0(y) = \int_{-\infty}^{\infty} g(y - x)v_T(x)\,dx$$

and

$$f_0(y) = \frac{d}{dy}\int_{-\infty}^{\infty} F(y - x)v_T(x)\,dx.$$

Their characteristic functions are clearly $\phi_G\phi_V$ and $\phi_F\phi_V$, respectively. Fourier inversion yields

$$f_0(y) - g_0(y) = \frac{1}{2\pi}\int_{-T}^{T} e^{-ity}[\phi_F(t) - \phi_G(t)]\phi_V(t)\,dt. \qquad (5.4.21)$$

We claim this is equivalent to

$$\int_{-\infty}^{\infty} [F(y - x) - G(y - x)]v_T(x)\,dx$$
$$= \frac{1}{2\pi}\int_{-T}^{T} e^{-ity}\frac{\phi_F(t) - \phi_G(t)}{-it}\phi_V(t)\,dt. \qquad (5.4.22)$$

Since $F$ is a distribution function with mean 0, $\phi_F(0) = 1$ and $\phi_F'(0) = 0$. The same is true for $G$. Hence (l'Hospital), $[\phi_F(t) - \phi_G(t)]/t$ is continuous and vanishes at $t = 0$ and thus the right hand side of Eq. (5.4.22) is well-defined. Differentiation of Eq. (5.4.22) yields Eq. (5.4.21). Hence the two sides of Eq. (5.4.22) differ only by a constant. Clearly, the left hand side of Eq. (5.4.22) approaches 0 as $y \to \infty$. The right hand side of Eq. (5.4.22) approaches 0 as $y \to \infty$ by the Riemann–Lebesgue Lemma of classical analysis. Thus Eq. (5.4.22) holds. Combining Eqs. (5.4.22) and (5.4.20) yields Eq. (5.4.18), as desired. ∎

Let $S_n' = \sum_{i=1}^{n} \zeta_i$ for $n \geq 1$.

**Lemma 5.4.10.**

$$\left|E\exp\left(\frac{itS_k'}{\sigma_n}\right) - \prod_{j=1}^{k} E\exp\left(\frac{it\zeta_j}{\sigma_n}\right)\right| \leq 4k\phi(q) = O(n^{-1/4})$$

uniformly in $t$.

**Remark.** $\prod_{j=1}^{k} E \exp(it\zeta_i/\sigma_n)$ would be the characteristic function of $S_k'$ if the $\zeta_j$ were independent.

**Proof of Lemma 5.4.10.** Fix $t$ real and $n \geq 1$.

$$E \exp\left(\frac{itS_{k-1}'}{\sigma_n}\right) \exp\left(\frac{it\zeta_k}{\sigma_n}\right) - E \exp\left(\frac{it\zeta_k}{\sigma_n}\right) E \exp\left(\frac{itS_{k-1}'}{\sigma_n}\right) \leq 4\phi(q)$$

by Lemma 5.4.5.

$$\left| E \exp\left(\frac{it\zeta_k}{\sigma_n}\right) E \exp\left(\frac{it\zeta_{k-1}}{\sigma_n}\right) \exp\left(\frac{itS_{k-2}'}{\sigma_n}\right) \right.$$
$$\left. - E \exp\left(\frac{it\zeta_k}{\sigma_n}\right) E \exp\left(\frac{it\zeta_{k-1}}{\sigma_n}\right) E \exp\left(\frac{itS_{k-2}'}{\sigma_n}\right) \right| \leq 4\phi(q)$$

by Lemma 5.4.5 and the fact that $| E \exp(it\zeta_k/\sigma_n) | \leq 1$. Hence

$$\left| E \exp\left(\frac{itS_k'}{\sigma_n}\right) - E \exp\left(\frac{it\zeta_k}{\sigma_n}\right) E \exp\left(\frac{it\zeta_{k-1}}{\sigma_n}\right) E \exp\left(\frac{itS_{k-2}'}{\sigma_n}\right) \right| \leq 2[4\phi(q)].$$

Proceeding inductively, and noting that $k\phi(q) = O(n^{-1/4})$ we obtain the desired result. ∎

**Lemma 5.4.11.** Let $\phi_2$ be the characteristic function of $\sum_{i=1}^{k} Z_i/\sigma_n$, where the $Z_i$ are independent identically distributed with the distribution of $\zeta_1$. Then

$$| \phi_2(t) - \exp(-t^2/2) | = O(n^{-\varepsilon/2})$$

uniformly for $| t | \leq n^{\varepsilon/4}$.

**Proof.** Let $\phi_{21}$ be the characteristic function of $Z_1/\sigma_n$ Taylor series expansion yields

$$\phi_{21}(t) = 1 - \frac{EZ_1^2 t^2}{2\sigma_n^2} + \frac{\Theta EZ_1^3 t^3}{6\sigma_n^3}$$

where $\Theta$ is a complex number satisfying $| \Theta | \leq 1$, not necessarily the same on each appearance.

$$\frac{EZ_1^2 t^2}{2\sigma_n^2} = O(n^{-1/4}).$$

Using Lemma 5.4.8,

$$\frac{EZ_1^3 t^3}{6\sigma_n^3} = O(n^{-1/4}).$$

Thus

$$\frac{EZ_1^2 t^2}{2\sigma_n^2} + \frac{\Theta E |Z_1|^3 t^3}{6\sigma_n^3} \to 0 \quad \text{as} \quad n \to \infty$$

uniformly for $|t| \leq n^{\varepsilon/4}$. If $|x| \leq \frac{1}{2}$, then $\log(1-x) = -x + \Theta |x|^2$ for some $|\Theta| \leq 1$ is easily shown by series expansion. Thus

$$\log \phi_{21}(t) = -\frac{EZ_1^2 t^2}{2\sigma_n^2} + \frac{\Theta E |Z_1|^3 t^3}{6\sigma_n^3}$$

$$+ \Theta \left| -\frac{EZ_1^2 t^2}{2\sigma_n^2} + \frac{\Theta E |Z_1|^3 t^3}{6\sigma_n^3} \right|^2$$

for $n$ sufficiently large. It follows easily from Lemmas 5.4.7 and 5.4.8 that

$$\log \phi_{21}(t) = -\frac{EZ_1^2 t^2}{2\sigma_n^2} + \frac{\Theta E |Z_1|^3 t^3}{2\sigma_n^3}.$$

Hence

$$\frac{t^2}{2} + \log \phi_2(t) = k \log \phi_{21}(t) + \frac{t^2}{2}$$

$$= \frac{t^2}{2} - \frac{kEZ_1^2 t^2}{2\sigma_n^2} + \frac{\Theta k E |Z_1|^3 t^3}{2\sigma_n^3}.$$

Thus, using Lemma 5.4.7 and Lemma 5.4.8,

$$\log \phi_2(t) + \frac{t^2}{2} \leq \frac{t^2}{2} \left| 1 - \frac{kEZ_1^2}{2\sigma_n^2} \right| + \frac{kE |Z_1|^3 t^3}{2\sigma_n^3} = O(n^{-\varepsilon/2}).$$

Since $|e^u - 1| \leq |u| e^{|u|}$ for all $u$,

$$|\phi_2(t) \exp(t^2/2) - 1| = O(n^{-\varepsilon/2}).$$

Therefore

$$|\phi_2(t) - \exp(-t^2/2)| = O(n^{-\varepsilon/2}). \quad \blacksquare$$

**Lemma 5.4.12.**

$$\left| P[S_k'/\sigma_n < x] - \int_{-\infty}^x (2\pi)^{-1/2} \exp(-y^2/2)\, dy \right| = O(n^{-\varepsilon/4})$$

for each $n \geq 1$ and real $x$.

**Proof.** Let $F_0$ be the distribution function of a normal random variable $U$ with mean 0 and variance 1. Let $F_1$ be the distribution function of $S_k'/\sigma_n$ and $F_2$ be the distribution function of $\sum_{i=1}^k Z_i/\sigma_n$, the $Z_i$ independent with each $Z_i$ having the same distribution as $\zeta_1$. Let $\phi_0$, $\phi_1$, and $\phi_2$ be, respectively, the characteristic functions of $U$, $S_k'/\sigma_n$, and $\sum_{i=1}^k Z_i/\sigma_n$. Then

$$\sup_x |F_1(x) - F_0(x)| = O\left(\int_{-n^{\varepsilon/4}}^{n^{\varepsilon/4}} \left|\frac{\phi_1(t) - \phi_0(t)}{t}\right| dt + n^{-\varepsilon/4}\right) \quad (5.4.23)$$

follows from Esseen's lemma (Lemma 5.4.9). $EU^2 = 1$. $E(\sum_{i=1}^k Z_i/\sigma_n)^2 = O(1)$ by Lemma 5.4.7. $E(S_k')^2/\sigma_n^2 = O(1)$ by Eq. (5.4.11). If $\phi$ is a characteristic function of a random variable $T$ with $ET = 0$ and $ET^2 < \infty$, then

$$|\phi(t) - 1| \leq t^2 ET^2/2$$

is easily shown by Taylor series expansion for all real $t$. It follows that

$$|\phi_1(t) - \phi_0(t)| \leq C|t| \quad \text{and} \quad |\phi_2(t) - \phi_0(t)| \leq C|t| \quad (5.4.24)$$

for all $|t| \leq 1$, some constant $C$, and all $n \geq 1$. It follows by Lemma 5.4.11 that

$$\phi_2(t) - \phi_0(t) = O(n^{-\varepsilon/2}) \quad \text{for} \quad |t| \leq n^{\varepsilon/4}. \quad (5.4.25)$$

$$\int_{-n^{\varepsilon/4}}^{n^{\varepsilon/4}} \left|\frac{\phi_1(t) - \phi_0(t)}{t}\right| dt$$
$$\leq \int_{-n^{\varepsilon/4}}^{n^{\varepsilon/4}} \left|\frac{\phi_1(t) - \phi_2(t)}{t}\right| dt + \int_{-n^{\varepsilon/4}}^{n^{\varepsilon/4}} \left|\frac{\phi_2(t) - \phi_0(t)}{t}\right| dt.$$

$$\int_{-n^{\varepsilon/4}}^{n^{\varepsilon/4}} \left|\frac{\phi_2(t) - \phi_0(t)}{t}\right| dt$$
$$= \int_{|t|<n^{-1/4}} \left|\frac{\phi_2(t) - \phi_0(t)}{t}\right| dt + \int_{n^{-1/4}\leq|t|\leq n^{\varepsilon/4}} \left|\frac{\phi_2(t) - \phi_0(t)}{t}\right| dt$$
$$= O(n^{-\varepsilon/4}),$$

using Eqs. (5.4.24) and (5.4.25).

Similarly, using Lemma 5.4.10 in the role of Eq. (5.4.25),

$$\int_{-n^{\varepsilon/4}}^{n^{\varepsilon/4}} \left|\frac{\phi_1(t) - \phi_2(t)}{t}\right| dt = O(n^{-1/8}).$$

Combining,

$$\int_{-n^{\varepsilon/4}}^{n^{\varepsilon/4}} \left| \frac{\phi_1(t) - \phi_0(t)}{t} \right| dt = O(n^{-\varepsilon/4}).$$

Using Eq. (5.4.23), the desired result follows. ∎

We need now to substitute $S_n$ for $S_k'$ in the statement of Lemma 5.4.12.

**Lemma 5.4.13.**

$$\left| P[S_n/\sigma_n \le x] - (2\pi)^{-1/2} \int_{-\infty}^{x} \exp(-y^2/2)\, dy \right| = O(n^{-\varepsilon/4}).$$

**Proof.** Let $A = [|\, S_k''\,|/\sigma_n \le n^{-\varepsilon/2}]$.

$$| P[S_k'/\sigma_n \le x] - P[S_n/\sigma_n \le x] |$$
$$\le | P[S_k'/\sigma_n \le x, A] - P[(S_k' + S_k'')/\sigma_n \le x, A] |$$
$$+ P[|\, S_k''\,|/\sigma_n > n^{-\varepsilon/2}].$$

$$| P[S_k'/\sigma_n \le x, A] - P[(S_k' + S_k'')/\sigma_n \le x, A] |$$
$$\le \max\{(P[S_k'/\sigma_n \le x + n^{-\varepsilon/2}, A] - P[S_k'/\sigma_n \le x, A]),$$
$$(P[S_k'/\sigma_n \le x, A] - P[S_k'/\sigma_n \le x - n^{-\varepsilon/2}, A])\}$$
$$\le O(n^{-\varepsilon/4}) + \int_{-n^{-\varepsilon/2}}^{n^{-\varepsilon/2}} \exp(-y^2/2)\, dy = O(n^{-\varepsilon/4})$$

by Lemma 5.4.12.

$$P[|\, S_k''\,|/\sigma_n > n^{-\varepsilon/2}] = O(n^{-\varepsilon})$$

by Eq. (5.4.10). Combining this with the preceding statement establishes the result using Lemma 5.4.12. ∎

**Lemma 5.4.14.** For $|\,\delta\,| < 1$ and $\alpha > 0$,

$$(\log \sigma_n^2)^{-(1+\delta)^2(1+\alpha)} < P[S_n > (1+\delta)(2\sigma_n^2 \log_2 \sigma_n^2)^{1/2}] < (\log \sigma_n^2)^{-(1+\delta)^2(1-\alpha)}$$

for $n$ sufficiently large.

**Proof.** By Lemma 5.4.13,

$$\left[ P[S_n/\sigma_n > (1+\delta)(2 \log_2 s_n^2)^{1/2}] - \int_{(1+\delta)(2\log_2\sigma_n^2)^{1/2}}^{\infty} (2\pi)^{-1/2} \exp(-y^2/2)\, dy \right]$$
$$\le Cn^{-\varepsilon/4}.$$

By Lemma 5.1.1,

$$(\log \sigma_n^2)^{-(1+\delta)^2(1+\alpha)} \leq \int_{(1+\delta)(2\log_2 \sigma_n^2)^{1/2}}^{\infty} (2\pi)^{-1/2} \exp(-y^2/2) \, dy$$

$$\leq (\log \sigma_n^2)^{-(1+\delta)^2}$$

for each $\alpha > 0$. Since $\sigma_n^2 = n\sigma^2(1 + o(1))$, the desired result follows. ∎

The inequality of Lemma 5.4.14 is central to the proof of Reznik's law of the iterated logarithm for mixing sequences. A second central inequality is the following maximal inequality.

**Lemma 5.4.15.**   For all $x > 0$ and all $n \geq 1$,

$$P[\max_{j \leq n} S_j > x] \leq 2P[S_n > x - 2\sigma_n] + Cn^{-\alpha}$$

for some $\alpha > 0$ and $C > 0$.

**Proof.**   Fix $x > 0$ and $\alpha > 0$. It suffices to prove the result for $n$ sufficiently large by choosing $C$ sufficiently large. Let

$$A_j = \{S_i \leq x, i < j, S_j > x\}$$

for $1 \leq j \leq n$.

$$P[\max_{j \leq n} S_j > x] = \sum_{j=1}^{n} P(A_j) \leq \sum{}' P(A_j) + Cn^{-\alpha},$$

where $\sum'$ denotes the sum over all $j$ such that $P(A_j) > Cn^{-(1+\alpha)}$. It suffices to show that

$$\sum{}' P(A_j) \leq 2P[S_n > x - 2\sigma_n].$$

$$[S_n > x - 2\sigma_n, A_j] \supset [S_n - S_j \geq -2\sigma_n, A_j]$$
$$\supset [|\, S_n - S_j\,| \leq 2\sigma_n, A_j].$$

Thus, summing on $j$,

$$P[S_n > x - 2\sigma_n] \geq \sum_{j=1}^{n} P[|\, S_n - S_j\,| \leq 2\sigma_n, A_j].$$

$$P[|\, S_n - S_j\,| \leq 2\sigma_n, A_j] = P[|\, S_n - S_j\,| \leq 2\sigma_n \,|\, A_j] P(A_j).$$

It thus suffices to prove that $P[|\, S_n - S_j\,| > 2\sigma_n \,|\, A_j] \leq \frac{1}{2}$ provided that $P(A_j) > Cn^{-(1+\alpha)}$ and $n$ is sufficiently large. Choose an integer $n_0$ such that $\phi(n_0) \leq 10^{-5}$ and choose $0 < \eta < 0.5$. Fix $j \geq 1$ for which $P(A_j)$

$> Cn^{-(1+\alpha)}$. We consider two cases. First, suppose $n - j > n_0$. Then

$$P[|\,S_n - S_j\,| > 2\sigma_n \mid A_j] \leq P\left[\left|\sum_{i=j+1}^{j+n_0} X_i\right| > \eta\sigma_n \mid A_j\right]$$

$$+P\left[\left|\sum_{i=j+1+n_0}^{n} X_i\right| > (2 - \eta)\sigma_n \mid A_j\right]. \quad (5.4.26)$$

Using Chebyshev's inequality,

$$P\left[\left|\sum_{i=j+1}^{j+n_0} X_i\right| > \eta\sigma_n \mid A_j\right] \leq P\left[\left|\sum_{i=j+1}^{j+n_0} X_i\right| > \eta\sigma_n\right]\Big/P(A_j)$$

$$\leq E\left|\sum_{i=1}^{n_0} X_i\right|^3 (P(A_j)\eta^3\sigma_n^3)^{-1}$$

$$\leq E\left|\sum_{i=1}^{n_0} X_i\right|^3 C^{-1}\eta^{-3}n^{-1/2+\alpha}\sigma^{-3}(1 + o(1))$$

$$\leq 10^{-5}$$

for $n$ sufficiently large, choosing $\alpha < \frac{1}{2}$. Here we have used the fact that $P(A_j) > Cn^{-(1+\alpha)}$.

$$P\left[\left|\sum_{i=j+1+n_0}^{n} X_i\right| > (2 - \eta)\sigma_n \mid A_j\right] \leq P\left[\left|\sum_{i=j+1+n_0}^{n} X_i\right| > 3\sigma_n/2 \mid A_j\right]$$

$$\leq \phi(n_0) + P\left[\left|\sum_{i=j+1+n_0}^{n} X_i\right| > 3\sigma_n/2\right]$$

by the definition of $\phi$ and the fact that $\eta < \frac{1}{2}$. By the Chebyshev inequality,

$$P\left[\left|\sum_{i=j+1+n_0}^{n} X_i\right| > 3\sigma_n/2\right] < 4\sigma_{n-j-n_0}^2/(9\sigma_n^2)$$

for all $j$ such that $n - j > n_0$. Thus,

$$P[|\,S_n - S_j\,| > 2\sigma_n \mid A_j] \leq (2)10^{-5} + \tfrac{4}{9} \leq \tfrac{1}{2}.$$

Now, considering the second case, suppose $n - j < n_0$.

$$P[|\,S_n - S_j\,| > 2\sigma_n \mid A_j] \leq P[|\,S_n - S_j\,| > 2\sigma_n]/P(A_j)$$

$$\leq E\left|\sum_{i=1}^{n-j+1} X_i\right|^3 (8P(A_j)\sigma_n^3)^{-1} \leq 1/2$$

for $n$ sufficiently large, using the fact that $P(A_j) > Cn^{-(1+\alpha)}$, $\alpha < \frac{1}{2}$, and $n - j < n_0$. ∎

At last we can prove Reznik's law of the iterated logarithm. Note the similarity of the proof to that of Theorem 5.1.1.

**Proof of Theorem 5.4.4.** Fix $\delta > 0$. Let $n_j = \varrho^j$ for $j \geq 1$ and some $\varrho > 1$. We first show that

$$P[\max_{n \leq n_{j+1}} S_n/(2\sigma_n^2 \log_2 \sigma_n^2)^{1/2} \geq 1 + \delta \text{ i.o.}] = 0.$$

By the Borel–Cantelli lemma, it suffices to show that

$$\sum_{j=1}^{\infty} P[\max_{n \leq n_{j+1}} S_n/(2\sigma_{n_j}^2 \log_2 \sigma_{n_j}^2)^{1/2} > 1 + \delta] < \infty.$$

By Lemma 5.4.15, it suffices to show that

$$\sum_{j=1}^{\infty} P[S_{n_{j+1}} > (1 + \delta)(2\sigma_{n_j}^2 \log_2 \sigma_{n_j}^2)^{1/2} - 2\sigma_{n_{j+1}}] < \infty$$

and hence that

$$\sum_{j=1}^{\infty} P[S_{n_j} > (1 + \delta')(2\sigma_{n_j}^2 \log_2 \sigma_{n_j}^2)^{1/2}] < \infty$$

for $0 < \delta' < \delta$, provided $\varrho$ is chosen sufficiently close to 1. But this follows immediately from Lemma 5.4.14 by choosing $\alpha > 0$ such that $(1 + \delta')^2(1 - \alpha) > 1$ and the fact

$$(\log \sigma_{n_j}^2)^{-(1+\delta')^2(1-\alpha)} \sim Cj^{-(1+\delta')^2(1-\alpha)}.$$

Hence

$$\limsup S_n/(2\sigma_n^2 \log_2 \sigma_n^2)^{1/2} \leq 1 \qquad \text{a.s.}$$

Since $\sigma_n^2 \sim n\sigma^2$,

$$\limsup S_n/(2n\sigma^2 \log_2 n\sigma^2)^{1/2} \leq 1 \qquad \text{a.s.}$$

To complete the proof we must show

$$\limsup S_n/(2n\sigma^2 \log_2 n\sigma^2)^{1/2} \geq 1 \qquad \text{a.s.}$$

Fix $N > 1$ and $\delta > 0$. Let

$$C_n = [S_{N^n} - S_{N^{n-1}+N^{n/2}} > (1 - \delta)\alpha(N^n - N^{n-1} - N^{n/2})],$$

where $\alpha(n) = (2n\sigma^2 \log_2 n\sigma^2)^{1/2}$. By Lemma 5.4.14 and the fact that $\sigma_n{}^2 = \sigma^2 n(1 + o(1))$,

$$P[C_n] \geq \tfrac{1}{2}[\log(N^n - N^{n-1} - N^{n/2})]^{-(1-\delta)^2(1+\alpha)}$$

for $n$ sufficiently large. Choose $\alpha > 0$ such that $(1 - \delta)^2(1 + \alpha) < 1$. For sufficiently large $n$ and $N$,

$$N^n - N^{n-1} - N^{n/2} \geq N^n/2.$$

Thus $\sum_{n=1}^{\infty} P[C_n] = \infty$. We now derive a converse Borel–Cantelli lemma in order to conclude $P[C_n \text{ i.o.}] = 1$. It suffices to show

$$P\left[\bigcap_{m=n}^{n+k} C_m{}^c\right] \to 0 \qquad \text{as} \quad k \to \infty \quad \text{and then} \quad n \to \infty.$$

$$\left|P\left[\bigcap_{m=n}^{n+k} C_m{}^c\right] - \prod_{m=n}^{n+k} P[C_m{}^c]\right|$$

$$= \left|P\left[\bigcap_{m=n}^{n+k} C_m{}^c\right] - P\left[\bigcap_{m=n}^{n+k-1} C_m{}^c\right]P[C_{n+k}^c] + P\left[\bigcap_{m=n}^{n+k-1} C_m{}^c\right]P[C_{n+k}^c]\right.$$

$$\left. - P\left[\bigcap_{m=n}^{n+k-2} C_m{}^c\right]P[C_{n+k-1}^c]P[C_{n+k}^c] \cdots - \prod_{m=n}^{n+k} P[C_m{}^c]\right|$$

$$\leq \phi(N^{(n+k)/2}) + \phi(N^{(n+k-1)/2}) + \cdots + \phi(N^{(n+1)/2}) \to 0$$

as $k \to \infty$ and then $n \to \infty$ by the mixing assumption. Thus it suffices to show

$$\prod_{m=n}^{n+k} P[C_m{}^k] \to 0 \qquad \text{as} \quad k \to \infty \quad \text{and then} \quad n \to \infty,$$

thus reducing the proof to the proof of the converse Borel–Cantelli lemma in the case of independence

$$\prod_{m=n}^{n+k} P[C_m{}^c] = \prod_{m=n}^{n+k} (1 - P[C_m]) \leq \exp\left(-\sum_{m=n}^{n+k} P[C_m]\right) \to 0$$

as $k \to \infty$, and then $n \to \infty$. Thus $P[C_n \text{ i.o.}] = 1$. The rest of the proof now follows that of the classical Kolmogorov law of the iterated logarithm. Let

$$B_n = [S_{N^{n-1}+N^{n/2}} > -2\alpha(N^{n-1} - N^{n/2})].$$

Using the conclusion of the first half of the proof

$$P[B_n \cap C_n \text{ i.o.}] = 1.$$

$$B_n \cap C_n \subset [S_{N^n} > (1 - \delta)\alpha(N^n - N^{n-1} - N^{n/2}) - 2\alpha(N^{n-1} + N^{n/2})].$$

It is easily seen that choosing $N$ sufficiently large implies for arbitrary $\delta' > \delta$ that

$$[B_n \cap C_n \text{ i.o.}] \subset [S_{N^n} > (1 - \delta')\alpha(N^n) \text{ i.o.}],$$

thus completing the proof. ∎

***Exercise 5.4.2.*** Let events $\{C_n, n \geq 1\}$ be given. Let $\mathscr{C}_n$ be the $\sigma$ field generated by $C_n$ and $\mathscr{C}_{1,n}$ the $\sigma$ field generated by $\{C_1, \ldots, C_n\}$ for each $n \geq 1$. Suppose for all $A \in \mathscr{C}_{n+1}$, $B \in \mathscr{C}_{1,n}$ and all $n \geq 1$ that

$$| P[AB] - P[A]P[B] | \leq \phi_n P[A].$$

Suppose $\sum_{n=1}^\infty \phi_n < \infty$ and $\sum_{n=1}^\infty P[C_n] = \infty$. Prove $P[C_n \text{ i.o.}] = 1$. ∎

Laws of the iterated logarithm for strictly stationary sequences satisfying mixing conditions were first established by Philipp [1967] and then by Iosifescu [1968] and Reznik [1968]. Philipp [1969] has also analyzed the more difficult situation of nonstationary sequences satisfying mixing conditions. In Philipp [1971], these results are applied to number theory. More recently, Heyde and Scott [1973] proved the law of the iterated logarithm for $\phi$-mixing sequences under the assumption that $EX_1^2 < \infty$ instead of $E | X_1 |^{2+\delta} < \infty$ for some $\delta > 0$ as assumed by Reznik. Their method is totally different and quite interesting. The main idea (due to Gordin [1969]) is the representation of a mixing sequence as the sum of martingale difference sequences and a small remainder part.

Let $\{X_n, n \geq 1\}$ be a stationary Markov sequence with discrete state space and $f$ be a real valued function defined on the state space of the Markov sequence. In his work, Chung [1960, Sections 14–16], proves various strong limit theorems including the law of the iterated logarithm for $\{f(X_n), n \geq 1\}$ under the additional assumption that $\{X_n, n \geq 1\}$ is a recurrent Markov chain with $Ef^2(X_1) < \infty$ and with certain regularity conditions satisfied. The trick (often called Doeblin's method or Doeblin's trick) is to reduce the results to the case of independence by means of renewal theory. Fix $i \in S$, the state space. Let $\tau_\nu$ be the $\nu$th time that $X_n = i$ for $\nu \geq 1$. Suppose $P[X_1 = i] = 1$. Then it is not difficult to show that

$\{Z_\nu, \nu \geq 1\}$ defined by $Z_\nu = \sum_{n=\tau_\nu}^{\tau_{\nu+1}-1} f(X_n)$ is a sequence of independent identically distributed random variables. Thus, provided second moments are finite, by the Hartman–Wintner law of the iterated logarithm

$$\limsup_{i=1}^{n} (Z_i - EZ_1)/[2(\text{var } Z_1)n \log_2 n]^{1/2} = 1 \qquad \text{a.s.}$$

It seems plausible that this would imply

$$\limsup_{i=1}^{n} [f(X_i) - Ef(X_1)]/(2 \text{ var } f(X_1)n \log_2 n)^{1/2} = 1 \qquad \text{a.s.}$$

Indeed it does. For details, one should consult Chung's book referred to above.

Results can be obtained for nondiscrete state spaces also. Let $\{X_n, n \geq 1\}$ be a stationary Markov sequence and $f$ be a real valued function defined on the state space of the Markov sequence. Suppose $\sigma^2 = Ef^2(X_1) > 0$ and $E|f(X_1)|^{2+\delta} < \infty$ for some $\delta > 0$. Let $T_n = \sum_{i=1}^{n} f(X_i)$. Provided $\{X_n, n \geq 1\}$ satisfies Doeblin's condition (see Definition 3.6.7 and Doob [1953, p. 192]) as well as certain regularity conditions, it follows that $\{X_n, n \geq 1\}$ satisfies the hypotheses of Reznik and hence that

$$\limsup(T_n - ET_n)/(2n\sigma^2 \log_2 n\sigma^2)^{1/2} = 1 \qquad \text{a.s.}$$

There is another interesting direction in which the law of the iterated logarithm can be generalized. $\limsup S_n/(2s_n^2 \log_2 s_n^2)^{1/2} = 1$ a.s. is equivalent to the upper class result

$$P[S_n > (1 + \varepsilon)(2s_n^2 \log_2 s_n^2)^{1/2} \text{ i.o.}] = 0$$

for each $\varepsilon > 0$ and the lower class result

$$P[S_n > (1 - \varepsilon)(2s_n^2 \log_2 s_n^2)^{1/2} \text{ i.o.}] = 1$$

for each $\varepsilon > 0$. Following Feller, $(1 + \varepsilon)(2 \log_2 s_n^2)^{1/2}$ is said to be in the upper class of $S_n/s_n$ for all $\varepsilon > 0$ and $(1 - \varepsilon)(2 \log_2 s_n^2)^{1/2}$ to be in the lower class of $S_n/s_n$ for all $\varepsilon > 0$. This raises the question of deciding whether an arbitrary positive sequence $\{\phi_n, n \geq 1\}$ such that $\phi_n \uparrow \infty$ is in the upper or lower class for a given $\{X_i, i \geq 1\}$. Feller [1943, 1946b] gives remarkably complete answers to this interesting question. The computations are correspondingly involved. Suppose $\{X_i, i \geq 1\}$ is a sequence of independent identically distributed symmetric random variables with $EX_1^2 \log_2^+ |X_1|$ $< \infty$ and $0 < \phi_n \uparrow \infty$. Then it is shown that $\phi_n$ belongs to the upper class if and only if

$$\sum_{i=1}^{\infty} (\phi_i/i) \exp(-\phi_i^2/2) < \infty.$$

Feller [1943] also gives analogous results for certain classes of bounded independent $X_i$ which are not necessarily identically distributed. Feller's results depend on delicate estimates of the rate of convergence of $P[S_n > s_n x]$ in the central limit theorem for large values of $x$. Strassen [1965] extends the upper–lower class results to the martingale case by means of an invariance principle for martingales analogous to Theorem 5.3.2. Recently Feller [1968b] has shown that $\lim \sup S_n/a_n = 1$ a.s. is implied in some cases for $\{X_i, i \geq 1\}$ independent identically distributed with $EX_1 = 0$ but $EX_1^2 = \infty$. Of course, by Theorem 5.3.5 $\{a_n, n \geq 1\}$ will not equal $\{(2nK \log_2 n)^{1/2}, n \geq 1\}$. This paper of Feller's contains some fascinating remarks on the asymptotic fluctuation behavior of $S_n$ as compared with $S_n - \max_{i \leq n} X_i$.

Heyde [1969] and Kesten [1972] bring a closely related problem to those discussed in the preceding paragraph into clear focus. Let $\{X_i, i \geq 1\}$ be a sequence of independent identically distributed random variables, assumed to be symmetric to avoid questions of centering constants. Then $\{X_i, i \geq 1\}$ is said to exhibit behavior of iterated logarithm type if

$$-\infty < \lim \inf S_n/a_n < \lim \sup S_n/a_n < \infty \qquad \text{a.s.}$$

for some $0 < a_n \uparrow \infty$. Whether $\{X_i, i \geq 1\}$ exhibits such behavior is not really a moment problem. Freedman (see Strassen [1966]) gives an example where $E|X_1| = \infty$ and yet there exists $0 < a_n \uparrow \infty$ such that

$$\lim \sup S_n/a_n = 1 \qquad \text{a.s.};$$

that is, $\{X_i, i \geq 1\}$ exhibits behavior of iterated logarithm type. Indeed, Feller [1968b] gives such an example where $E|X_1|^\alpha = \infty$ for all $\alpha > 0$. The following rather striking result is true.

**Theorem 5.4.5.** $\{X_i, i \geq 1\}$ exhibits behavior of iterated logarithm type if and only if

$$\lim_{x \to \infty} \inf x^2 P[|X_1| > x]/EX_1^2 I(|X_1| \leq x) = 0. \qquad (5.4.27)$$

**Remark.** Equation (5.4.27) is precisely the analytic condition which characterizes $X_1$ as belonging to the domain of partial attraction of the normal law. That is, there exists $a_n \uparrow \infty$, integers $n_k \uparrow \infty$ such that

$$P[S_{n_k}/a_k \leq x] \to (2\pi)^{-1/2} \int_{-\infty}^{x} \exp(-y^2/2) \, dy$$

for each $x \in R$.

**Proof.** The sufficiency is due to Kesten [1972] and lies very deep. Thus we omit the proof. The necessity, which we prove, is due to Heyde [1969a]. (It was discovered independently by Steiger.) Assume Eq. (5.4.27) fails. Hence there exists $C > 0$ such that

$$x^2 P[|X_1| > x] > CEX_1^2 I[|X_1| \leq x] \tag{5.4.28}$$

for all $x$ sufficiently large. Thus, for all $x$ sufficiently large and $\eta > 1$

$$x^2\{P[|X_1| > x] - P[|X_1| > \eta x]\} \leq EX_1^2 I[x < |X_1| \leq \eta x]$$
$$\leq x^2\eta^2 C^{-1} P[|X_1| > \eta x].$$

Let $a_n \uparrow \infty$. Then for $n$ sufficiently large and $\eta > 1$,

$$(1 + C^{-1}\eta^2)^{-1} P[|X_1| > a_n] \leq P[|X_1| > \eta a_n]$$
$$\leq P[|X_1| > a_n].$$

On the other hand, if $\eta < 1$,

$$P[|X_1| > a_n] \leq P[|X_1| > \eta a_n]$$
$$\leq (1 + C^{-1}\eta^{-2}) P[|X_1| > a_n].$$

Thus, either $\sum_{n=1}^{\infty} P[|X_n| > \eta a_n] < \infty$ for all $\eta > 0$ or $\sum_{n=1}^{\infty} P[|X_n| > \eta a_n] = \infty$ for all $\eta > 0$. (Note the connection with Lemma 3.2.4.)

Thus, either $\limsup |X_n|/a_n = \infty$ a.s. or $X_n/a_n \to 0$ a.s. $\limsup |X_n|/a_n = \infty$ a.s. implies $\limsup |S_n|/a_n = \infty$ a.s., not behavior of the iterated logarithm type. Suppose the other case, namely that $X_n/a_n \to 0$ a.s. Let $X_n' = X_n I(|X_n| \leq a_n)$ for $n \geq 1$. By Eq. (5.4.28),

$$E(X_n')^2 \leq C^{-1} a_n^2 P[|X_1| > a_n].$$

Thus, $\sum_{n=1}^{\infty} E(X_n')^2/a_n^2 < \infty$, thus implying $\sum_{i=1}^{n} X_i'/a_n \to 0$ a.s. Clearly $P[X_n' \neq X_n \text{ i.o.}] = 0$ and hence, $\sum_{i=1}^{n} X_i/a_n \to 0$ a.s., not behavior of the iterated logarithm type. ∎

**Exercise 5.4.3.** Suppose that $X_1$ does not belong to the domain of partial attraction of the normal distribution. Let $0 < a_n \uparrow \infty$. Prove that $S_n/a_n \to 0$ a.s. if and only if

$$\sum_{n=1}^{\infty} P[|X_1| > a_n] < \infty. \quad \blacksquare$$

Two recent papers of Feller's [1969, 1970] explore the question of when $\lim \sup S_n/a_n = 1$ a.s. for independent $X_i$ with zero means but not necessarily satisfying the classical assumptions (such as $|X_i| \leq K_i s_i (\log_2 s_i^2)^{-1}$ a.s. for all $i \geq 1$ with $K_i \to 0$). In the same vein, Egorov [1969] shows that $X_i$ independent with mean 0 and $|X_i| \leq K s_i (\log_2 s_i^2)^{-1}$ for all $i \geq 1$ implies that

$$\lim \sup S_n/(2s_n^2 \log_2 s_n^2)^{1/2} = C \qquad \text{a.s.}$$

for some $0 < C < \infty$.

In the spirit of Chapter 4, laws of the iterated logarithm exist for weighted sums of random variables. Strassen's Example 5.3.1 and Azuma's Theorem 4.2.3 are of this type. The first such result was due to Gal [1951] (the upper class result) and Stackelberg [1964] (the lower class result). They showed that $\{X_i, i \geq 1\}$ independent with $P[X_i = 1] = P[X_i = -1] = \frac{1}{2}$ for $i \geq 1$ implies

$$\lim \sup \sum_{i=1}^{n} (1 - i/n) X_i / [(2/3)n \log_2 n]^{1/2} = 1 \qquad \text{a.s.}$$

Gaposhkin [1965] generalized the Gal–Stackelberg result to independent uniformly bounded random variables with mean zero and equal variance and considered weights of the form $(1 - i/n)^\alpha$ for any $\alpha > 0$. Recently Tomkins [1971a] and Chow and Lai [1973] have extended these results to still more general classes of weights.

Another direction is the study of the law of the iterated logarithm for lacunary trigonometric series; that is, series of the form

$$\sum_{i=1}^{\infty} (a_i \cos n_i x + b_i \sin n_i x)$$

with integers $n_{i+1}/n_i \geq C > 1$ for all $i \geq 1$. Salem and Zygmund [1950] proved the upper half of the law of the iterated logarithm under the assumption

$$B_n = \left( \sum_{i=1}^{n} a_i^2 + b_i^2 \right)^{1/2} \to \infty$$

and

$$\max_{i \leq n} (a_i^2 + b_i^2)^{1/2} = o(B_n (\log_2 B_n)^{-1/2}). \tag{5.4.29}$$

Erdös and Gal [1965] established the full law of the iterated logarithm for series of the form $\sum_{k=1}^{\infty} \exp(in_k x)$. Finally, Weiss [1959] established the lower half of the law of the iterated logarithm as well, assuming Eq. (5.4.29).

Khintchine [1924] proved the first law of the iterated logarithm in 1924 for $X_i$ independent with

$$P[X_i = 1] = P[X_i = -1] = \tfrac{1}{2} \quad \text{for} \quad i \geq 1.$$

For a result as classical as the law of the iterated logarithm, there has been a striking amount of activity in recent years as this chapter indicates.

CHAPTER **6**

# Recurrence of $\{S_n, n \ge 1\}$ and Related Results

### 6.1. Recurrence of $\{S_n, n \ge 1\}$ in the Case of Independent Identically Distributed $X_i$

Throughout Section 6.1, $\{X_i, i \ge 1\}$ denotes a sequence of independent identically distributed random variables. As is traditional, we refer to $\{S_n, n \ge 1\}$ as a random walk. When $X_1$ is nondegenerate, $S_n$ is almost surely divergent by the results of Chapter 2. That is,

$$P[(S_n \to \infty) \cup (S_n \to -\infty) \cup S_n \text{ oscillates})] = 1.$$

We have barely studied the various modes of divergence of $S_n$. What conditions imply $S_n \to \infty$ a.s.; is it possible for $\limsup S_n < \infty$ and $\liminf S_n > -\infty$ a.s. to hold and yet $S_n$ be a.s. divergent; etc.? Satisfying answers exist to these and closely related questions.

**Theorem 6.1.1.** Either

(i)   $S_n \to \infty$ a.s.,

(ii)  $S_n \to -\infty$ a.s.,

(iii) $S_n = 0$ a.s. for each $n \ge 1$, or

(iv)  $\limsup S_n = \infty$ and $\liminf S_n = -\infty$ a.s.

*Proof.* Each of (i)–(iv) occurs with probability one or zero by the Hewitt–Savage 0–1 law. (Of course, the Kolmogorov 0–1 law suffices for (i), (ii), and (iv) while (iii) is trivial to analyze. Thus the Hewitt–Savage law is not really needed here.)

Suppose the events (i), (ii), and (iii) each have probability zero. Since $P[S_n \to -\infty] = 0$,

$$P[\limsup S_n > -\infty] = 1.$$

331

There then exists $-\infty < c \leq \infty$ such that $P[\limsup S_n = c] = 1$. Otherwise, there would exist a finite $b$ such that $0 < P[\limsup S_n > b] < 1$, an impossibility by the Hewitt–Savage 0–1 law. Since $(X_1, X_2, \ldots)$ has the same distribution as $(X_2, X_3, \ldots)$, it follows that

$$\limsup S_n = \limsup(S_n - X_1) = c \qquad \text{a.s.}$$

Thus $c = c - X_1$ a.s. Thus either $c = \infty$ or $P[X_1 = 0] = 1$. $P[X_1 = 0] = 1$ if and only if $S_n = 0$ a.s. for every $n \geq 1$. Since (iii) is assumed to have probability 0, it follows that $c = \infty$, that is, $\limsup S_n = \infty$ a.s. Replacing each $X_i$ by $-X_i$ in the above argument then yields $\liminf S_n = -\infty$ a.s. ∎

Note that the thrust of the proof was the showing that

$$P[-\infty < \liminf S_n < \limsup S_n < \infty] = 0.$$

We have isolated the four possible modes of divergence and have shown that one such mode must occur with probability one. What conditions are sufficient for each of these modes of divergence? Clearly $E \mid X_1 \mid < \infty$ and $EX_1 \neq 0$ implies by the strong law that $S_n \to \infty$ a.s. when $EX_1 > 0$ and $S_n \to -\infty$ a.s. when $EX_1 < 0$. What happens when $EX_1 = 0$? If $EX_1^2 < \infty$, we know by the Hartman–Wintner law of the iterated logarithm that $\limsup S_n = \infty$ a.s. and $\liminf S_n = -\infty$ a.s. Naively, one might suspect that the symmetry imposed by the condition $EX_1 = 0$ would imply that $\limsup S_n = \infty$ and $\liminf S_n = -\infty$ a.s. provided that $P[X_1 = 0] < 1$. It is possible to construct $\{X_i, i \geq 1\}$ such that $EX_1 = 0$ and

$$P[S_n > n/(2 \log n)] \to 1.$$

(See Exercise 15, Chapter 10 of Feller [1966].) This might make one suspect (also naively) that $S_n \to \infty$ a.s. is possible when $EX_1 = 0$.

The study of recurrence provides us with an answer to this rather difficult question.

**Definition 6.1.1.** Let $\{T_n, n \geq 1\}$ be a stochastic sequence.

(i)   A number $x$ is a recurrent state if

$$P[T_n \in (x - \varepsilon, x + \varepsilon) \text{ i.o.}] = 1$$

for every $\varepsilon > 0$. Intuitively, $x$ is recurrent if $T_n$ comes close to $x$ infinitely often with probability one.

(ii)   A number $x$ is a possible state if, for each $\varepsilon > 0$, there exists an integer $n \geq 1$ such that

$$P[T_n \in (x - \varepsilon, x + \varepsilon)] > 0.$$

(Intuitively, $x$ is possible if at some time $n$, $T_n$ may be close to $x$.)   ∎

**Remark.**   Note by the Hewitt–Savage 0–1 law, $x$ not recurrent implies

$$P[S_n \in (x - \varepsilon, x + \varepsilon) \text{ i.o.}] = 0$$

for all sufficiently small $\varepsilon > 0$.

**Lemma 6.1.1.**   Let $R_1$ be the real line with the usual topology. Let $L_d = \{0, \pm d, \pm 2d, \dots\}$ for each $d > 0$ and $L_0 = R_1$, and $L_\infty = \{0\}$. Then a subset $G$ of $R_1$ is a topologically closed subgroup (under addition) of $R_1$ if and only if $G = L_d$ for some $d$, $\infty \geq d \geq 0$.

**Exercise 6.1.1.**   Prove Lemma 6.1.1.   ∎

We distinguish between two types of random variables and two resulting kinds of random walks.

**Definition 6.1.2.**   A random variable $X$ is said to be distributed on the lattice $L_d = (0, \pm d, \pm 2d, \dots)$ for some $\infty > d > 0$ if

$$\sum_{i=0}^{\infty} P[|X| = id] = 1 \quad \text{and} \quad \sum_{i=0}^{\infty} P[|X| = id'] < 1$$

for all $d' > d$. $X$ is said to be distributed on $L_\infty = \{0\}$ if $P[X = 0] = 1$. If $X$ is distributed on a lattice for some $\infty \geq d > 0$, then $X$ is said to be of lattice type.   ∎

**Exercise 6.1.2.**   Let $a \neq 0$ and $b \neq 0$ be given. Suppose $0 < P[X = a] < 1$ and $P[X = b] = 1 - P[X = a]$. Prove that $X$ is of lattice type if and only if $a/b$ is rational. Supposing that $a/b$ is rational, find the lattice $L_d$ on which $X$ is distributed.   ∎

If $X$ is not of lattice type we say that $X$ is distributed on $L_0 = R_1$. Note that if $X_1$ is distributed on $L_d$, then $S_n$ is distributed on $L_d$ for each $n \geq 1$. In this situation, we call $L_d$ the range of the random walk $\{S_n, n \geq 1\}$. Clearly random walks with range $L_d$ for some $d > 0$ (lattice type random walks) will behave very differently than random walks with range $L_0$ (nonlattice type random walks). However, for our purposes, lattice and nonlattice type random walks can be analyzed together.

**Example 6.1.1.**   Let $P[X_1 = 1] = p$ and $P[X_1 = -1] = q \equiv 1 - p$. We consider the random walk $\{S_n, n \geq 1\}$. Clearly it is distributed on the lattice $L_1$. $p > \frac{1}{2}$ implies $S_n \to \infty$ a.s. and $p < \frac{1}{2}$ implies $S_n \to -\infty$ a.s. By the law of the iterated logarithm applied to $\{X_i, i \geq 1\}$ and then to $\{-X_i, i \geq 1\}$ it follows that $p = \frac{1}{2}$ implies $\limsup S_n = \infty$ a.s. and $\liminf S_n = -\infty$ a.s. We have in previous chapters considered a gambling interpretation for $\{S_n, n \geq 1\}$.

Another interesting interpretation considers $\{S_n, n \geq 1\}$ as a model for the returns in a two candidate election. Consider an infinite population voting sequentially in a two candidate election. Let $X_i = 1$ if the $i$th person votes for Candidate I and $X_i = -1$ if the $i$th person votes for Candidate II. Then $S_n$ represents the lead that Candidate I has after $n$ people have voted. If $p = \frac{1}{2}$, it follows from the law of the iterated logarithm that the lead changes from Candidate I to Candidate II infinitely often with probability one. Indeed, $P[S_n = k \text{ i.o. in } n] = 1$ for each integer $k$ follows from the law of the iterated logarithm; that is, all states of the range of $\{S_n, n \geq 1\}$ are recurrent. Of course if $p < \frac{1}{2}$ there will be a last time Candidate I is ahead, if he ever is, with probability one. Indeed, for each $k \geq 1$, there will be a last time Candidate I is behind by $k$ votes or less.  ∎

**Exercise 6.1.3.**   Let $L_d$ be the range of $\{S_n, n \geq 1\}$, for some $\infty \geq d \geq 0$.

(i)   Let $x \in L_d$ and $0 \leq d < \infty$. Show by example that $x$ is not necessarily possible.

(ii)   Let $x$ be possible. Prove that $x \in L_d$.  ∎

**Theorem 6.1.2.**   [Chung and Fuchs, 1951]. Let $L_d$ be the range of $\{S_n, n \geq 1\}$ for some $\infty \geq d \geq 0$. Then either no members of $L_d$ are recurrent or all members of $L_d$ are recurrent.

**Proof.**   Let $G$ denote the recurrent states of $L_d$. Suppose $G$ is nonempty. We first show that $G$ is topologically closed and that $G$ is a group. Choose $x_n \in G$ such that $x_n \to x \in R_1$. Fix $\varepsilon > 0$ and consider the interval $(x - \varepsilon, x + \varepsilon)$. Choose $n \geq 1$ such that $(x_n - n^{-1}, x_n + n^{-1}) \subset (x - \varepsilon, x + \varepsilon)$. Then

$$P[S_m \in (x - \varepsilon, x + \varepsilon) \text{ i.o.}]$$
$$\geq P[S_m \in (x_n - n^{-1}, x_n + n^{-1}) \text{ i.o. in } m] = 1.$$

Thus $x$ is recurrent and therefore $G$ is closed.

In order to show that $G$ is a group, it suffices to show that $x$ recurrent and $y$ recurrent implies $x - y$ recurrent since $G$ nonempty and closed under subtraction implies that $G$ is a group. Because we will use it, we prove the more general statement that $x$ recurrent and $y$ possible implies $x - y$ recurrent. Choose $\varepsilon > 0$ and then choose $n \geq 1$ such that $P[|\,S_n - y\,| < \varepsilon] > 0$. Since $x$ is recurrent, it follows that

$$1 = P[|\,S_m - x\,| < \varepsilon \text{ i.o.} \,|\, |\,S_n - y\,| < \varepsilon]$$
$$= P[|\,S_m - x\,| < \varepsilon \text{ i.o.}, |\,S_n - y\,| < \varepsilon \,|\, |\,S_n - y\,| < \varepsilon]$$
$$\leq P[|\,S_m - S_n - (x - y)\,| < 2\varepsilon \text{ i.o. in } m \,|\, |\,S_n - y\,| < \varepsilon]$$
$$= P[|\,S_m - S_n - (x - y)\,| < 2\varepsilon \text{ i.o. in } m]$$
$$= P[|\,S_m - (x - y)\,| < 2\varepsilon \text{ i.o. in } m].$$

The last two equalities hold since the $X_i$ are independent identically distributed. Thus $x - y$ is recurrent and $G$ is a group.

Since $G$ is closed and $G$ is a group, $G = L_{d'}$ for some $\infty \geq d' \geq 0$ by Lemma 6.1.1. There are three cases.

If $d' = 0$, $G = R_1$. Thus the random walk is nonlattice and the range $L_d = R_1$.

If $d' = \infty$, then $x = 0$ is the only recurrent state. Recalling that 0 is recurrent and $y$ possible implies $0 - y$ recurrent, it follows that 0 is the only possible state of $\{S_n, n \geq 1\}$. Thus $P[X_1 = 0] = 1$ and $L_d = \{0\}$.

Suppose (the third case) $0 < d' < \infty$. By Exercise 6.1.3 (ii), $L_{d'} = G \subset L_d$. Thus $0 \leq d \leq d'$. Suppose $d < d'$. It follows that there exists a possible state $y \notin L_{d'}$. But 0 being recurrent implies that $0 - y$ and hence $0 - (-y) = y$ is recurrent; this is a contradiction. Thus $L_{d'} = L_d$. ∎

Variations of the simple renewal argument used above to show that

$$P[|\,S_m - x\,| < \varepsilon \text{ i.o.} \,|\, |\,S_n - y\,| < \varepsilon] = P[|\,S_m - (x - y)\,| < 2\varepsilon \text{ i.o. in } m]$$

will be used repeatedly in Chapter 6.

**Definition 6.1.3.** A random walk is called recurrent or transient according as the states of the range are recurrent or transient. ∎

**Corollary 6.1.1.** If $\{S_n, n \geq 1\}$ is recurrent and $P[X_1 = 0] < 1$, then $\limsup S_n = \infty$ and $\liminf S_n = -\infty$ a.s.

**Proof.** Immediate from Theorem 6.1.2. ∎

**Theorem 6.1.3.**   [Chung and Fuchs, 1951].   $\{S_n, n \geq 1\}$ is recurrent if and only if there exists a finite open interval $J$ such that

$$\sum_{n=1}^{\infty} P[S_n \in J] = \infty.$$

Moreover, $\sum_{n=1}^{\infty} P[S_n \in J] = \infty$ for such an interval implies that $\sum_{n=1}^{\infty} P[S_n \in J'] = \infty$ for every open interval $J'$ such that $J'$ contains at least one possible state.

**Proof.**   If $\sum_{n=1}^{\infty} P[S_n \in J] < \infty$ for all finite open intervals $J$, then by the Borel–Cantelli lemma, there can be no recurrent states.

Suppose $\sum_{n=1}^{\infty} P[S_n \in J] = \infty$ for some finite open interval $J$. Fix $\varepsilon > 0$. Since $J$ is finite, there exists $x \in R_1$ and $I = (x - \varepsilon, x + \varepsilon)$ such that

$$\sum_{n=1}^{\infty} P[S_n \in I] = \infty. \tag{6.1.1}$$

Let

$$A_n = [S_n \in I, S_{n+j} \notin I \quad \text{for all} \quad j \geq 1]$$

for $n \geq 1$ and let $A_0 = [S_j \notin I \text{ for } j \geq 1]$. $A_n$ is the event that the last visit to $I$ occurs at time $n$. Fix $n \geq 1$.

$$A_n \supset [S_n \in I, \mid S_{n+j} - S_n \mid \geq 2\varepsilon \quad \text{for all} \quad j \geq 1].$$

Then

$$P[S_n \in I \quad \text{for at most finitely many } n]$$

$$= \sum_{n=0}^{\infty} P(A_n)$$

$$\geq \sum_{n=1}^{\infty} P[S_n \in I] P[\mid S_{n+j} - S_n \mid \geq 2\varepsilon \text{ for all } j \geq 1]$$

$$= P[\mid S_j \mid \geq 2\varepsilon \text{ for all } j \geq 1] \sum_{n=1}^{\infty} P[S_n \in I],$$

since the $X_i$ are independent identically distributed. Using Eq. (6.1.1),

$$P[\mid S_j \mid \geq 2\varepsilon \text{ for all } j \geq 1] = 0 \tag{6.1.2}$$

for each $\varepsilon > 0$.

Thus, loosely speaking, 0 is visited with probability one. Intuitively, once 0 is visited, the sequence $\{S_n, n \geq 1\}$ starts from scratch and once again 0 should be visited with probability one, etc. To make this renewal argument rigorous, fix $\varepsilon > 0$ and let $I_0 = (-\varepsilon, \varepsilon)$,

$$B_n = [S_n \in I_0, S_{n+j} \notin I_0 \text{ for all } j \geq 1]$$

for $n \geq 1$, and

$$B_0 = [S_j \notin I_0 \text{ for all } j \geq 1].$$

$$1 - P[S_n \in I_0 \text{ i.o.}] = \sum_{n=0}^{\infty} P(B_n)$$

since the $B_n$ are disjoint. By Eq. (6.1.2), $P(B_0) = 0$. Thus to show that $P[S_n \in I_0 \text{ i.o.}] = 1$, it suffices to show $P(B_n) = 0$ for arbitrary $n \geq 1$. Fix $n \geq 1$. Let

$$B_{n,\delta} = [S_n \in (-\delta, \delta), S_{n+j} \notin I_0 \text{ for all } j \geq 1]$$

for $\varepsilon > \delta > 0$.

$$P[B_{n,\delta} \triangle B_n] \to 0 \qquad \text{as} \quad \delta \uparrow \varepsilon.$$

Hence it suffices to show that $P[B_{n,\delta}] = 0$ for fixed $\varepsilon > \delta > 0$

$$B_{n,\delta} \subset [S_n \in (-\delta, \delta), |S_{n+j} - S_n| > \varepsilon - \delta \text{ for all } j \geq 1].$$

Thus

$$P(B_{n,\delta}) \leq P[S_n \in (-\delta, \delta), |S_{n+j} - S_n| > \varepsilon - \delta \text{ for all } j \geq 1]$$
$$= P[S_n \in (-\delta, \delta)]P[|S_j| > \varepsilon - \delta \text{ for all } j \geq 1] = 0$$

by Eq. (6.1.2). Thus $P[S_n \in I_0 \text{ i.o.}] = 1$ for all $\varepsilon > 0$; that is, 0 is recurrent. By Theorem 6.1.2, all states in the range of $\{S_n, n \geq 1\}$ are recurrent or else none are. Thus $\{S_n, n \geq 1\}$ is recurrent. Hence $\sum_{n=1}^{\infty} P[S_n \in J'] = \infty$ for every interval containing at least one possible state by the Borel–Cantelli lemma. ∎

Recall that $\{S_n, n \geq 1\}$ recurrent and $P[X_1 = 0] < 1$ implies by Corollary 6.1.1 that $\limsup S_n = \infty$ and $\liminf S_n = -\infty$ a.s. It is interesting to ask whether the converse is true:

**Example 6.1.2.** We exhibit a random walk $\{S_n, n \geq 1\}$ such that $\limsup S_n = \infty$ and $\liminf S_n = -\infty$ a.s. and yet $\{S_n, n \geq 1\}$ is transient (that is, $|S_n| \to \infty$ a.s.). The example depends on the theory of stable distributions. Stable distributions play a central role in the study of weak convergence. $X_1$ is said to be a symmetric stable law if $X_1$ is symmetric and for every $c_1 > 0$, $c_2 > 0$ there exists a corresponding $c > 0$ such that $c_1 X_1 + c_2 X_2$ and $c X_1$ have the same distribution. $X_1$ normal with $EX_1^2 = 1$, $EX_1 = 0$, and $c = (c_1^2 + c_2^2)^{1/2}$ provides one example. According to the general theory, there exist symmetric stable random variables corresponding to each $\beta \geq \frac{1}{2}$ such that $\sum_{i=1}^{n} X_i / n^\beta$ has the same distribution as $X_1$ for each $n \geq 1$. Such random variables possess continuous densities $f$ with $f(0) \neq 0$ (see Feller [1966, pp. 548–549]). Fix $\beta > 1$ and let $X_1$ be a symmetric stable random variable corresponding to $\beta$ with density $f(\cdot)$.

Since $X_1$ is symmetric and nondegenerate, it follows immediately from Theorem 6.1.1 that

$$\limsup S_n = \infty \quad \text{and} \quad \liminf S_n = -\infty \quad \text{a.s.}$$

In order to prove that $\{S_n, n \geq 1\}$ is transient, it suffices to show that $\sum_{n=1}^{\infty} P[|S_n| \leq 1] < \infty$ by Theorem 6.1.3.

$$P[|S_n| \leq 1] = P[|S_n|/n^\beta \leq n^{-\beta}] = P[|X_1| \leq n^{-\beta}]$$

$$= \int_{-n^{-\beta}}^{n^{-\beta}} f(x) \, dx \sim 2n^{-\beta} f(0)$$

since $f$ is continuous. Hence $\sum_{n=1}^{\infty} P[|S_n| \leq 1] < \infty$ since $\sum_{n=1}^{\infty} n^{-\beta} < \infty$. ∎

Next we will see that $EX_1 = 0$ implies recurrence thereby answering our initial question affirmatively of whether $EX_1 = 0$ and $P[X_1 = 0] < 1$ implies $\limsup S_n = \infty$ and $\liminf S_n = -\infty$ a.s.

**Theorem 6.1.4.** [Chung and Fuchs, 1951]. $EX_1 = 0$ implies that $\{S_n, n \geq 1\}$ is recurrent.

*Proof.* [Chung and Ornstein, 1962]. By Theorem 6.1.3, it suffices to show that $\sum_{n=1}^{\infty} P[|S_n| \leq 1] = \infty$. Let $J$ be any interval of length 1.

Let the (possibly infinite valued) random variable $N$ be the number of times $S_n \in J$ for $n \geq 1$.

$$N = \sum_{n=1}^{\infty} I(S_n \in J).$$

Thus

$$EN = \sum_{n=1}^{\infty} P[S_n \in J]. \tag{6.1.3}$$

Let the stopping rule $t$ be the smallest $n \geq 1$ such that $S_n \in J$ if such an $n$ exists, otherwise let $t = \infty$.

$$EN = \sum_{n=1}^{\infty} \int_{t=n} N \, dP.$$

$t = n$ implies

$$\sum_{k=1}^{\infty} I(S_k \in J) \leq 1 + \sum_{k=n+1}^{\infty} I(S_k \in J)$$

$$= 1 + \sum_{k=n+1}^{\infty} I[(S_k - S_n) + S_n \in J]$$

$$\leq 1 + \sum_{k=n+1}^{\infty} I(S_k - S_n \in [-1, 1])$$

since $J$ has length 1 and $t = n$ implies $S_n \in J$. Thus

$$EN \leq \sum_{n=1}^{\infty} \int_{\Omega} I(t = n)\left\{1 + \sum_{k=n+1}^{\infty} I(S_k - S_n \in [-1, 1])\right\} dP$$

$$= \sum_{n=1}^{\infty} P[t = n]\left(1 + \sum_{k=1}^{\infty} P\{S_k \in [-1, 1]\}\right)$$

$$\leq 1 + \sum_{k=1}^{\infty} P\{S_k \in [-1, 1]\}.$$

Here we have used the fact that $S_k - S_n \in [-1, 1]$ and $t = n$ are independent for each $k \geq n$ and that $S_k - S_n$ is distributed as $S_{k-n}$ for each $k > n$. Thus, using Eq. (6.1.3),

$$\sum_{n=1}^{\infty} P[S_n \in J] \leq 1 + \sum_{k=1}^{\infty} P\{S_k \in [-1, 1]\}. \tag{6.1.4}$$

Let $M$ be a positive integer.

$$\sum_{n=1}^{\infty} P[-M < S_n \leq M] = \sum_{k=-M}^{M-1} \sum_{n=1}^{\infty} P\{S_n \in (k, k+1]\}$$

$$\leq 2M\left(1 + \sum_{n=1}^{\infty} P[|S_n| \leq 1]\right)$$

by Eq. (6.1.4). Thus

$$\limsup_{M \to \infty} (2M)^{-1} \sum_{n=1}^{\infty} P[|S_n| < M] \leq 1 + \sum_{n=1}^{\infty} P[|S_n| \leq 1]. \tag{6.1.5}$$

To complete the proof, we show that the left hand side of Eq. (6.1.5) equals $+\infty$. $EX_1 = 0$ implies $S_n/n \to 0$ a.s. implies $S_n/n \to 0$ in probability. Fix $\varepsilon > 0$. There exists $n_0$ such that $n \geq n_0$ implies $P[|S_n| < \varepsilon n] > \frac{1}{2}$. Hence for $M > \varepsilon n$ and $n \geq n_0$, $P[|S_n| < M] > \frac{1}{2}$. Hence

$$\limsup_{M \to \infty} (2M)^{-1} \sum_{n=1}^{\infty} P[|S_n| < M] \geq \limsup_{M \to \infty} (2M)^{-1}(1/2)[(M/\varepsilon) - 1 - n_0]$$

$$= (4\varepsilon)^{-1}.$$

Since $\varepsilon > 0$ is arbitrary,

$$\limsup_{M \to \infty} (2M)^{-1} \sum_{n=1}^{\infty} P[|S_n| < M] = \infty. \quad \blacksquare$$

Theorem 6.1.4 and the strong law of large numbers give us a description of the recurrence behavior of $\{S_n, n \geq 1\}$ when $E|X_1| < \infty$. What can be said when $E|X_1| = \infty$? Just as the strong law was used, by Exercise 3.2.4, $E(X_1^+) = \infty$ and $E(X_1^-) < \infty$ implies that $\lim S_n/n = \infty$ a.s. and hence that $\lim S_n = \infty$ a.s. Under what conditions can we assert that $\limsup S_n = \infty$ and $\liminf S_n = -\infty$ a.s.? Of course, $X_1$ symmetric with $P[X_1 = 0] < 1$ is trivially such a condition by Theorem 6.1.1. A deeper analysis shows that

$$\liminf S_n = -\infty \quad \text{a.s.}$$

if and only if

$$\sum_{n=1}^{\infty} n^{-1} P[S_n < 0] = \infty$$

and hence that

$$\limsup S_n = \infty \quad \text{and} \quad \liminf S_n = -\infty \quad \text{a.s.}$$

if and only if

$$\sum_{n=1}^{\infty} n^{-1} P[S_n > 0] = \infty \quad \text{and} \quad \sum_{n=1}^{\infty} n^{-1} P[S_n < 0] = \infty.$$

Before proving this result we need two preparatory lemmas.

A combinatoric result which we isolate in Lemma 6.1.2 lies at the heart of the proof. Let $\{x_1, x_2, \ldots, x_n\}$ be $n$ numbers, $n$ fixed. Consider the partial sums $s_0 = 0$, $s_1 = x_1, \ldots, s_n = \sum_{i=1}^{n} x_i$. $\nu > 0$ is said to be a ladder index if $s_\nu > \max_{0 \leq i < \nu} s_i$; that is, if $s_\nu$ exceeds all preceding partial sums. Consider the $n$ cyclic reorderings: $(x_1, x_2, \ldots, x_n)$, $(x_2, x_3, \ldots, x_n, x_1)$, $\ldots$, $(x_n, x_1, \ldots, x_{n-1})$. Number these reorderings $0, 1, \ldots, n-1$, respectively. The partial sums $\{s_k^{(\nu)}, k \geq 1\}$ for the $\nu$th reordering are given by

$$s_k^{(\nu)} = \begin{cases} s_{\nu+k} - s_\nu & \text{for} \quad 0 \leq k \leq n - \nu \\ s_n - s_\nu + s_{k-n+\nu} & \text{for} \quad n - \nu + 1 \leq k \leq n. \end{cases} \qquad (6.1.6)$$

**Lemma 6.1.2.**  [Feller, 1966, p. 394].  Suppose $s_n > 0$. Let $r$ be the number of cyclic reorderings in which $n$ is a ladder index. Then $r \geq 1$. Moreover, for each of these $r$ reorderings, there are exactly $r$ ladder indices.

**Remark.**  Note that a hypothesis like $s_n > 0$ is needed since $x_i < 0$ for all $i \leq n$ implies that $r = 0$. Note $x_i > 0$ for all $i \leq n$ implies $r = n$. One should construct other examples for small $n$ to get some feel for the lemma.

**Proof.**  Let $\nu$ be the smallest positive integer such that $s_\nu > \max_{0 \leq k < \nu} s_i$, $s_\nu \geq \max_{0 \leq k \leq n} s_k$. Such a $\nu$ exists since $s_n > 0$. By Eq. (6.1.6), the last partial sum of the $\nu$th reordering is strictly maximal. Hence $n$ is a ladder index for the $\nu$th reordering. Thus $r \geq 1$.

Suppose, without loss of generality, that $n$ is a ladder index for the given ordering $(x_1, x_2, \ldots, x_n)$. To prove the lemma, we must show that the number of reorderings $r \geq 1$ for which $n$ is a ladder index equals the number of ladder indices for the given ordering. We establish a $1 : 1$ correspondence in two steps. Correspond the given order ($\nu = 0$) with the last ladder index of the given order.

Suppose that $n$ is a ladder index for the $\nu$th reordering for some $1 \leq \nu \leq n - 1$; that is,

$$s_n = s_n^{(\nu)} > \max_{0 \leq k \leq n-1} s_k^{(\nu)}.$$

Thus

$$s_n > s_n - s_\nu + s_{k-n+\nu}$$

for $n - 1 \geq k \geq n - \nu$, by Eq. (6.1.6). That is, $s_\nu > s_j$ for $0 \leq j < \nu$. Hence $\nu$ is a ladder index of the given ordering.

Suppose $\nu$ is a ladder index of the given ordering for some $1 \leq \nu \leq n-1$. Consider the $\nu$th reordering.

$$s_k^{(\nu)} = s_{\nu+k} - s_\nu < s_{\nu+k} \leq s_n$$

for $1 \leq k \leq n - v$ since $s_v > 0$ and $s_n \geq \max_{1 \leq i \leq n} s_i$.

$$s_k^{(v)} = s_n - s_v + s_{k-n+v} < s_n$$

for $n - v + 1 \leq k < n$ since $v$ is a ladder index for the given ordering. Thus

$$S_n = s_n^{(v)} > \max_{0 \leq i < n} s_i^{(v)}.$$

That is, the $v$th reordering has $n$ as a ladder index.

We have thus established the desired $1 : 1$ correspondence between ladder indices of the given ordering and reorderings with $n$ as a ladder index. Since the given ordering was arbitrarily chosen among those orderings having $n$ as a ladder index, the result is established. ∎

Now we prove the basic analytical lemma.

**Lemma 6.1.3.** [Sparre-Anderson, 1954]. Let $p_n = P[S_i \leq 0$ for all $0 \leq i < n$, $S_n > 0]$ for each $n \geq 1$ where $S_0 = 0$. Let the generating function $p(\cdot) : [0, 1] \to [0, 1]$ be defined by $p(s) = \sum_{n=1}^{\infty} p_n s^n$. Then

$$\log\left(\frac{1}{1 - p(s)}\right) = \sum_{n=1}^{\infty} n^{-1} P[S_n > 0] s^n \tag{6.1.7}$$

for $0 \leq s \leq 1$.

*Remark.* Let $S_0 = 0$. $p(\cdot)$ is the generating function of the first ladder index or the entrance time of $\{S_n, n \geq 0\}$ into $(0, \infty)$. Lemma 6.1.3 is rather striking in that it says that the distribution of this entrance time depends on the distribution of $X_1$ only through the values of $\{P[S_n > 0], n \geq 1\}$. For example, the distribution is the same for all continuous symmetric random variables $X_1$. Lemma 6.1.3 will be very useful.

*Exercise 6.1.4.* Let $Y_1$ be a positive integer valued random variable (with $P[Y_1 = \infty] > 0$ possible, contrary to our usual convention). Let $a_i = P[Y_1 = a_i]$ for $i \geq 1$. Let the generating function $a(\cdot) : [0, 1] \to [0, 1]$ be defined by $a(s) = \sum_{i=1}^{\infty} a_i s^i$. For fixed $r \geq 1$, let $\{Y_i, 1 \leq i \leq r\}$ be a sequence of independent random variables, each with the given distribution of $Y_1$. Let $T_r = \sum_{i=1}^{r} Y_i$ and $a_i^{(r)} = P[T_r = i]$ for $1 \leq i \leq r$. Let $a^{(r)}(\cdot)$ be the generating function of $T_r$. Show that

$$a^{(r)}(s) = [a(s)]^r$$

for $0 \leq s \leq 1$. ∎

**_Proof of Lemma 6.1.3._**  [Feller, 1966, pp. 393–397].  Fix $n \geq 1$. We apply the combinatorial Lemma 6.1.2 to $(X_1, X_2, \ldots, X_n)$ and its $n$ cyclic permutations with the obvious correspondence of notation. Fix $n \geq r \geq 1$ and define a sequence of random variables $\{I_r^{(\nu)}, 0 \leq \nu < n\}$ by $I_r^{(\nu)} = 1$ if $n$ is the $r$th ladder index of $\{S_0^{(\nu)}, S_1^{(\nu)}, \ldots, S_n^{(\nu)}\}$; $= 0$ otherwise. Fix $n > \nu \geq 0$. Let $T_j^{(\nu)}$ be the value of the $j$th ladder index of $\{S_i^{(\nu)}, 0 \leq i \leq n\}$ for $1 \leq j \leq n$ when such ladder indices exist; otherwise, let $T_j^{(\nu)} = \infty$ for $1 \leq j \leq n$.

Now,

$$EI_r^{(\nu)} = P[I_r^{(\nu)} = 1] = P[T_r^{(\nu)} = n].$$

Since the $X_i$ are independent identically distributed, $I_r^{(0)}, I_r^{(1)}, \ldots$ and $I_r^{(n-1)}$ each have the same distribution. Thus

$$P[T_r^{(\nu)} = n] = n^{-1} E\left[\sum_{j=0}^{n-1} I_r^{(j)}\right].$$

On $S_n \leq 0$, $\sum_{j=0}^{n-1} I_r^{(j)} = 0$. On $S_n > 0$, $\sum_{j=0}^{n-1} I_r^{(j)} = r$ or $= 0$ by Lemma 6.1.2. Thus

$$P[T_r^{(\nu)} = n] = n^{-1} r P\left[\sum_{j=0}^{n-1} I_r^{(j)} = r\right].$$

Summing over $r$,

$$\sum_{r=1}^{n} P[T_r^{(\nu)} = n]/r = n^{-1} \sum_{r=1}^{n} P\left[\sum_{j=0}^{n-1} I_r^{(j)} = r\right]$$

$$= n^{-1} P[S_n > 0] \tag{6.1.8}$$

since

$$\bigcup_{r=1}^{n} \left(\sum_{j=0}^{n-1} I_r^{(j)} = r\right) = [S_n > 0]$$

by Lemma 6.1.2. By the definition of $T_1^{(\nu)}$, $p_i = P[T_1^{(\nu)} = i]$ for $i \leq n$.

$$T_r^{(\nu)} = T_1^{(\nu)} + (T_2^{(\nu)} - T_1^{(\nu)}) + \cdots + (T_r^{(\nu)} - T_{r-1}^{(\nu)}).$$

It is intuitively clear and easy to show that $T_r^{(\nu)}$ is a sum of $r$ independent random variables, each with the same distribution as $T_1^{(\nu)}$. That is, $P[T_r^{(\nu)} = n]$ is the coefficient of $s^n$ in $[p(\cdot)]^r$. Multiplying by $s^n$ where

$0 \leq s \leq 1$ in Eq. (6.1.8) and summing on $n$,

$$\sum_{n=1}^{\infty} s^n \sum_{r=1}^{n} r^{-1} P[T_r^{(\nu)} = n] = \sum_{n=1}^{\infty} n^{-1} P[S_n > 0] s^n.$$

Also

$$\sum_{n=1}^{\infty} s^n \sum_{r=1}^{n} r^{-1} P[T_r^{(\nu)} = n] = \sum_{r=1}^{\infty} r^{-1} \sum_{n=1}^{\infty} P[T_r^{(\nu)} = n] s^n$$

$$= \sum_{r=1}^{\infty} r^{-1} [p(s)]^r = \log\left(\frac{1}{1 - p(s)}\right),$$

thus completing the proof. ∎

Now we can state and prove Spitzer's [1956] important result. We assume $X_1$ is nondegenerate.

**Theorem 6.1.5.**   (i)   $S_n \to -\infty$ a.s. if and only if

$$\sum_{n=1}^{\infty} n^{-1} P[S_n > 0] < \infty.$$

(ii)   $\limsup S_n = \infty$ and $\liminf S_n = -\infty$ a.s. if and only if

$$\sum_{n=1}^{\infty} n^{-1} P[S_n > 0] = \infty \qquad \text{and} \qquad \sum_{n=1}^{\infty} n^{-1} P[S_n < 0] = \infty.$$

(iii)   (i) and (ii) hold when $>$ and $<$ are replaced by $\geq 0$ and $\leq 0$, respectively.

**Proof.**   (ii) follows immediately from (i) and Theorem 6.1.1. The analysis required to prove (i) establishes (iii) with only minor changes. Thus we only prove (i). Let $p(\cdot)$ be the generating function of the first ladder index of $\{S_n, n \geq 0\}$ as in Lemma 6.1.3. For each $r \geq 1$ let $T_r$ be the value of the $r$th ladder index of $\{S_n, n \geq 0\}$ when the $r$th ladder index exists, otherwise, let $T_r = \infty$.

Suppose $S_n \to -\infty$ a.s. Thus $P[T_n = \infty] \to 1$. $T_n$ is the sum of $n$ independent random variables, each with generating function $p(\cdot)$. $P[T_1 = \infty] > 0$ must hold since $P[T_1 = \infty] = 0$ implies that $P[T_n = \infty] = 0$ for all $n \geq 1$. Thus

$$p(1) = \sum_{n=1}^{\infty} P[T_1 = n] = P[T_1 < \infty] < 1.$$

Thus, by Eq. (6.1.7) of Lemma 6.1.3,

$$\sum_{n=1}^{\infty} n^{-1} P[S_n > 0] < \infty,$$

as desired.

Suppose $P[S_n \to -\infty] < 1$. By Theorem 6.1.1, $\limsup S_n = \infty$ a.s. Hence $P[T_1 = \infty] = 0$; that is, $p(1) = 1$ and by Eq. (6.1.7), $\sum_{n=1}^{\infty} n^{-1} \times P[S_n > 0] = \infty$. ∎

In a recent paper, Williamson [1970] has applied Spitzer's result to the analysis of random walks with $E|X_1| = \infty$. A function $L(\cdot)$ defined on $R^+$ is said to be slowly varying at $\infty$ if $L(x) \geq 0$ for all $x \geq 0$ and $L(tx)/L(x) \to 1$ as $x \to \infty$ for each $t > 0$. Let $X_1$ have distribution function $F$. Let $L(\cdot)$ be slowly varying. Suppose for $x < 0$ that

$$F(x) = L(-x)/|x|^{\alpha}$$

for some $0 < \alpha < 1$. Then according to Williamson, $S_n \to -\infty$ a.s. if and only if

$$E[(X_1^+)^{\alpha}/L(X_1^+)] < \infty.$$

Clearly this result can be stated with hypotheses on both tails of $F$ to yield a necessary and sufficient set of moment conditions for $\limsup S_n = \infty$ and $\liminf S_n = -\infty$ a.s. We will not prove these results.

Besides asking whether a random walk is recurrent, one might ask how much time $S_n$ spends in various subsets of $L_d$. For each finite interval $I$ such that $I \cap L_d \neq \varnothing$, let $\|I\| = $ the length of $I$ if $d = 0$; $= $ the number of points of $L_d$ in $I$ if $d > 0$. For each $n \geq 1$, let $N_n(I) = $ the number of visits to $I$ by $S_1, S_2, \ldots, S_n$. Suppose $\{S_n, n \geq 1\}$ is recurrent. Then, of course, $N_n(I) \to \infty$ a.s. Using the pointwise ergodic theorem, Harris and Robbins [1953] show that

$$N_n(I_1)/N_n(I_2) \to \|I_1\|/\|I_2\|$$

for every pair of intervals $I_1$ and $I_2$. The result is somewhat striking in that one might expect the limit of $N_n(I_1)/N_n(I_2)$ to depend on the distribution of $X_1$. Following Breiman [1968, p. 60], we only give a proof for the lattice case.

**Lemma 6.1.4.**  Let $t$ be a stopping rule with respect to $\{S_n, n \geq 1\}$ (that is, $(t = n) \in \mathscr{B}(S_i, i \leq n)$ for all $n \geq 1$). Let $S_n' = S_{t+n} - S_t$ for $n \geq 1$ define $(S_n', n \geq 1)$. Then $\{S_n', n \geq 1\}$ and $\{S_n, n \geq 1\}$ have the same distributions and $\mathscr{B}(S_n, n \leq t)$ and $\mathscr{B}(S_n', n \geq 1)$ are independent $\sigma$ fields. (Recall that $\mathscr{B}(S_n, n \leq t)$ consists of all events $A \in \mathscr{F}$ for which $A \cap (t = n) \in \mathscr{B}(S_i, i \leq n)$ for each $n \geq 1$.)

**Proof.**  Choose $A \in \mathscr{B}(S_n, n \leq t)$ and $C \in \mathscr{C}_\infty$, the Borel sets of $R_\infty$.

$$P[(S_1', S_2', \ldots) \in C, A]$$

$$= \sum_{n=1}^{\infty} P[S_1', S_2', \ldots) \in C, t = n, A]$$

$$= \sum_{n=1}^{\infty} P[S_{n+1} - S_n, S_{n+2} - S_n, \ldots) \in C, t = n, A]$$

$$= \sum_{n=1}^{\infty} P[S_{n+1} - S_n, S_{n+2} - S_n, \ldots) \in C]P[t = n, A]$$

by the fact that $(t = n) \cap A \in \mathscr{B}(S_i, i \leq n)$ and hence is independent of $[(S_{n+1} - S_n, S_{n+2} - S_n, \ldots)] \in C$.

$$\sum_{n=1}^{\infty} P[(S_{n+1} - S_n, S_{n+2} - S_n, \ldots) \in C]P[t = n, A]$$

$$= P[(S_1, S_2, \ldots) \in C] \sum_{n=1}^{\infty} P[t = n, A]$$

$$= P[(S_1, S_2, \ldots) \in C]P(A).$$

Thus

$$P[(S_1', S_2', \ldots) \in C, A] = P[(S_1, S_2, \ldots) \in C]P(A).$$

Taking $A = \Omega$ shows that $\{S_n', n \geq 1\}$ and $\{S_n, n \geq 1\}$ have the same distribution. The desired independence follows. ∎

Suppose $\{S_n, n \geq 1\}$ is recurrent with range $L_d$ for some $d > 0$. Let

$t_1$  be the smallest $n \geq 1$ such that $S_n = 0$,

$t_2$  be the smallest $n > t_1$ such that $S_n = 0$,

$\vdots$

$t_k$  be the smallest $n > t_{k-1}$ such that $S_n = 0$,

$\vdots$

Let $\tau_1 = t_1, \tau_2 = t_2 - t_1, \ldots, \tau_k = t_k - t_{k-1}, \ldots$ . Supposing $\{S_n, n \geq 1\}$ recurrent one would suspect that $\{\tau_i, i \geq 1\}$ is a sequence of independent identically distributed random variables.

**Exercise 6.1.5.**   Use Lemma 6.1.4 to prove the above statement.   ∎

**Theorem 6.1.6.**   Suppose (for ease of notation) that $\{S_n, n \geq 1\}$ has range $L_d$ with $d = 1$ and that $\{S_n, n \geq 1\}$ is recurrent. Then

$$N_n(j)/N_n(i) \to 1 \qquad \text{a.s.}$$

for all pairs of integers $(i, j)$. (This clearly implies $N_n(I_1)/N_n(I_2) \to \| I_1 \|/ \| I_2 \|$ a.s. for arbitrary $d > 0$ and finite intervals $I_1$ and $I_2$.)

**Proof.**   The proof depends in part on the strong law of large numbers. Let $\{\tau_i, i \geq 1\}$ and $\{t_i, i \geq 1\}$ be as defined above. For each $j \neq 0$ let $v_1(j)$ be the number of times $\{S_n, n \geq 1\}$ hits $j$ before the first hitting of zero, $v_2(j)$ be the number of times $\{S_n, n \geq 1\}$ hits $j$ between the first and second hittings to zero, etc. Using the recurrence of $\{S_n, n \geq 1\}$ it follows that $\{v_i(j), i \geq 1\}$ is a sequence of independent identically distributed random variables for each $j \neq 0$ similarly to the way in which $\{\tau_i, i \geq 1\}$ was shown to be a sequence of independent identically distributed random variables. Let $\pi(j) = E v_1(j)$ for $j \neq 0$ and $\pi(0) = 1$. Clearly $\pi(j) > 0$ for all integers $j$. Suppose (as will be proved shortly) that $\pi(j) < \infty$ for all integers $j$.

Fix integers $j \neq 0$, $k \neq 0$. By the strong law,

$$\sum_{i=1}^{n} v_i(l)/n \to \pi(l) \qquad \text{a.s.}$$

for every $l \neq 0$. Thus

$$\frac{\sum_{i=1}^{n} v_i(j)}{\sum_{i=1}^{n} v_i(k)} = \frac{\sum_{i=1}^{n} v_i(j)/n}{\sum_{i=1}^{n} v_i(k)/n} \to \frac{\pi(j)}{\pi(k)} \qquad \text{a.s.}$$

But $\sum_{i=1}^{n} v_i(l) = N_{t_n}(l)$ for all $l \neq 0$. Moreover $N_{t_n}(0) = n$. Thus (allowing $j = 0$ or $k = 0$ possibly)

$$N_{t_n}(j)/N_{t_n}(k) \to \pi(j)/\pi(k) \qquad \text{a.s. as } n \to \infty. \tag{6.1.9}$$

$$\max_{0 \leq m \leq \tau_{n+1}} \left| \frac{N_{t_n+m}(j)}{N_{t_n+m}(k)} - \frac{N_{t_n}(j)}{N_{t_n}(k)} \right|$$

$$\leq \frac{N_{t_{n+1}}(j) - N_{t_n}(j)}{N_{t_n}(k)} + \frac{N_{t_n}(j)[N_{t_{n+1}}(k) - N_{t_n}(k)]}{N_{t_n}^2(k)}.$$

$$\frac{N_{t_{n+1}}(j) - N_{t_n}(j)}{N_{t_n}(k)} = \frac{v_{n+1}(j)}{n} \frac{n}{N_{t_n}(k)}.$$

$\nu_{n+1}(j)/n \to 0$ a.s. since $\{\nu_i(j), i \geq 1\}$ is a sequence of independent identically distributed random variables with finite mean. $N_{t_n}(k)/n \to \pi(k)$ a.s. for every integer $k$ as we have seen. Thus

$$\frac{N_{t_{n+1}}(j) - N_{t_n}(j)}{N_{t_n}(k)} \to 0 \qquad \text{a.s.}$$

as $n \to \infty$. Similarly

$$\frac{N_{t_n}(j)[N_{t_{n+1}}(k) - N_{t_n}(k)]}{N_{t_n}^2(k)} \to 0 \qquad \text{a.s.}$$

can be shown. Thus

$$\max_{0 \leq m \leq \tau_{n+1}} \left| \frac{N_{t_n+m}(j)}{N_{t_n+m}(k)} - \frac{N_{t_n}(j)}{N_{t_n}(k)} \right| \to 0 \qquad \text{as} \quad n \to \infty.$$

Combining this with Eq. (6.1.9), it follows that

$$N_n(j)/N_n(k) \to \pi(j)/\pi(k) \qquad \text{a.s.} \tag{6.1.10}$$

as $n \to \infty$ for all integers $j, k$.

It remains to show the surprising fact that $\pi(j) = 1$ for all integers $j$. Given that we are in state $j \neq 0$, the probability of hitting zero is one by the recurrence of $\{S_n, n \geq 1\}$. Let $r_j$ be the probability of hitting $j$ without hitting 0 first. Given that we have just hit $j$, let $p_j$ be the conditional probability of hitting zero before a visit to $j$ again. Then $\nu_1(j) = m$ consists of a visit to $j$ before hitting 0, then $m - 1$ failures to visit 0 before $j$ is hit again, and then a successful hitting of 0 before $j$ is hit again. These $m + 1$ trials are independent. Thus

$$E\nu_1(j) = \sum_{m=1}^{\infty} m(1 - p_j)^{m-1}p_j r_j < \infty$$

for all $j \neq 0$, a fact asserted earlier without proof.

Fix integers $j, k$. We have seen that $N_n(j)/N_n(k) \to \pi(j)/\pi(k)$ a.s. This is under the assumption that the random walk starts at 0; that is, $P[S_0 = 0] = 1$. If one were to suppose that the random walk was started at $k$, the limit for this new random walk is then $\pi(j - k)/\pi(0)$. But the limit should not be influenced by the starting behavior. Thus

$$\pi(j)/\pi(k) = \pi(j - k)/\pi(0)$$

is to be expected. To give a rigorous argument, let $t$ be the smallest $n \geq 1$ such that $S_n = k$.

$$N_{t+n}(j)/N_{t+n}(k) \to \pi(j)/\pi(k) \qquad \text{a.s.}$$

as $n \to \infty$. $N_{t+n}(k)$ is the number of times $S_{i+t} - S_t = 0$ for $0 \leq i \leq n$. $N_{t+n}(j)$ is the number of times $S_{i+t} - S_t = j - k$ for $1 \leq i \leq n$ plus the number of times $S_i = j$ for $1 \leq i \leq t$. Recall that $\{S_{n+t} - S_n, n \geq 1\}$ has the same distribution as $\{S_n, n \geq 1\}$. Thus

$$N_{t+n}(j)/N_{t+n}(k) \to \pi(j-k)/\pi(0) \qquad \text{a.s. as} \quad n \to \infty;$$

that is,

$$\pi(j)/\pi(k) = \pi(j-k)/\pi(0) = \pi(j-k).$$

Thus

$$\pi(j) = \pi(j-k)\pi(k)$$

for all integers $j, k$. The only solutions to this well known functional equation (the integer version of $f(x+y) = f(x)f(y)$) are $\pi(j) = r^j$ for some real number $r$. $\pi(j) > 0$ clearly holds for all real $j$; thus $r \neq 0$.

To complete the argument we resort to a combinatorial trick. Fix $j \neq 0$. Consider a sequence of $n$ states of $\{S_n, n \geq 1\}$ terminating at 0; that is, $m_1, m_2, \ldots, m_{n-1}, 0$. Since the $X_i$ are independent identically distributed,

$$P[X_1 = m_1, X_2 = m_2 - m_1, \ldots, X_n = 0 - m_{n-1}]$$
$$= P[X_1 = 0 - m_{n-1}, X_2 = m_{n-1} - m_{n-2}, \ldots, X_n = m_1].$$

Thus

$$P[S_1 = m_1, S_2 = m_2, \ldots, S_{n-1} = m_{n-1}, S_n = 0]$$
$$= P[S_1 = -m_{n-1}, S_2 = -m_{n-2}, \ldots, S_2 = -m_1, S_1 = 0].$$

But

$$[S_1 = m_1, S_2 = m_2, \ldots, S_{n-1} = m_{n-1}, S_n = 0]$$

is a path which hits $j$ $i$ times before returning to 0 if and only if

$$[S_1 = -m_{n-1}, S_2 = -m_{n-2}, \ldots, S_{n-1} = -m_1, S_n = 0]$$

is a path which hits $-j$ $i$ times before returning to 0. Thus

$$\pi(j) = Ev_1(j) = Ev_1(-j) = \pi(-j)$$

for all $j \neq 0$. Since $\pi(j) = r^j$ and $r \neq 0$, it follows that $r = 1$. ∎

If $\{S_n, n \geq 1\}$ is transient, then one cannot obtain a theorem of the form Theorem 6.1.6 since $N_n(i)$ converges a.s. to a finite limit as $n \to \infty$. Recall that our purpose in stating Theorem 6.1.6 was to study how much time $S_n$ spends in various subsets of its range $L_d$. Suppose $d > 0$. For each $n \geq 1$, let $U_n$ be the number of *distinct* states visited by $(S_1, S_2, \ldots, S_n)$. $U_n$ will be referred to as the range of $(S_1, S_2, \ldots, S_n)$. Note that $0 \leq U_n/n \leq 1$ and that $U_n/n$ close to one indicates that very few states have been revisited and that $U_n/n$ close to zero indicates that many states have been revisited. Thus $U_n/n$, as is $N_n(I_1)/N_n(I_2)$, is a measure of how $S_1, S_2, \ldots, S_n$ spends its time in various states.

**Theorem 6.1.7.** (Kesten, Spitzer, and Whitman; see Spitzer [1964, p. 35]). Let $\{S_n, n \geq 1\}$ be a random walk with range $L_d$ for some $d > 0$.

(i)  $EU_n/n \to P[S_n \neq 0 \text{ for all } n \geq 1]$

and

(ii)  $U_n/n \to P[S_n \neq 0 \text{ for all } n \geq 1]$ a.s.

**Remark.** Note that $P[S_n \neq 0 \text{ for all } n \geq 1] = P[\text{no visit to } 0]$.

**Proof.** For each $k \geq 1$, let $W_k = 1$ if $S_j \neq S_k$ for all $1 \leq j < k$; $= 0$ otherwise.

$$U_n = \sum_{k=1}^{n} W_k \qquad \text{for} \quad n \geq 1. \tag{6.1.11}$$

$$EW_k = P[S_k - S_{k-1} \neq 0, S_k - S_{k-2} \neq 0, \ldots, S_k - S_1 \neq 0]$$
$$= P[X_k \neq 0, X_k + X_{k-1} \neq 0, \ldots, X_k + X_{k-1} + \cdots + X_2 \neq 0].$$

Since $(X_k, X_{k-1}, \ldots, X_2)$ and $(X_1, X_2, \ldots, X_{k-1})$ have the same distribution (once again a combinatorial trick is helpful),

$$P[S_1 \neq 0, S_2 \neq 0, \ldots, S_{k-1} \neq 0]$$
$$= P[X_1 \neq 0, X_1 + X_2 \neq 0, \ldots, X_1 + X_2 + \cdots + X_{k-1} \neq 0]$$
$$= P[X_k \neq 0, X_k + X_{k-1} \neq 0, \ldots, X_k + X_{k-1} + \cdots + X_2 \neq 0] = EW_k.$$

But

$$P[S_1 \neq 0, S_2 \neq 0, \ldots, S_{k-1} \neq 0] \to P[\text{no return to } 0].$$

Thus $EW_k \to P[\text{no return to } 0]$. Hence using Eq. (6.1.11) the weaker statement

$$EU_n/n = \sum_{k=1}^{n} EW_k/n \to P[S_n \neq 0 \text{ for all } n \geq 1]$$

holds.

The proof of (ii) is carried out in two parts. First

$$\limsup U_n/n \leq P[\text{no return to } 0] \qquad \text{a.s.}$$

is proved. Fix a positive integer $m$ and let $N_k$ be the number of distinct points visited by the partial sums with indices from $(k-1)m + 1$ to $km$; that is, $N_k$ is the number of distinct points visited by $\{S_{(k-1)m+1}, S_{(k-1)m+2}, \ldots, S_{km}\}$. Note that $N_k$ depends only on the $X_i$ with indices from $(k-1)m +2$ to $km$. Thus the $N_k$ are independent. The $N_k$ are easily seen to be identically distributed. $|N_k| \leq m$ for all $k \geq 1$. $U_{mn} \leq \sum_{i=1}^{n} N_i$ for all $n \geq 1$. Thus, using the strong law,

$$\limsup_{n \to \infty} U_{mn}/(mn) \leq \limsup_{n \to \infty} \sum_{i=1}^{n} N_i/(mn) = EN_1/m = EU_m/m.$$

For $mn \leq j < m(n+1)$, $U_j \leq U_{mn} + m$. Combining,

$$\limsup U_j/j \leq EU_m/m \qquad \text{a.s.}$$

Letting $m \to \infty$ and using (i), it follows that

$$\limsup U_j/j \leq P[S_n \neq 0 \text{ for all } n \geq 1] \qquad \text{a.s.}$$

It suffices to show that $\liminf U_j/j \geq P[S_n \neq 0 \text{ for all } n \geq 1]$ a.s. to complete the proof. This is accomplished by an appeal to the pointwise ergodic theorem. Let $S_0 = 0$. For $k \geq 1$, let $V_k = 1$ if $S_j \neq S_k$ for all $j \geq k + 1$; $= 0$ otherwise. Thus $V_k = 1$ if the last visit to a state occurs at time $k$. $\sum_{i=1}^{n} V_i$ equals the number of states visited for the last time in the first $n$ trials. $U_n$ is the number of states visited at least once in the first $n$ trials. Thus

$$U_n \geq \sum_{i=1}^{n} V_i \qquad \text{a.s.} \tag{6.1.12}$$

Let

$$\phi(X_1, X_2, \ldots) = \begin{cases} 1 & \text{if } S_k \neq 0 \text{ for all } k \geq 1 \\ 0 & \text{otherwise} \end{cases}$$

define $\phi : R_\infty \to R_1$. Note that

$$V_k = \begin{cases} 1 & \text{if } X_{k+1} + \cdots + X_{k+j} \neq 0 \qquad \text{for all } j \geq 1 \\ 0 & \text{otherwise.} \end{cases}$$

Thus $V_k = \phi(X_{k+1}, X_{k+2}, \ldots)$. $\{V_k, k \geq 0\}$ is a stationary sequence by Theorem 3.5.3. $\{V_k, k \geq 0\}$ is ergodic since $\{X_k, k \geq 1\}$ is ergodic. Thus

$$\sum_{k=1}^{n} V_k/n \to EV_0 = P[S_n \neq 0 \text{ for all } n \geq 1] \qquad \text{a.s.}$$

by the pointwise ergodic theorem. Combining with Eq. (6.1.12),

$$\liminf U_n/n \geq P[S_n \neq 0 \text{ for all } n \geq 1] \qquad \text{a.s.} \quad \blacksquare$$

**Remark.**  Note that the proof of Theorem 6.1.7 depends in an essential way on the pointwise ergodic theorem.

## 6.2.  Generalizations of the Results of Section 6.1 to Other Stochastic Structures

A random walk is of course a special case of a Markov sequence with stationary transition probabilities. The study of recurrence as presented in Section 6.1 extends easily to Markov sequences with stationary transition probabilities and discrete state space. Dropping the restriction that the state space is discrete appears difficult and has not been successfully carried out as of the writing of this book. Throughout Section 6.2, $\{T_n, n \geq 0\}$ denotes a Markov sequence with stationary transition probabilities and (unless otherwise specified) state space $\{0, \pm 1, \ldots, \pm n, \ldots\} \equiv N$.

**Definition 6.2.1.**  (i)  $j \in N$ is said to be Markov recurrent if

$$P[T_n = j \text{ i.o.} \mid T_0 = j] = 1.$$

(ii)  $j \in N$ is said to be Markov transient if

$$P[T_n = j \text{ i.o.} \mid T_0 = j] = 0. \quad \blacksquare$$

Note that Markov recurrence and Markov transience depend only on the transition probability unlike recurrence as defined in Definition 6.1.1. The initial distribution plays no role in determining whether a state is Markov recurrent or Markov transient. It is intuitively clear that $P[T_n = j$

for some $n \geq 1] = 1$ implies that Markov recurrence of $j$ is equivalent to $P[T_n = j \text{ i.o.}] = 1$.

**Theorem 6.2.1.** (i) If $P[T_n = j \text{ i.o.}] = 1$ (i.e., recalling Definition 6.1.1, $j$ is recurrent), then $j$ is Markov recurrent. If $j$ is Markov transient then $P[T_n = j \text{ i.o.}] = 0$ (i.e., recalling Definition 6.1.1, $j$ is transient).

(ii) Let $P[T_n = j \text{ for some } n \geq 0] = 1$. Then $j$ is Markov recurrent if and only if $P[T_n = j \text{ i.o.}] = 1$.

(iii) Let $P[T_n = j \text{ for some } n \geq 0] > 0$. Then $j$ is Markov transient if and only if $P[T_n = j \text{ i.o.}] = 0$.

**Proof.** (ii) Let $t$ be the first $n \geq 0$ such that $T_n = j$. $P[t < \infty] = 1$ by hypothesis.

Let $C \in \mathscr{C}_\infty$, the Borel sets of $R_\infty$.

$$P[(T_{t+1}, T_{t+2}, \ldots) \in C] = \sum_{n=0}^{\infty} P[(T_{n+1}, T_{n+2}, \ldots) \in C, T_n = j, t = n]$$

$$= \sum_{n=0}^{\infty} P[(T_{n+1}, T_{n+2}, \ldots) \in C \mid T_n = j] P[t = n]$$

$$= P[(T_1, T_2, \ldots) \in C \mid T_0 = j]$$

using the assumption of stationary transition probabilities. Thus

$$P[T_n = j \text{ i.o.}] = P[T_{t+n} = j \text{ i.o.}]$$
$$= P[T_n = j \text{ i.o.} \mid T_0 = j].$$

(i) and (iii) follow similarly and are left as exercises. ∎

**Exercise 6.2.1.** (i) Construct a Markov sequence such that $0 < P[T_n = 0 \text{ i.o.}] < 1$.

(ii) Construct a Markov sequence such that $P[T_n = 0 \text{ i.o.}] = 0$ and $0$ is Markov recurrent. ∎

**Theorem 6.2.2.** $j \in N$ is Markov recurrent if and only if

$$\sum_{n=1}^{\infty} P[T_n = j \mid T_0 = j] = \infty. \tag{6.2.1}$$

**Proof.** $j$ Markov recurrent implies

$$\sum_{n=1}^{\infty} P[T_n = j \mid T_0 = j] = \infty$$

by the definition of Markov recurrence and the Borel–Cantelli lemma.

Suppose Eq. (6.2.1) holds. Fix $n \geq 1$. Let $A_n = [T_n = j, T_{n+i} \neq j$ for all $i \geq 1]$ and $A_0 = [T_i \neq j$ for all $i \geq 1]$.

$$[T_n = j \text{ i.o.}]^c = \bigcup_{i=0}^{\infty} A_i.$$

Thus

$$1 - P[T_n = j \text{ i.o.} \mid T_0 = j] = \sum_{n=0}^{\infty} P[A_n \mid T_0 = j]. \tag{6.2.2}$$

$$P[A_n \mid T_0 = j] = P[T_{n+i} \neq j \text{ for all } i \geq 1 \mid T_n = j]P[T_n = j \mid T_0 = j]$$
$$= P[T_i \neq j \text{ for all } i \geq 1 \mid T_0 = j]P[T_n = j \mid T_0 = j]$$

since $\{T_n, n \geq 0\}$ is a Markov sequence with stationary transition probabilities. Thus

$$P[A_n \mid T_0 = j] = P[A_0 \mid T_0 = j]P[T_n = j \mid T_0 = j]. \tag{6.2.3}$$

Thus

$$1 \geq \sum_{n=0}^{\infty} P[A_n \mid T_0 = j] = P[A_0 \mid T_0 = j] \sum_{n=0}^{\infty} P[T_n = j \mid T_0 = j].$$

Thus, by Eq. (6.2.1), $P[A_0 \mid T_0 = j] = 0$. By Eq. (6.2.3), $P[A_n \mid T_0 = j] = 0$ for all $n \geq 1$. By Eq. (6.2.2),

$$P[T_n = j \text{ i.o.} \mid T_0 = j] = 1. \quad \blacksquare$$

**Corollary 6.2.1.** Let $j \in N$. Either $j$ is Markov recurrent or $j$ is Markov transient (a 0–1 law).

**Proof.** Suppose $j$ is not Markov recurrent. By Theorem 6.2.1,

$$\sum_{n=1}^{\infty} P[T_n = j \mid T_0 = j] < \infty.$$

Thus, by the Borel–Cantelli lemma, $P[T_n = j \text{ i.o.} \mid T_0 = j] = 0$. $\quad \blacksquare$

**Definition 6.2.2.** (i) A state $k \in N$ is said to lead to a state $j \in N$ if there exists $n \geq 1$ such that

$$P[T_n = j \mid T_0 = k] > 0.$$

(ii) If $k$ leads to $j$ and $j$ leads to $k$, $j$ and $k$ are said to communicate. $\quad \blacksquare$

**Theorem 6.2.3.** If $j$ and $k$ communicate then $j$ and $k$ are simultaneously Markov recurrent or Markov transient.

**Proof.** Suppose $k$ is Markov recurrent. Since $j$ and $k$ communicate, there exists $n_0$ and $n_1$ such that

$$a \equiv P[T_{n_0} = k \mid T_0 = j] > 0 \qquad \text{and} \qquad b \equiv P[T_{n_1} = j \mid T_0 = k] > 0.$$

Applying the Markov property, for each $m \geq 1$,

$$P[T_{n_0+m+n_1} = j \mid T_0 = j]$$
$$\geq P[T_{n_0} = k, T_{n_0+m} = k, T_{n_0+m+n_1} = j \mid T_0 = j]$$
$$= P[T_{n_0+m+n_1} = j \mid T_{n_0+m} = k]P[T_{n_0+m} = k \mid T_{n_0} = k]P[T_{n_0} = k \mid T_0 = j]$$
$$= baP[T_m = k \mid T_0 = k].$$

Summing on $m$ and applying Theorem 6.2.2, it follows that $j$ is Markov recurrent. The argument is symmetric in $j$ and $k$. Thus $j$ is Markov recurrent if and only if $k$ is Markov recurrent. Then, by Corollary 6.2.1 $j$ is Markov transient if and only if $k$ is Markov transient. ∎

We thus obtain an analog of Theorem 6.1.2:

**Corollary 6.2.2.** Suppose that the state space $N$ is a "communicating class"; that is, $i$ and $j$ communicate for every $i, j \in N$.

(i)   Then either all states of $N$ are Markov recurrent or all states of $N$ are Markov transient.

(ii)   Moreover, either all states of $N$ are recurrent or all states are transient.

**Proof.** (i)   Immediate from Corollary 6.2.1 and Theorem 6.2.3.

(ii)   If all states of $N$ are transient, we have finished. Suppose there exists a state $j$ which is not transient; that is, $P[T_n = j \text{ i.o.}] > 0$. Thus $P[T_n = j \text{ i.o.} \mid T_0 = j] > 0$. Thus, by Corollary 6.2.1, $j$ is Markov recurrent. By Corollary 6.2.2 (i), all states of $N$ are thus Markov recurrent. Fix $j \in N$.

$$P[T_n = j \text{ i.o.}] = \sum_{k=-\infty}^{\infty} P[T_n = j \text{ i.o.} \mid T_0 = k]P[T_0 = k].$$

To complete the proof, it thus suffices to show that $P[T_n = j \text{ i.o.} \mid T_0 \in k]$ $= 1$ for arbitrary $k \in N$. Let $t$ be the smallest $m \geq 1$ such that $T_m = j$ if such an $n$ exists; otherwise let $t = \infty$.

$$P[T_n = j \text{ i.o. } | \ T_0 = k] \geq \sum_{m=1}^{\infty} P[T_{n+t} = j \text{ i.o., } t = m \ | \ T_0 = k]$$

$$= \sum_{m=1}^{\infty} P[T_{n+m} = j \text{ i.o. } | \ T_m = j]P[t = m \ | \ T_0 = k]$$

$$= P[t < \infty \ | \ T_0 = k].$$

Thus it suffices to show that $P[t < \infty \ | \ T_0 = k] = 1$.

The proof is completed by a renewal argument which we state intuitively: Each time the state $k$ is hit, there is a fixed nonzero probability $\alpha$ of $j$ being hit before $k$ is hit again. Each time $k$ is hit, this process is renewed; that is, again $j$ has probability $\alpha$ of being hit before $k$. It is thus intuitively clear that $P[t < \infty \ | \ T_0 = k] = 1$ since $k$ is Markov recurrent. It is left as an exercise to make this argument rigorous. ∎

**Example 6.2.1.** Let the state space of $\{T_n, n \geq 0\}$ be $N = \{0, 1, \ldots\}$. Let the transition probability be given by

$$P[T_1 = k \ | \ T_0 = 0] = 2^{-k} \qquad \text{for} \quad k \geq 1,$$
$$P[T_1 = 0 \ | \ T_0 = j] = 1 \qquad \text{for} \quad j \geq 1,$$

and

$$P[T_1 = k \ | \ T_0 = j] = 0 \qquad \text{for} \quad j \neq 0, k \neq 0.$$

Clearly all states of $N$ communicate.

$$\sum_{n=1}^{\infty} P[T_n = 0 \ | \ T_0 = 0] = \sum_{n=1}^{\infty} P[T_{2n} = 0 \ | \ T_0 = 0] = \infty.$$

Thus, as was obvious, 0 is Markov recurrent. Using Theorem 6.2.3, all states of $\{T_n, n \geq 0\}$ are Markov recurrent. ∎

Some work has been done on recurrence for independent nonidentically distributed random variables. Let $\{S_n, n \geq 1\}$ be the partial sums of independent $X_i$. Orey [1966] has given conditions under which all states of $\{S_n, n \geq 1\}$ must be either recurrent or transient. Recently Mineka and Silverman [1970] used Orey's results to obtain sufficient conditions for the recurrence of $\{S_n, n \geq 1\}$.

Another interesting topic is the question of the behavior of higher dimensional random walks. Let $\{Y_n, n \geq 1\}$ be a sequence of independent identically distributed vectors taking values in $R_k$ for some $k \geq 1$. The proofs of Theorems 6.1.2 and 6.1.3 go through in the $k$-dimensional case.

Thus either all possible states are recurrent or none are. Moreover, the random walk is recurrent if and only if there exists an open interval $J$ of $R_k$ such that

$$\sum_{n=1}^{\infty} P[S_n \in J] = \infty.$$

The following example which we present as an exercise introduces the topic.

**Exercise 6.2.2.** (Polya's [1921] classical work on recurrence of simple symmetric random walk in $R_k$ for $k \geq 1$.)

(i) Let the $X_i$ be independent with $P[X_i = 1] = P[X_i = -1] = \frac{1}{2}$ for $i \geq 1$. Prove that $\{S_n, n \geq 1\}$ is recurrent (a fact we know of course since $EX_1 = 0$) by showing that $\sum_{n=1}^{\infty} P[S_n = 0] = \infty$. Hint: $S_n = 0$ if and only if $n$ is even and $n/2$ $X_i = +1$ and $n/2$ $X_i = -1$.

(ii) Let $Y_i$ be independent two-dimensional random variables with $P[Y_i = (1, 0)] = P[Y_i = (0, 1)] = P[Y_i = (0, -1)] = P[Y_i = (-1, 0)] = \frac{1}{4}$ for $i \geq 1$. Let $T_n = \sum_{i=1}^{n} Y_i$. $\{T_n, n \geq 1\}$ is the simple ("simple" because of the range of the $Y_i$) symmetric ("symmetric" since all probabilities are $\frac{1}{4}$) random walk in $R_2$. Prove that $\{T_n, n \geq 1\}$ is recurrent by showing that $\sum_{n=1}^{\infty} P[T_n = (0, 0)] = \infty$. Hint: $T_n = (0, 0)$ if and only if $n$ is even and for some even $k \leq n$ and there are $k/2$ $Y_i$ equal $(1, 0)$, $k/2$ $Y_i$ equal $(-1, 0)$, $(n - k)/2$ $Y_i$ equal $(0, 1)$, and $(n - k)/2$ $Y_i$ equal $(0, -1)$.

(iii) Let $\{T_n, n \geq 1\}$ be a simple symmetric random walk in $R_k$ for some $k \geq 3$. (For example, $P[Y_i = (1, 0, 0)] = P[Y_i = (0, 1, 0)] = P[Y_i = (0, 0, 1)] = P[Y_i = (-1, 0, 0)] = P[Y_i = (0, -1, 0)] = P[Y_i = (0, 0, -1)] = \frac{1}{6}$ and $T_n = \sum_{i=1}^{n} Y_i$ when $k = 3$.) It can be proved that $\{T_n, n \geq 1\}$ is transient for $k \geq 3$. (The computations are messier than for (ii) but similar—see Bailey [1970, p. 36].) ∎

Let $\{Y_i, i \geq 1\}$ be a sequence of independent identically distributed two-dimensional random vectors with distribution function $F$. Let $T_n = \sum_{i=1}^{n} Y_i$ for $n \geq 1$. Chung and Fuchs [1951] show that

$$\iint\limits_{R_2} x \, dF(x, y) = 0, \qquad \iint\limits_{R_2} y \, dF(x, y) = 0$$

and

$$\iint\limits_{R_2} (x^2 + y^2) \, dF(x, y) < \infty$$

implies that $\{T_n, n \geq 1\}$ is recurrent. Generalizing Exercise 6.2.2 (iii), Chung and Fuchs show that all three-dimensional random walks are transient. By three-dimensional, it is meant that $P_{Y_1}$ defined on the Borel sets of $R_3$ does not assign mass one to some two-dimensional subset of $R_3$.

The notion of the range of a random walk being a lattice $L_d$ for some $0 \leq d \leq \infty$ extends to more than one dimension in the obvious way. This leads to generalizations of and results related to Theorem 6.1.7: Dvoretzky and Erdös [1951] first showed that

$$(U_n - EU_n)/n \to 0 \qquad \text{a.s.}$$

for simple (probability $2^{-k}$ assigned to each of the $k$ nearest neighbors) $k$-dimensional random walk with $k \geq 2$. For all transient random walks with $d > 0$ it is known that

$$EU_n/n \to P[\text{no visit to } (0, 0, \ldots, 0)].$$

Thus the Dvoretzky and Erdös result implies that

$$U_n/n \to P[\text{no visit to } (0, 0, \ldots, 0)] \qquad \text{a.s.}$$

for all simple $k$-dimensional random walks with $k \geq 3$. More recently, Kesten, Spitzer, and Whitman (see Spitzer [1964, p. 38]) have shown that

$$U_n/n \to P[\text{no visit to } (0, 0, \ldots, 0)] \qquad \text{a.s.}$$

for all $k \geq 1$ and $d > 0$, thus generalizing the previous result. Very recently Jain and Pruitt [1972a] have shown that

$$U_n/n \to 1 \qquad \text{a.s.}$$

for all recurrent random walks with $d > 0$. Jain and Pruitt [1972a] even derive a law of the iterated logarithm for $\{U_n, n \geq 1\}$ when $k \geq 4$ and $d > 0$.

## 6.3.  Recurrence of $\{S_n/n^\alpha, n \geq 1\}$ in the Case of Independent Identically Distributed $X_i$

Throughout Section 6.3, $\{X_i, i \geq 1\}$ denotes a sequence of independent identically distributed random variables. Let $\alpha > 0$. Many of the questions of Section 6.1 can be answered for $\{S_n/n^\alpha, n \geq 1\}$ also. First we look at the question of when

$$\limsup S_n/n^\alpha = \infty \qquad \text{and} \qquad \liminf S_n/n^\alpha = -\infty \quad \text{a.s.}$$

By the Hartman–Wintner law of the iterated logarithm, we already know that $EX_1 = 0$ and $EX_1^2 < \infty$ implies that

$$\limsup S_n/n^\alpha = \infty \qquad \text{and} \qquad \liminf S_n/n^\alpha = -\infty \quad \text{a.s.}$$

for all $0 < \alpha \leq \frac{1}{2}$. Stone [1966] shows that $EX_1 = 0$ alone implies that

$$\limsup S_n/n^{1/2} = \infty \qquad \text{and} \qquad \liminf S_n/n^{1/2} = -\infty \quad \text{a.s.}$$

Of course, trivially

$$\limsup S_n/n^{1/2} = \infty \qquad \text{and} \qquad \liminf S_n/n^{1/2} = -\infty \quad \text{a.s.}$$

implies

$$\limsup S_n/n^\alpha = \infty \qquad \text{and} \qquad \liminf S_n/n^\alpha = -\infty \quad \text{a.s.}$$

for each $0 < \alpha < \frac{1}{2}$. Stone's method of proof is clever and of definite independent interest.

**Lemma 6.3.1.** Let $F$ be the distribution function of a nondegenerate random variable with mean 0. Then there exists distribution functions $G$ and $H$ and constant $0 < \alpha < 1$ such that

$$F = \alpha G + (1 - \alpha)H$$

(that is, $F$ is a "mixture" of $G$ and $H$) with $H$ being the distribution of a nondegenerate random variable with a mean of 0 and *finite* variance.

**Remark.** Let $X$ have distribution function $F$, $Y$ have distribution function $G$, and $Z$ have distribution $H$. Then $F = \alpha G + (1 - \alpha)H$ has a probabilistic interpretation. Probabilistically, it is the same if one observes $X$ (that is, carry out an experiment where the observation has distribution function $F$) or if one flips a coin with probability of heads $\alpha$, observing $Y$ if the coin is heads and observing $Z$ if the coin is tails.

**Proof of Lemma 6.3.1.** Let $X$ be a random variable with distribution function $F$. Choose $a < b \in R_1$ and $0 \leq c_1 \leq 1, 0 \leq c_2 \leq 1$ such that

$$E[XI(a < X < b)] + c_1 a P[X = a] + c_2 b P[X_2 = b] = 0$$

and

$$0 < 1 - \alpha \equiv c_1 P[X = a] + P[a < X < b] + c_2 P[X = b] < 1.$$

Define a distribution function $H$ by

$$(1 - \alpha)H(x) = \begin{cases} 0 & \text{if } x < a \\ c_1 P[X = a] & \text{if } x = a \\ c_1 P[X = a] + P[a < X \leq x] & \text{if } a < x < b \\ c_1 P[X = a] + P[a < X < b] \\ \quad + c_2 P[X = b] & \text{if } x \geq b. \end{cases}$$

Clearly

$$\int_{-\infty}^{\infty} x \, dH(x) = 0 \quad \text{and} \quad 0 < \int_{-\infty}^{\infty} x^2 \, dH(x) < \infty.$$

Define a distribution function $G$ by

$$\alpha G(x) = \begin{cases} P[X \leq x] & \text{if } x < a \\ P[X < a] + (1 - c_1)P[X = a] & \text{if } a \leq x < b \\ P[X < a] + (1 - c_1)P[X = a] \\ \quad + (1 - c_2)P[X = b] + P[b < X \leq x] & \text{if } x \geq b. \end{cases}$$

Clearly $F = \alpha G + (1 - \alpha)H$.  ∎

**Theorem 6.3.1.**  If $EX_1 = 0$ and $X_1$ is nondegenerate, then

$$\limsup S_n/n^{1/2} = \infty \quad \text{and} \quad \liminf S_n/n^{1/2} = -\infty \quad \text{a.s.}$$

**Proof.**  Let $F$ be the distribution of $X_1$. Let $G$, $H$, and $\alpha$ be as in the statement of Lemma 6.3.1. Let

$$\{\xi_1, Y_1, Z_1, \xi_2, Y_2, Z_2, \ldots, \xi_n, Y_n, Z_n, \ldots\}$$

be independent with the $Y_i$ having distribution function $G$ and the $Z_i$ having distribution function $H$ and $P[\xi_i = 1] = \alpha$, $P[\xi_i = 0] = 1 - \alpha$ for all $i \geq 1$. Let $T_n = \sum_{i=1}^{n} Y_i$ and $U_n = \sum_{i=1}^{n} Z_i$ for $n \geq 1$ and $T_0 = U_0 = 0$. Let $J(n) = \sum_{i=1}^{n} \xi_i$ and $K(n) = n - J(n)$ for $n \geq 1$. Let $V_n = T_{J(n)} + U_{K(n)}$ for $n \geq 1$.

Since $F = \alpha G + (1 - \alpha)H$, it is clear that $\{V_n, n \geq 1\}$ and $\{S_n, n \geq 1\}$ have the same distributions. The additional structure of $\{V_n, n \geq 1\}$ will allow us to prove the theorem.

Fix $N > 0$. If $P[\limsup V_n/n^{1/2} \geq N] > 0$ then $P[\limsup V_n/n^{1/2} \geq N] = 1$ by the Kolmogorov 0–1 law.

$$P[\limsup V_n/n^{1/2} \geq N] \geq P\left[ \bigcap_{m=1}^{\infty} (V_n/n^{1/2} \geq N \text{ for some } n \geq m) \right]$$

$$= \lim_{m \to \infty} P[V_n/n^{1/2} \geq N \text{ for some } n \geq m].$$

Hence, to prove that $\limsup S_n/n^{1/2} = \infty$ a.s., it suffices to show that

$$\lim_{m \to \infty} P[V_n/n^{1/2} \geq N \text{ for some } n \geq m] > 0.$$

Stone's main idea is both intuitive and clever. $\{T_n, n \geq 1\}$ is a recurrent random walk and $\{U_n, n \geq 1\}$ obeys the central limit theorem. Thus $T_n/n^{1/2} \approx 0$ i.o. and $\limsup U_n/n^{1/2} = \infty$ a.s.

$$V_n/n^{1/2} = T_{J(n)}/n^{1/2} + U_{K(n)}/n^{1/2}.$$

If $J(n)/n$ and $K(n)/n$ are asymptotically well behaved, it should follow that $\limsup V_n/n^{1/2} = \infty$ a.s.

Fix $m \geq 1$ and $M > 0$. Let $\tau(m)$ be the smallest $n \geq m$ such that

$$| T_{J(n)} | \leq M \quad \text{and} \quad K(n) \geq n(1 - \alpha)/2.$$

$\tau(m)$ is finite almost surely since $K(n)/n \to 1 - \alpha$ a.s. and since $\{T_n, n \geq 1\}$ is recurrent by virtue of the fact that $EY_1 = EX_1 - EZ_1 = 0$.

$P[V_n/n^{1/2} \geq N \text{ for some } n \geq m]$

$$\geq P[U_{K[\tau(m)]}/(\tau(m))^{1/2} \geq M/(\tau(m))^{1/2} + N]$$

$$\geq P\{U_{K[\tau(m)]}/(K[\tau(m)])^{1/2} \geq M/(K[\tau(m)])^{1/2} + N2^{1/2}(1 - \alpha)^{-1/2}\}.$$

$$= \sum_{i=1}^{\infty} P\{U_i/i^{1/2} \geq M/i^{1/2} + N2^{1/2}(1 - \alpha)^{-1/2} \mid K(\tau(m)) = i\}$$

$$\times P[K(\tau(m)) = i]$$

$$= \sum_{i=1}^{\infty} P[U_i/i^{1/2} \geq M/i^{1/2} + N2^{1/2}(1 - \alpha)^{-1/2}]P[K(\tau(m)) = i]$$

by independence. By the central limit theorem,

$$\lim_{i \to \infty} P[U_i/i^{1/2} \geq M/i^{1/2} + N2^{1/2}(1 - \alpha)^{-1/2}] > 0.$$

Combining and using standard summability theory,

$$\limsup_{m \to \infty} P[V_n/n^{1/2} \geq N \text{ for some } n \geq m] > 0.$$

$\liminf S_n/n^{1/2} = -\infty$ a.s. follows in the same way. ∎

In a recent paper Stone [1969] uses Spitzer's characterization of $S_n \to -\infty$ a.s. (Theorem 6.1.5) in conjunction with a Fourier–Stieltjes transform analysis to show that if $X_1$ is nondegenerate, then either

$$S_n/n^{1/2} \to -\infty \quad \text{a.s.} \quad \text{or} \quad \limsup S_n/n^{1/2} = \infty \quad \text{a.s.}$$

This clearly extends Theorem 6.3.1 since $EX_1 = 0$ implies $\{S_n, n \geq 1\}$ is recurrent which implies

$$P[S_n/n^{1/2} \to -\infty] = 0.$$

The result is also closely related to Theorem 3.2.7.

Further $\{S_n, n \geq 1\}$ recurrent and $X_1$ nondegenerate implies

$$\limsup S_n/n^{1/2} = \infty \quad \text{and} \quad \liminf S_n/n^{1/2} = -\infty \quad \text{a.s.}$$

which follows from this result of Stone's. Indeed $X_1$ nondegenerate with

$$P[S_n \geq 0 \text{ i.o.}] > 0 \quad \text{and} \quad P[S_n \leq 0 \text{ i.o.}] > 0$$

implies

$$\limsup S_n/n^{1/2} = \infty \quad \text{and} \quad \liminf S_n/n^{1/2} = -\infty \quad \text{a.s.}$$

by Stone's result.

Spitzer's result (Theorem 6.1.5) can be used to quickly obtain a characterization of when $\limsup S_n/n = \infty$ a.s.:

**Theorem 6.3.2.** [Katz, 1968]. (i) $\limsup S_n/n = \infty$ a.s. if and only if

$$\sum_{n=1}^{\infty} n^{-1} P[S_n > nM] = \infty$$

for all $M > 0$.

(ii) $\limsup S_n/n = \infty$ and $\liminf S_n/n = -\infty$ a.s. if and only if

$$\sum_{n=1}^{\infty} n^{-1} P[S_n > nM] = \infty \quad \text{and} \quad \sum_{n=1}^{\infty} n^{-1} P[S_n < -nM] = \infty$$

for all $M > 0$.

**Proof.** (ii) follows immediately from (i).
Suppose

$$\sum_{n=1}^{\infty} n^{-1} P[S_n > nM] = \infty$$

for all $M > 0$. Fix $M > 0$.

$$\sum_{n=1}^{\infty} n^{-1} P\left[\sum_{i=1}^{n} (X_i - M) > 0\right] = \infty.$$

By Theorem 6.1.5,

$$\limsup_{n \to \infty} (S_n - nM) = \limsup_{n \to \infty} \sum_{i=1}^{n} (X_i - M) = \infty \qquad \text{a.s.}$$

Hence

$$\limsup S_n/n \geq M \qquad \text{a.s.}$$

Since $M$ is arbitrary,

$$\limsup S_n/n = \infty \qquad \text{a.s.}$$

Unwinding this argument proves the other half of (i). ∎

**Exercise 6.3.1.** [Spitzer, 1956]. Prove that $S_n/n \to c$ a.s. if and only if $\sum_{n=1}^{\infty} n^{-1} P[|S_n/n - c| > \varepsilon] < \infty$ for every $\varepsilon > 0$. ∎

Katz [1968] uses Theorem 6.3.2 to show that $S_n/n \to 0$ in probability but not almost surely implies $\limsup S_n/n = \infty$ and $\liminf S_n/n = -\infty$ a.s.

One could ask about the recurrence of $S_n/n^\alpha$ at $a$ for various $\alpha > 0$ and $a \in R_1$. Binmore and Katz characterize the recurrence of $S_n/n^\alpha$ at $a$:

**Theorem 6.3.3.** (See Kesten [1970]). Fix $D > 1$, $\varepsilon > 0$, $-\infty < a < b < \infty$ and $\alpha > 0$. Then

(i)  $P[S_n/n^\alpha \in (a, b) \text{ i.o.}] = 1$ implies

(ii)  $\sum_{i=0}^{\infty} P[S_n/n^\alpha \in (a, b) \text{ for some } n \in [D^i, D^{i+1})] = \infty$ implies

(iii)  $P[S_n/n^\alpha \in (a - \varepsilon, b + \varepsilon) \text{ i.o.}] = 1$.

**Proof.** The proof is very much in the spirit of the proof of Theorem 6.1.3. (i) implying (ii) is immediate by the Borel–Cantelli lemma.

Assume (ii) and let $I = (a, b)$. Let $j$ be a positive integer. Splitting the sum in (ii) into $j$ parts, it follows that there exists an integer $0 \leq k < j$ for which

$$\sum_{i=0}^{\infty} P[S_n/n^\alpha \in I \text{ for some } n \in [D^{ij+k}, D^{ij+k+1})] = \infty. \qquad (6.3.1)$$

Fix such an integer $k$ and define events

$$E_i = \{S_n/n^\alpha \in I \text{ for some } n \in [D^{ij+k}, D^{ij+k+1}), S_n/n^\alpha \notin I \text{ for all } n \geq D^{(i+1)j+k}\}$$

for $i \geq 1$. Fix $i$. Let $\mathscr{C}_1$ be the Borel sets of $R_1$.

$$P(E_i) = \sum_{D^{ij+k} \leq m < D^{ij+k+1}} \int_{u \in I} P[S_n/n^\alpha \notin I \text{ for}$$
$$n \geq D^{(i+1)j+k} \mid S_m/m^\alpha = u] \, dP_m(u), \qquad (6.3.2)$$

where

$$P_m(A) = P[S_n/n^\alpha \notin I \text{ for } D^{ij+k} \leq n < m, S_m/m^\alpha \in A]$$

for $m \geq 1$ and $A \in \mathscr{C}_1$ defines $P_m(\cdot)$. The Markov property of $\{S_n, n \geq 1\}$ was used here to conclude

$$P[S_n/n^\alpha \notin I \text{ for } n \geq D^{(i+1)j+k} \mid S_n/n^\alpha \notin I \text{ for } D^{ij+k} \leq n < m, S_m/m^\alpha \in A]$$
$$= P[S_n/n^\alpha \notin I \text{ for } n \geq D^{(i+1)j+k} \mid S_m/m^\alpha \in A].$$

Since $\{X_l, l \geq 1\}$ is a sequence of independent identically distributed random variables,

$$P[S_n/n^\alpha \notin I \text{ for } n \geq D^{(i+1)j+k} \mid S_m/m^\alpha = u]$$
$$= P[S_{n-m}/(n-m)^\alpha \notin (n-m)^{-\alpha}(n^\alpha I - m^\alpha u) \text{ for } n - m \geq D^{(i+1)j+k} - m].$$
$$(6.3.3)$$

Note that $m \leq D^{ij+k+1}$ and $n \geq D^{(i+1)j+k}$ in Eq. (6.3.2). It follows that

$$n - m \geq D^{(i+1)j+k} - m \geq (D^{j-1} - 1)D^{ij+k+1} \geq (D^{j-1} - 1)m. \qquad (6.3.4)$$

Moreover, $n/(n-m) = (1 - m/n)^{-1} \leq (1 - D^{1-j})^{-1}$. Thus, $n \leq (1 - D^{1-j})^{-1} \times (n - m)$.

Thus, choosing $j$ sufficiently large and supposing $u \in I$, it follows that

$$| (n - m)^{-\alpha}(n^\alpha a - m^\alpha u) - a | \leq | a | \left| \frac{n^\alpha}{(n - m)^\alpha} - 1 \right|$$
$$+ (| a | + | b |) \frac{m^\alpha}{(n - m)^\alpha} < \varepsilon$$

and

$$| (n - m)^{-\alpha}(n^\alpha b - m^\alpha u) - b | < \varepsilon.$$

Combining the last two inequalities, we obtain

$$(n - m)^{-\alpha}(n^\alpha I - m^\alpha u) \subset (a - \varepsilon, b + \varepsilon).$$

Thus by Eq. (6.3.3) and Eq. (6.3.4)

$$P[S_n/n^\alpha \notin I \text{ for } n \geq D^{(i+1)j+k} \mid S_m/m^\alpha = u]$$
$$\geq P[S_l/l^\alpha \notin (a - \varepsilon, b + \varepsilon) \text{ for } l \geq D^{j-1} - 1].$$

Substituting this into Eq. (6.3.2) yields

$$P(E_i) \geq \sum_{D^{ij+k} \leq m < D^{ij+k+1}} P[S_n/n^\alpha \notin I \text{ for } D^{ij+k} \leq n < m, S_m/m^\alpha \in I]$$
$$\times P[S_l/l^\alpha \notin (a - \varepsilon, b + \varepsilon) \text{ for } l \geq D^{j-1} - 1]$$
$$= P\{S_n/n^\alpha \in I \text{ for some } n \in [D^{ij+k}, D^{ij+k+1})\}$$
$$\times P[S_l/l^\alpha \notin (a - \varepsilon, b + \varepsilon) \text{ for } l \geq D^{j-1} - 1].$$

Since the $E_i$ are disjoint,

$$1 \geq \sum_{i=0}^\infty P(E_i) \geq \sum_{i=0}^\infty P\{S_n/n^\alpha \in I \text{ for some } n \in [D^{ij+k}, D^{ij+k+1})\}$$
$$\times P[S_l/l^\alpha \notin (a - \varepsilon, b + \varepsilon) \text{ for } l \geq D^{j-1} - 1].$$

Recalling Eq. (6.3.1), this implies that

$$P[S_l/l^\alpha \notin (a - \varepsilon, b + \varepsilon) \text{ for } l \geq D^{j-1} - 1] = 0$$

for all $j$ sufficiently large, thus implying (iii). ∎

**Corollary 6.3.1.** Fix $D > 1$ and $\alpha > 0$. $b$ is a recurrent state of $\{S_n/n^\alpha, n \geq 1\}$ if and only if for all $\varepsilon > 0$,

$$\sum_{i=0}^\infty P\{| S_n/n^\alpha - b | < \varepsilon \text{ for some } n \in [D^i, D^{i+1})\} = \infty.$$

**Proof.** Immediate from Theorem 6.3.3. ∎

**Corollary 6.3.2.** If for some $\alpha > 0$,

$$\sum_{n=1}^\infty n^{-1}P[| S_n/n^\alpha - b | < \varepsilon] = \infty$$

for some $b \in R_1$ and every $\varepsilon > 0$, then $S_n/n^\alpha$ is recurrent at $b$.

**Proof.** Fix $i \geq 1$.

$$\sum_{D^i \leq n < D^{i+1}} n^{-1} P[|\, S_n/n^\alpha - b\,| < \varepsilon]$$

$$\leq (D^{i+1} - D^i) D^{-i}$$

$$\times P\{|\, S_n/n^\alpha - b\,| < \varepsilon \quad \text{for some} \quad n \in [D^i, D^{i+1})\}.$$

Apply Corollary 6.3.1.   ∎

Building on these initial results of Binmore and Katz and stimulated by Stone's work, Kesten [1970] discovered some remarkable results about the recurrence behavior of $\{S_n/n^\alpha, n \geq 1\}$. As usual $X_1$ is assumed to be nondegenerate. If $\alpha < \frac{1}{2}$, Kesten showed that either no finite real numbers are recurrent or all real numbers (including $+\infty$ and $-\infty$) are recurrent. If $\alpha = \frac{1}{2}$, then either no finite real numbers are recurrent or there is at least a half line $[-\infty, a]$ or $[a, \infty]$ of recurrent points. Perhaps most remarkable of all, for any closed set $C$ of $R_1 \cup (-\infty) \cup (\infty)$ which includes $+\infty$ and $-\infty$, there exists a random walk $\{S_n, n \geq 1\}$ such that the recurrent points of $S_n/n$ are exactly $C$. Related to Theorem 6.1.5 and Theorem 6.3.2, it is shown that if $EX_1^+ = \infty$, then

$$\sum_{n=1}^{\infty} n^{-1} P[S_n > 0] = \infty$$

if and only if

$$\limsup S_n/n = \infty \qquad \text{a.s.}$$

# References

Alexitz, G. (1961). "Convergence Problems of Orthogonal Series." Pergamon, Oxford.

Ash, R. B. (1972). "Real Analysis and Probability." Academic Press, New York.

Austin, D. G. (1966). A sample function property of martingales. *Ann. Math. Statist.* **37**, 1396–1397.

Azuma, K. (1967). Weighted sums of certain dependent random variables. *Tôhoku Math. J.* **19**, 357–367.

Baez-Duarte, L. (1971). An a.e. divergent martingale that converges in probability. *J. Math. Anal. Appl.* **36**, 149–150.

Bailey, N. T. J. (1970). "The Elements of Stochastic Processes with Applications to the Natural Sciences." Wiley, New York.

Basu, A. K. (1973). A note on Strassen's version of the law of the iterated logarithm. *Proc. Amer. Math. Soc.* **41**, 596–601.

Baxter, G. (1964). An ergodic theorem with weighted averages. *J. Math. Mech.* **13**, 481–488.

Baxter, G. (1965). A general ergodic theorem with weighted averages. *J. Math. Mech.* **14**, 277–288.

Berman, S. M. (1965). Sign-invariant random variables and stochastic processes with sign-invariant increments. *Trans. Amer. Math. Soc.* **119**, 216–243.

Billingsley, P. (1968). "Convergence of Probability Measures." Wiley, New York.

Birkoff, G. D. (1931). Proof of the ergodic theorem. *Proc. Nat. Acad. Sci. U.S.A.* **17**, 656–660.

Blum, J. R., Hanson, D. L., and Koopmans, L. H. (1963). On the strong law of large numbers for a class of stochastic processes. *Z. Wahrscheinlichkeitstheorie und Verw. Gebiete* **2**, 1–11.

Borel, E. (1909). Sur les probabilites denombrables et leurs applications arithmetiques. *Rend. Circ. Mat. Palermo* **26**, 247–271.

Breiman, L. (1967). On the tail behavior of sums of independent random variables. *Z. Wahrscheinlichkeitstheorie und Verw. Gebiete* **9**, 20–25.

Breiman, L. (1968). "Probability." Addison-Wesley, Reading, Massachusetts.

Burkholder, D. L. (1964). Maximal inequalities as necessary conditions for almost everywhere convergence. *Z. Wahrscheinlichkeitstheorie und Verw. Gebiete* **3**, 75–88.

Burkholder, D. L. (1966). Martingale transforms. *Ann. Math. Statist.* **37**, 1497–1504.

Burkholder, D. L. (1973). Distribution function inequalities for martingales. *Ann. of Prob.* **1**, 1–19.

Burkholder, D. L., and Gundy, R. F. (1970). Extrapolation and interpolation of quasi-linear operators on martingales. *Acta Math.* **124**, 248–304.

Carleson, L. (1966). On convergence and growth of partial sums of Fourier series. *Acta Math.* **116**(1-2), 135–137.

Chover, J. (1967). On Strassen's version of the log log law. *Z. Wahrscheinlichkeitstheorie und Verw. Gebiete* **1**, 340–346.

Chow, Y. S. (1960). A martingale inequality and the law of large numbers. *Proc. Amer. Math. Soc.* **11**, 107–111.

Chow, Y. S. (1963). Convergence theorems of martingales. *Z. Wahrscheinlichkeitstheorie und Verw. Gebiete* **1**, 340–346.

Chow, Y. S. (1965). Local convergence of martingales and the law of large numbers. *Ann. Math. Statist.* **36**, 552–558.

Chow, Y. S. (1966). Some convergence theorems for independent random variables. *Ann. Math. Statist.* **37**, 1482–1493.

Chow, Y. S. (1967). On a strong law of large numbers for martingales. *Ann. Math. Statist.* **38**, 610–611.

Chow, Y. S. (1968). Convergence of sums of squares of martingale differences. *Ann. Math. Statist.* **39**, 123–133.

Chow, Y. S. (1969). Martingale extensions of a theorem of Marcinkiewicz and Zygmund. *Ann. Math. Statist.* **40**, 427–433.

Chow, Y. S., and Robbins, H. (1961). On sums of independent random variables with infinite moments and "fair" games. *Proc. Nat. Acad. Sci. U.S.A.* **47**, 330–335.

Chow, Y. S., and Teicher, H. (1971). Almost certain summability of independent, identically distributed random variables. *Ann. Math. Statist.* **42**, 401–404.

Chow, Y. S. and Lai, T. L. (1973). Limiting behavior of weighted sums of independent random variables. *Ann. of Prob.* **1**, 810–824.

Chung, K.-L. (1947). Note on some strong laws of large numbers. *Amer. J. Math.* **69**, 189–192.

Chung, K.-L. (1948). On the maximum partial sums of sequences of independent random variables. *Trans. Amer. Math. Soc.* **64**, 205–233.

Chung, K.-L. (1951). The strong law of large numbers. *Proc. Symp. Statist. and Probability, 2nd, Berkeley, 1951*, pp. 341–352.

Chung, K.-L. (1960). "Markov Chains with Stationary Transition Probabilities." Springer-Verlag, Berlin and New York.

Chung, K.-L. (1968). "A Course in Probability Theory." Harcourt, New York.

Chung, K.-L., and Erdös, P. (1947). On the lower limit of sums of independent random variables. *Ann. of Math.* **48**, 1003–1012.

Chung, K.-L., and Fuchs, W. H. J. (1951). On the distribution of values of sums of random variables. *Mem. Amer. Math. Soc.* **6**, 1–12.

Chung, K.-L., and Ornstein, D. (1962). On the recurrence of sums of random variables. *Bull. Amer. Math. Soc.* **68**, 30–32.

Darling, D. A., and Robbins, H. (1967). Confidence sequences for means, variance and median. *Proc. Nat. Acad. Sci. U.S.A.* **58**, 66–68.

Davis, B. (1968). Comparison tests for the convergence of martingales. *Ann. Math. Statist.* **39**, 2141–2144.

Davis, B. (1969). Divergence properties of some martingale transforms. *Ann. Math. Statist.* **40**, 1852–1854.

Davis, B. (1970). On the integrability of the martingale square function. *Israel J. Math.* **8**, 187–190.

Derman, C., and Robbins, H. (1955). The strong law of large numbers when the first moment does not exist. *Proc. Nat. Acad. Sci. U.S.A.* **41**, 586–587.

Dharmadhikari, S. W., and Jogdeo, K. (1969). Bounds on moments of sums of random variables. *Ann. Math. Statist.* **40**, 1506–1509.

Doob, J. L. (1953). "Stochastic Processes." Wiley, New York.

Dvoretzky, A. (1949). On the strong stability of a sequence of events. *Ann. Math. Statist.* **20**, 296–299.

Dvoretzky, A., and Erdös, P. (1951). Some problems on random walk in space. *Proc. Symp. Math. Statist. and Probability, Berkeley,* **2**, pp. 353–367.

Egorov, V. A. (1969). On the law of the iterated logarithm, *Theor. Probability Appl.* **14**, 693–699.

Erdös, P. (1949). On a theorem of Hsu and Robbins. *Ann. Math. Statist.* **20**, 286–291.

Erdös, P., and Gal, I. S. (1955). On the law of the iterated logarithm. *Nederl. Akad. Wetensch. Proc. Ser. A* **58**, 65–84.

Feller, W. (1943). The general form of the so-called law of the iterated logarithm. *Trans. Amer. Math. Soc.* **54**, 373–402.

Feller, W. (1946a). A limit theorem for random variables with infinite moments. *Amer. J. Math.* **68**, 257–262.

Feller, W. (1946b). The law of the iterated logarithm for identically distributed random variables. *Ann. of Math.* [2], **47**, 631–638.

Feller, W. (1966). "An Introduction to Probability Theory and Its Applications," Vol. II. Wiley, New York.

Feller, W. (1968a). "An Introduction to Probability Theory and Its Applications," Vol. I. 3rd ed. Wiley, New York.

Feller, W. (1968b). An extension of the law of the iterated logarithm to variables without variance. *J. Math. Mech.* **18**, 343–355.

Feller, W. (1969). General analogues to the law of the iterated logarithm. *Z. Wahrscheinlichkeitstheorie und Verw. Gebiete* **14**, 21–26.

Feller, W. (1970). On the oscillations of sums of independent random variables. *Ann. of Math.* **91**, 402–418.

Franck, W. E., and Hanson, D. L. (1966). Some results giving rates of convergence in the law of large numbers for weighted sums of independent random variables. *Trans. Amer. Math. Soc.* **124**, 347–359.

Gal, I. S. (1951). Sur la majoration des suites des fonctions. *Nederl. Akad. Wetensch. Proc. Ser. A* **54**, 243–251.

Gaposhkin, V. F. (1965). The law of the iterated logarithm for Cesaro's and Abel's methods of summation. *Theor. Probability Appl.* **10**, 411–420.

Garsia, A. (1965). A simple proof of Eberhard Hopf's maximal ergodic theorem. *J. Math. Mech.* **14**, 381–382.

Garsia, A. (1967). Combinatorial inequalities and convergence of some orthogonal series with an appendix by John A. R. Holbrook. *In* "Orthogonal Expansions and their Continuous Analogues" (D. T. Haimo, ed.), pp. 75–92. Southern Illinois Univ. Press, Carbondale, Illinois.

Garsia, A. (1970). "Topics in Almost Everywhere Convergence." Markham Publ., Chicago, Illinois.

Garsia, A., and Sawyer, S. (1965). On an Ergodic Theorem with Weighted Means. Unpublished manuscript.

Gilat, D. (1971). On the nonexistence of a three series condition for series of non-independent random variables. *Ann. Math. Statist.* **42**, 409–410.

Gordin, M. I. (1969). The central limit theorem for stationary processes *Dokl. Akad. Nauk SSSR* **188**, 1174–1176.

Gundy, R. F. (1966). Martingale theory and pointwise convergence of certain orthogonal series. *Trans. Amer. Math. Soc.* **124**, 228–248.

Gundy, R. F. (1967). The martingale version of a theorem of Marcinkiewicz and Zygmund. *Ann. Math. Statist.* **38**, 725–734.

Gundy, R. F. (1968). A decomposition for $L_1$ bounded martingales. *Ann. Math. Statist.* **39**, 134–138.

Hájek, J., and Rényi, A. (1955). Generalization of an inequality of Kolmogorov. *Acta Math. Acad. Sci. Hungar.* **6**, 281–283.

Halmos, P. R. (1950). "Measure Theory." Van Nostrand-Reinhold, Princeton, New Jersey.

Halmos, P. R. (1956). Lectures on ergodic theory. Mathematical Society of Japan, No. 3, 1956.

Hardy, G. H. (1967). "Divergent Series." Oxford Univ. Press, London and New York.

Hardy, G. H., Littlewood, J. E., and Polya, G. (1964). "Inequalities", 2nd ed. Cambridge Univ. Press, London.

Harris, T. E., and Robbins, H. (1953). Ergodic theory of Markov chains admitting an infinite invariant measure. *Proc. Nat. Acad. Sci. U.S.A.* **39**, 862–864.

Hartman, P., and Wintner, A. (1941). On the law of the iterated logarithm. *Amer. J. Math.* **63**, 169–176.

Hewitt, E., and Savage, L. J. (1955). Symmetric measures on Cartesian products. *Trans. Amer. Math. Soc.* **80**, 470–501.

Heyde, C. C. (1968a). On almost sure convergence for sums of independent random variables. *Sankhyā Ser. A* **30**, 353–358.

Heyde, C. C. (1968b). On the converse to the iterated logarithm law. *J. Appl. Probability* **5**, 210–215.

Heyde, C. C. (1969a). A note concerning behavior of iterated logarithm type. *Proc. Amer. Math. Soc.* **23**, 85–90.

Heyde, C. C. (1969b). Some properties of metrics in a study of convergence to normality. *Z. Wahrscheinlichkeitstheorie und Verw. Gebiete* **11**, 181–192.

Heyde, C. C. and Scott, D. J. (1973). Invariance principles for the law of the iterated logarithm for martingales and processes with stationary increments, *Ann. of Prob.* **1**, 428–437.

Hill, J. D. (1951). The Borel property of summability methods. *Pacific J. Math.* **1**, 399–409.

Hsu, P. L., and Robbins, H. (1947). Complete convergence and the law of large numbers. *Proc. Nat. Acad. Sci. U.S.A.* **33**, 25–31.

Ibragimov, I. A. (1962). Some limit theorems for stationary processes. *Theor. Probability Appl.* **7**, 349–382.

Iosifescu, M. (1968). La loi du logarithme itere pour une classe de variables aletoires dependent. *Teor. Verojanost. i Premenen* **13**, 315–325.

Jain, N. C., and Pruitt, W. E. (1972a). "The Range of Random Walk." *Proc. Symp. Statist. and Probability, Sixth, Berkeley, 1972*, pp. 31–50.

Jain, N. C., and Pruitt, W. E. (1972b). The law of the iterated logarithm for the range of random walk. *Ann. Math. Statist.* **43**, 1962–1967.

Jain, N. C., Jogder, K., and Stout, W. F. (1973). Upper and lower functions for martingales with applications to mixing. *Ann. of Prob.* (to appear).

Jamison, B., Orey, S., and Pruitt, W. (1965). Convergence of weighted averages of independent random variables. *Z. Wahrscheinlichkeitstheorie und Verw. Gebiete* **4**, 40–44.

Jessen, B. (1934). The theory of integration in a space of an infinite number of dimensions. *Acta Math.* **63**, 249–323.

Jessen, B., and Wintner, A. (1935). Distribution functions and the Riemann Zeta function. *Trans. Amer. Math. Soc.* **38**, 48–88.

Katz, M. (1968). A note on the weak law of large numbers. *Ann. Math. Statist.* **39**, 1348–1349.

Kesten, H. (1970). The limit points of a random walk. *Ann. Math. Statist.* **41**, 1173–1205.

Kesten, H. (1972). Sums of independent random variables without moment conditions. *Ann. Math. Statist.* **43**, 701–732.

Khintchine, A. (1924). Über einen Satz der Wahrscheinlichkeitsrechnung. *Fund. Math.* **6**, 9–20.

Kolmogorov, A. (1928). Über die Summen durch den Zufall bestimmter unabhängiger Grössen. *Math. Ann.* **99**, 309–319.

Kolmogorov, A. (1929). Über das Gesetz des iterierten Logarithmus. *Math. Ann.* **101**, 126–135.

Kolmogorov, A. (1930). Sur la loi forte des grandes nombres. *C. R. Acad. Sci. Paris* **191**, 910–912.

Komlos, J. (1967). A generalization of a problem of Steinhaus. *Acta Math. Acad. Sci. Hungar.* **18**, 217–229.

Komlos, J., and Revesz, P. (1964). On the weighted averages in independent random variables. *Publ. Math. Inst. Hungar. Acad. Sci.* **9**, pp. 583–587.

Krickeberg, K. (1965). "Probability Theory." Addison-Wesley, Reading, Massachusetts.

Lévy, P. (1931). Sur les series dont les termes sont des variables eventuelles independantes. *Studia Math.* **3**, 119–155.

Lévy, P. (1954). "Theorie de l'Addition des Variables Aleatoires," 2nd ed. Gauthier-Villars, Paris.

Loève, M. (1951). On almost sure convergence. *Proc. Symp. Statist. and Probability, 2nd, Berkeley, 1951*, pp. 279–303.

Loève, M. (1963). "Probability Theory," 3rd ed. Van Nostrand-Reinhold, Princeton, New Jersey.

Marcinkiewicz, J., and Zygmund, A. (1937). Remarque sur la loi du logarithme itere. *Fund. Math.* **29**, 215–222.

Marcinkiewicz, J., and Zygmund, A. (1938). Quelques theoremes sur les fonctions independantes. *Studia Math.* **7**, 104–120.

Menchoff, D. (1923). Sur les series de fonctions orthogales I. *Fund. Math.* **4**, 82–105.

Meyer, P.-A. (1972). "Martingales and Stochastic Integrals." Springer-Verlag, Berlin and New York.

Miller, H. D. (1967). A note on sums of independent random variables with infinite first moments, *Ann. Math. Statist.* **38**, 751–758.

Mineka, J., and Silverman, S. (1970). A local limit theorem and recurrence conditions for sums of independent nonlattice random variables. *Ann. Math. Statist.* **41**, 592–600.

Nagaev, S. V. (1972). On sufficient and necessary conditions for the strong law of large numbers. (In Russian). *Theor. Probability. Appl.* **17**, 609–618.

Nelson, P. (1970). A class of orthogonal series related to martingales. *Ann. Math. Statist.* **41**, 1684–1694.

Orey, S. (1966). Tail events for sums of independent random variables. *J. Math. Mech.* **15**, 937–951.

Pakshirajan, R. P., and Sreehari, M. (1970). The law of the iterated logarithm for a Markov process. *Ann. Math. Statist.* **41**, 945–955.

Petrov, V. V. (1960). On the law of the iterated logarithm. *Uspehi Mat. Nauk* **15**, 189–193.

Petrov, V. V. (1966). On a relation between an estimate of the remainder in the central limit theorem and the law of the iterated logarithm. *Theor. Probability Appl.* **11**, 454–458.

Philipp, W. (1967). Das Gesetz vom iterierten Logarithmus für stark mischende stationare Prozesse. *Z. Wahrscheinlichkeitstheorie und Verw. Gebiete* **8**, 204–209.

Philipp, W. (1969). The law of the iterated logarithm for mixing stochastic processes. *Ann. Math. Statist.* **40**, 1985–1991.

Philipp, W. (1971). "Mixing Sequences of Random Variables and Probabilistic Number Theory," Amer. Math. Soc. Mem. No. 114. Amer. Math. Soc., Providence, Rhode Island.

Polya, G. (1921). Über eine Aufgabe der Wahrscheinlichkeitsrechnung betreffend die Irrfahrt in Strassennetz. *Math. Ann.* **89**, 149–160.

Prokhorov, Yu. V. (1949). On the strong law of large numbers. *Dokl. Akad. Nauk SSSR*, **69**, 607–610.

Prokhorov, Yu. V. (1959a). Some remarks on the strong law of large numbers. *Theor. Probability Appl.* **4**, 204–208.

Prokhorov, Yu. V. (1959b). An extremal problem in probability theory. *Theor. Probability Appl.* **4**, 201–203.

Pruitt, W. E. (1966). Summability of independent random variables. *J. Math. Mech.* **15**, 769–776.

Rademacher, H. (1922). Einige Satze uber Reihen von allgemeinen Orthogonalfunktion. *Math. Ann.* **87**, 112–138.

Revesz, P. (1968). "The Laws of Large Numbers." Academic Press, New York.

Revesz, P. (1972). The Law of the Iterated Logarithm for Multiplicative Systems. *Indiana Univ. Math J.* **21**, 557–564.

Reznik, M. Kh. (1968). The law of the iterated logarithm for some classes of stationary processes. *Theor. Probability Appl.* **8**, 606–621.

Riesz, F. (1945). Sur la theorie ergodique. *Comment. Math. Helv.* **17**, 221–239.

Riesz, F., and Sz.-Nagy, B. (1955). "Functional Analysis." Ungar, New York.

Robbins, H. (1953). On the equidistribution of sums of independent random variables. *Proc. Amer. Math. Soc.* **4**, 786–799.

Salem, R., and Zygmund, A. (1950). La loi du logarithmi pour les series trigonetriques lacunaires, *Bull. Sci. Math.* **74**, 209–224.

Serfling, R. J. (1969). Probability inequalities and convergence properties for sums of multiplicative random variables. Rep. M151. Florida State Dept. of Statist. Tallahassee, Florida.

Serfling, R. J. (1970a). Moment inequalities for maximum cumulative sum. *Ann. Math. Statist.* **41**, 1227–1234.

Serfling, R. J. (1970b). Convergence properties of $S_n$ under moment restrictions. *Ann. Math. Statist.* **41**, 1235–1248.

Shiryaev, A. N. (1963). On conditions for ergodicity of stationary processes in terms of higher order moments. *Theor. Probability Appl.* **8**, 436–439.

Skorokhod, A. V. (1957). Limit theorems for stochastic processes with independent increments. *Teor. Verojatnost. i Primenen.* **2**, 145–177.

Skorokhod, A. V. (1961). "Studies in the Theory of Random Processes." Addison-Wesley, Reading, Massachusetts.

Sparre-Anderson, E. (1954). On the fluctuations of sums of random variables II. *Math. Scand.* **2**, 195–223.

Spitzer, F. (1956). A combinatorial lemma and its applications to probability theory. *Trans. Amer. Math. Soc.* **82**, 323–339.

Spitzer, F. (1964). "Principles of Random Walk." Van Nostrand-Reinhold, Princeton, New Jersey.

Stackelberg, O. (1964). On the law of the iterated logarithm. *Nederl. Akad. Wetensch Proc. Ser. A* **67**, 48–67.

Steiger, W. L. (1972). A converse to the log-log law for martingales, Working paper 185, Centre de Recherches Mathematiques, U. de Montreal, Montreal, Canada.

Steiger, W. L., and Zaremba, S. K. (1972). The converse of the Hartman–Wintner theorem. *Z. Wahrscheinlichkeitstheorie und Verw. Gebiete* **22**, 193–194.

Stone, C. J. (1966). The growth of a recurrent random walk. *Ann. Math. Statist.* **37**, 1040–1041.

Stone, C. J. (1969). The growth of a random walk. *Ann. Math. Statist.* **36**, 2203–2206.

Stout, W. F. (1967). Some results on almost sure and complete convergence in the independent and martingale cases. Ph. D. Thesis, Purdue Univ., Lafayette, Indiana.

Stout, W. F. (1968). Some results on the complete and almost sure convergence of linear combinations of independent random variables and martingale differences. *Ann. Math. Statist.* **39**, 1549–1562.

Stout, W. F. (1970a). The Hartman–Wintner law of the iterated logarithm for martingales. *Ann. Math. Statist.* **41**, 2158–2160.

Stout, W. F. (1970b). A martingale analogue of Kolmogorov's law of the iterated logarithm. *Z. Wahrscheinlichkeitstheorie und Verw. Gebiete* **15**, 279–290.

Strassen, V. (1964). An invariance principle for the law of the iterated logarithm. *Z. Wahrscheinlichkeitstheorie und Verw. Gebiete* **3**, 211–226.

Strassen, V. (1965). Almost sure behavior of sums of independent random variables and martingales. *Proc. Symp. Math. Statist. and Probability, 5th, Berkeley, 1965,* **2**, pp. 315–344.

Strassen, V. (1966). A converse to the law of the iterated logarithm. *Z. Wahrscheinlichkeitstheorie und Verw. Gebiete* **4**, 265–268.

Takahashi, S. (1972). Notes on the law of the iterated logarithm. *Studia Sci. Math. Hung.* **7**, 21–24.

Talalyan, A. A., and Arutyunyan, F. G. (1965). On the convergence of Haar series to $+\infty$. *Mat. Sb.* **66**(103), 240–247.

Teicher, H. (1968). Some new conditions for the strong law. *Proc. Nat. Acad. Sci. U.S.A.* **59**, 705–707.

Tomkins, R. J. (1971a). An iterated logarithm theorem for some weighted averages of independent random variables. *Ann. Math. Statist.* **42**, 760–763.

Tomkins, R. J. (1971b). Some iterated logarithm results related to the central limit theorem. *Trans. Amer. Math. Soc.* **156**, 185–192.

Tomkins, R. J. (1972). A generalization of Kolmogorov's law of the iterated logarithm. *Proc. Amer. Math. Soc.* **32**, 268–274.

Varberg, D. E. (1966). Convergence of quadratic forms in independent random variables. *Ann. Math. Statist.* **37**, 567–576.

Weiss, M. (1959). The law of the iterated logarithm for lacunary trigonometric series. *Trans. Amer. Math. Soc.* **91**, 444–469.

Williamson, J. A. (1970). Fluctuations when $E(|X_1|) = \infty$. *Ann. Math. Statist.* **41**, 865–875.

Zygmund, A. (1959). "Trigonometric Series," 2nd ed. Cambridge Univ. Press, London and New York.

# Index

Numbers in italic type indicate pages where an author's work is fully referenced

A  4
B  5
C  6
D  7
E  8
F  9
G  0
H  1
I  2
J  3